Management of Nematode and Insect-Borne Plant Diseases

HAWORTH FOOD & AGRICULTURAL PRODUCTS PRESS™
Crop Science

The Lowland Maya Area: Three Millennia at the Human-Wildland Interface edited by A. Gómez-Pompa, M. F. Allen, S. Fedick, and J. J. Jiménez-Osornio

Biodiversity and Pest Management in Agroecosystems, Second Edition by Miguel A. Altieri and Clara I. Nicholls

Plant-Derived Antimycotics: Current Trends and Future Prospects edited by Mahendra Rai and Donatella Mares

Concise Encyclopedia of Temperate Tree Fruit edited by Tara Auxt Baugher and Suman Singha

Landscape Agroecology by Paul A. Wojtkowski

Concise Encyclopedia of Plant Pathology by P. Vidhyasekaran

Molecular Genetics and Breeding of Forest Trees edited by Sandeep Kumar and Matthias Fladung

Testing of Genetically Modified Organisms in Foods edited by Farid E. Ahmed

Fungal Disease Resistance in Plants: Biochemistry, Molecular Biology, and Genetic Engineering edited by Zamir K. Punja

Plant Functional Genomics edited by Dario Leister

Immunology in Plant Health and Its Impact on Food Safety by P. Narayanasamy

Abiotic Stresses: Plant Resistance Through Breeding and Molecular Approaches edited by M. Ashraf and P. J. C. Harris

Teaching in the Sciences: Learner-Centered Approaches edited by Catherine McLoughlin and Acram Taji

Handbook of Industrial Crops edited by V. L. Chopra and K. V. Peter

Durum Wheat Breeding: Current Approaches and Future Strategies edited by Conxita Royo, Miloudi M. Nachit, Natale Di Fonzo, José Luis Araus, Wolfgang H. Pfeiffer, and Gustavo A. Slafer

Handbook of Statistics for Teaching and Research in Plant and Crop Science by Usha Rani Palaniswamy and Kodiveri Muniyappa Palaniswamy

Handbook of Microbial Fertilizers edited by M. K. Rai

Eating and Healing: Traditional Food As Medicine edited by Andrea Pieroni and Lisa Leimar Price

Physiology of Crop Production by N. K. Fageria, V. C. Baligar, and R. B. Clark

Plant Conservation Genetics edited by Robert J. Henry

Introduction to Fruit Crops by Mark Rieger

Generations Gardening Together: Sourcebook for Intergenerational Therapeutic Horticulture by Jean M. Larson and Mary Hockenberry Meyer

Agriculture Sustainability: Principles, Processes, and Prospects by Saroja Raman

Introduction to Agroecology: Principles and Practice by Paul A. Wojtkowski

Handbook of Molecular Technologies in Crop Disease Management by P. Vidhyasekaran

Handbook of Precision Agriculture: Principles and Applications edited by Ancha Srinivasan

Dictionary of Plant Tissue Culture by Alan C. Cassells and Peter B. Gahan

Handbook of Potato Production, Improvement, and Postharvest Management edited by Jai Gopal and S. M. Paul Khurana

Carbon Sequestration in Soils of Latin America edited by Rattan Lal, Carlos C. Cerri, Martial Bernoux, Jorge Etchevers, and Eduardo Cerri

Genetically Engineered Crops: Interim Policies, Uncertain Legislation edited by Iain E. P. Taylor

Management of Nematode and Insect-Borne Plant Diseases edited by Geeta Saxena and K. G. Mukerji

Management of Nematode and Insect-Borne Plant Diseases

Geeta Saxena
K. G. Mukerji
Editors

The Haworth Press
Taylor & Francis Group
New York • London

For more information on this book or to order, visit
http://www.haworthpress.com/store/product.asp?sku=5754

or call 1-800-HAWORTH (800-429-6784) in the United States and Canada
or (607) 722-5857 outside the United States and Canada
or contact orders@HaworthPress.com

Published by

The Haworth Press, Taylor & Francis Group, 270 Madison Avenue, New York, NY 10016.

Cover design by Marylouise E. Doyle.

Library of Congress Cataloging-in-Publication Data

Management of nematode and insect-borne diseases/Geeta Saxena, K.G. Mukerji, editors.
p. cm.
ISBN: 978-1-56022-134-0 (hard : alk. paper)
ISBN: 978-1-56022-135-7 (soft : alk. paper)
1. Nematode diseases of plants—Control. 2. Insect pests—Control. I. Saxena, Geeta, M.Sc., Ph.D. II. Mukerji, K. G.

SB998.N4M32 2007
632'.6257-dc22 2007000472

CONTENTS

ABOUT THE EDITORS

Geeta Saxena, PhD, is Associate Professor of Botany at Swami Shraddhanand College, the University of Delhi, India. She received the UNESCO Young Scientist of the Year Award in 1989 in the field of Biotechnology-Mycology and Plant Pathology, and worked as a DAAD Fellow on fluorescence and electron microscopy, and X-ray microanalysis of nematophagous fungi at Freie Universitat in Berlin. Dr. Saxena was a visiting scientist on zoosporic fungi at the Institute of Cell and Molecular Biology at the University of Edinburgh in Scotland. She has extensively worked on taxonomy, ecology of nematophagous fungi, and its applications in biocontrol. She has had forty research papers published in the leading international journals. Dr. Saxena has also edited books entitled *Recent Developments in Biocontrol of Plant Diseases,* and *New Approaches in Microbial Ecology.*

K. G. Mukerji, PhD, is a retired Senior Professor at the University of Delhi, India. A distinguished mycologist and microbial ecologist, he has published more than 600 research papers. Dr. Mukerji is an editorial board member of several national and international journals, and is an honorary member of the research board of advisors for the American Biological Institute. He is co-author of *Taxonomy of Indian Myxomycetes, Index to Plant Diseases of India,* and *The Haustorium,* and editor of several books, including *Biocontrol of Plant Diseases, Fruit and Vegetable Diseases,* and *Recent Developments in Biocontrol of Plant Diseases.*

Management of Nematode and Insect-Borne Plant Diseases

doi:10.1300/5754_a

NOTES FOR PROFESSIONAL LIBRARIANS AND LIBRARY USERS

This is an original book title published by The Haworth Press, Taylor & Francis Group. Unless otherwise noted in specific chapters with attribution, materials in this book have not been previously published elsewhere in any format or language.

CONSERVATION AND PRESERVATION NOTES

All books published by The Haworth Press are printed on certified pH neutral, acid-free book grade paper. This paper meets the minimum requirements of American National Standard for Information Sciences-Permanence of Paper for Printed Material, ANSI Z39.48-1984.

DIGITAL OBJECT IDENTIFIER (DOI) LINKING

The Haworth Press is participating in reference linking for elements of our original books. (For more information on reference linking initiatives, please consult the CrossRef Web site at www.crossref.org.) When citing an element of this book such as a chapter, include the element's Digital Object Identifier (DOI) as the last item of the reference. A Digital Object Identifier is a persistent, authoritative, and unique identifier that a publisher assigns to each element of a book. Because of its persistence, DOIs will enable The Haworth Press and other publishers to link to the element referenced, and the link will not break over time. This will be a great resource in scholarly research.

CONTRIBUTORS

Ginny Antony, Plant Transformation Laboratory, Department of Plant Pathology, Kansas State University, Manhattan, KS, 66506. E-mail: ginns@ksu.edu.

Steven P. Arthurs, USDA-ARS, Yakima Agricultural Research Laboratory, Wapato, WA, 98951. E-mail: sarthurs@yarl.ars.usda.gov.

Antonio Bello, Departamento Agroecologia, CCMA, CSIC, Madrid, Spain. E-mail: antoon.ploeg@ucr.edu.

J. D. Dutcher, Entomology Department, University of Georgia, Coastal Plain Experiment Station, Tifton, GA, 31793-0748. E-mail: dutcher@tifton.uga.edu.

Daniel L. Frank, Department of Entomology and Nematology, University of Florida, Gainesville, FL, 32611. E-mail: oeliburd@mail.ifas.ufl.edu.

Wayne A. Gardner, Department of Entomology, University of Georgia, Griffin, GA, 30223-1797. E-mail: Wgardner@griffin.uga.edu.

Tony Grace, Molecular Genetics Laboratory, Department of Entomology, Kansas State University, Manhattan, KS, 66506. E-mail: tonygrac@ksu.edu.

Nicola Greco, Consiglio Nazionale delle Ricerche, Istituto per la Protezione delle Piante, Sezione di Bari, 70126 Bari, Italy. E-mail: n.greco@ba.ipp.cnr.it.

G. W. Krewer, Entomology Department, University of Georgia, Coastal Plain Experiment Station, Tifton, GA, 31793-0748. E-mail: dutcher@tifton.uga.edu.

doi:10.1300/5754_b

Lawrence A. Lacey, USDA-ARS, Yakima Agricultural Research Laboratory, Wapato, WA, 98951. E-mail: llacey@yarl.ars.usda.gov.

Franco Lamberti, Consiglio Nazionale delle Ricerche, Istituto per la Protezione delle Piante, Sezione di Bari, 70126 Bari, Italy.

Oscar E. Liburd, Department of Entomology and Nematology, University of Florida, Gainesville, FL, 32611. E-mail: oeliburd@mail.ifas.ufl.edu.

José-Antonio López-Pérez, Centro de Experimentación Agraria de Marchamalo, Marchamalo, Guadalajara, Spain.

Maria Nenmaura Gomes Pessoa, Universidade Federal do Ceará, Centro de Ciéncias Agrárias, Departmento de Fitotecnia, Fortaleza, Ceará, Brasil.

Antoon Ploeg, Department of Nematology, University of California, Riverside, CA, 92521-0415.

Philip A. Roberts, Department of Nematology, University of California, Riverside, CA, 92521-0415. E-mail: philip.roberts@ucr.edu.

Ericka Scocco, Department of Entomology, University of Georgia, Griffin, GA, 30223-1797.

Nandita Selvanathan, 289 Bowman Avenue, Winnipeg, Manitoba, Canada R2K 1P1. E-mail: selvanathan@shaw.ca.

Alberto Troccoli, Consiglio Nazionale delle Ricerche, Istituto per la Protezione delle Piante, Sezione di Bari, 70126 Bari, Italy. E-mail: a.troccoli@ba.ipp.cnr.it.

Preface

A host of organisms can be harmful to the productivity of an agricultural field if their presence is not managed. Plant diseases that curtail production are sometimes disastrous. For any one pathogen or for a range of natural pests—nematodes and insects—control is best instituted through a coordinated, planned, and often location-based approach. Management that may be good for individual farmers is not necessarily good for farmers in aggregate or society at large.

The aim of this book is to provide a detailed description of the management of diseases caused by nematodes and insects. Control measures usually pass through a period during which they represent the "new" or "in vogue" approach. The development of transgenic crop plants is one such measure currently receiving a great deal of interest and, of course, funding.

The majority of plant-parasitic nematodes are root feeders, completing their life cycles in the root zone, and are found in association with most plants. Some are endoparasitic, living and feeding within the tissue of roots, tubers, buds, seeds, and so on. Others are ectoparasitic, feeding externally through plant walls. A few species are highly host specific, but the majority have a wide host range. The most sustainable approach to nematode control integrates several tools and strategies, including cover crops, crop rotation, soil solarization, the least toxic pesticides, and plant varieties resistant to nematode damage.

Insect pests cause heavy crop losses, either through direct destruction of plants or through wounds produced by the insect itself that allow the infection and diffusion of other pathogens. Research on the use of natural insecticides has received a greater impetus in recent years because of growing consciousness of the ecological damage caused by chemical substances.

Management of Nematode and Insect-Borne Plant Diseases

doi:10.1300/5754_c

Biopesticides are rapidly substituting and replacing the conventional methods of controlling pests and disease-causing vectors. The attractiveness stems from their ecofriendliness and target specificity. Other advantages are the absence of insect resistance, the absence of residue build-up in the environment, and the potential impact of biotechnological research and development. Biological pest controls have a number of limitations, but it is hoped that biotechnology and a responsible environmental outlook will help us to change the limitations and make biological pest control agents products for the agriculture of today. Several major crop plants have been engineered with genes that make them resistant to pests. Biological control is a critical component of many integrated pest management programs, supplemented by the judicious use of cultural management practices.

This book contains ten chapters, four on the management of insect-borne diseases, five on plant-parasitic nematodes, and one on RNAi for the management of insect- and nematode-borne diseases.

Chapter 1 provides an in-depth description of the use of arthropod microbial control agents against the more common pests or pest complexes found in different types of tree fruits. This is followed by a chapter looking at reducing strains of insects transmitting the bacterium *Xylella fastidiosa* in fruit and nut crops. This chapter outlines the biology and control of the bacterium and its vectors. Chapter 3 highlights the use of living and synthetic mulches to reduce the population densities of aphids and whiteflies successfully while also delaying the onset and spread of associated insect-borne diseases. These mulches offer significant protection against pests and can reduce vector efficiencies. Chapter 4 describes genetically transforming microbial control agents to improve the performance of these agents in pest management systems. Chapter 5 gives an overview of the integrated use of different control options to manage plant-parasitic nematodes. Among these, the most feasible are crop rotations, especially for nematodes of narrow host range, soil solarization in hot areas, and the use of resistant cultivars and rootstocks for a number of crops. Chapter 6 focuses on the nematophagous fungi as potential biocontrol agents of plant-parasitic nematodes. It describes predatory, endoparasitic, and parasitic fungi of root-knot and cyst nematodes, their mode of trapping, and their application in nematode control. Chapter 7 emphasizes the use of biofumigation as a safe, sustainable, and economically viable strategy to manage plant-parasistic nematodes. Chapter 8 deals with Potato Early Dying complex—a disease associated with the nematode *Pratylenchus penetrans* and the fungus *Verticillium.* Management practices include resistant varieties, sanitation, crop rotation, and irrigation. Chapter 9 describes host-plant resistance to nematodes in vegetable crops in relation to resistance gene availability and the effective integration of resistance into vegetable cropping systems for nematode management. The final chapter deals with RNA interference, also known as RNA silencing or posttranscriptional gene silencing—a molecular method for the management of insect- and nematode-borne diseases. RNAi is a recently discovered phenomenon that responds to double-stranded RNA by silencing gene expression in a sequence-specific manner.

All chapters are new and have been specially written for this book. Indeed, probably no other publication has dealt with these topics in one place. Each chapter has been written by experienced scientists in the field. Our hope in assembling this variety of approaches in a single volume is that researchers and students working on the management of insect- and nematode-borne diseases can use it as their first reference for starting a new avenue of investigation.

We are grateful to all the authors for their contributions to this book and for accepting suggestions to produce the final balance of the book. The chapters are original, and some topics have been included for the first time in any book. We pay our deep respects to Dr. Franco Lamberti, one of our authors, who passed away. We are grateful to a large number of people for their assistance and encouragement during the preparation of this book. Special mention should be made of Mr. Vishal Saurav and Professor Ajay Kumar, who extended constant help from the initiation of this book. Thanks are also due to Professor Ashok K. Singh for technical advice about insects and nematodes during preparation of the chapters.

It is our hope that this book will be useful to all students and researchers in plant pathology, entomology, nematology, agriculture, and horticulture, and to fruit and vegetable specialists.

Chapter 1

Microbial Control of Insect Pests of Tree Fruit

Steven P. Arthurs
Lawrence A. Lacey

INTRODUCTION

Many tree fruit growers rely on chemical insecticides to eliminate arthropod pests or suppress their numbers below economically damaging thresholds. However, environmental and human health concerns have placed the continued reliance on broad-spectrum insecticides in orchards under increasing scrutiny. In the United States, the Food Quality Protection Act (FQPA) of 1996 amended the Federal Insecticide, Fungicide, and Rodenticide Act (FIFRA), and the Federal Food, Drug, and Cosmetic Act (FFDCA) fundamentally changed the way the Environmental Protection Agency (EPA) regulates pesticides. As the requirements stipulate a new safety standard, "reasonable certainty of no harm," that must be applied to all pesticides used on foods, the continued registration of certain insecticides cannot be guaranteed. Indeed, a number of products that tree fruit growers have come to rely on will be phased out over the coming years. Moreover, resistance to commonly used chemical pesticides continues to be documented among some important pests of orchards, further reducing insecticidal options for many orchard managers. It is clear that alternative

---

We are grateful for the comments and suggestions of Alan Knight, Wee Yee, and Allison Walston.

Management of Nematode and Insect-Borne Plant Diseases

doi:10.1300/5754_01

approaches to managing arthropod pests will be increasingly emphasized through public demands for pesticide-free food, by legislation, and owing to biological necessity.

Integrated pest management (IPM), which combines biological, cultural, physical, and chemical tools in a way that minimizes economic, health, and environmental risks, is a more sustainable approach to managing pests than pesticides alone. Owing to their specificity for insects and safety for vertebrates (Laird, Lacey and Davidson, 1990; Hokkanen and Hajek, 2004), microbial pathogens of insects are ideal candidates for incorporation into IPM. They may be used inundatively for rapid short-term control or to augment naturally occurring pathogens (augmentation), conserved or activated in nature (conservation), or introduced into pest populations as classical biological control agents aimed at establishing long-term control (inoculative release) (Lacey et al., 2001). The most common method in orchards is through inundative means (Tanada and Kaya, 1993; Kaya and Lacey, 2000). In this approach pathogens are often applied in a water-based spray with the same agrochemical equipment used to apply conventional pesticides. Entomopathogens that have received most commercial interest as microbial pesticides are those that can be mass produced using artificial media and stored as formulated products for extended periods of time. For these reasons, entomopathogens have a distinct advantage over arthropod biological control agents (i.e., predators and parasitoids) because they can be rapidly deployed over relatively large areas in response to pest outbreaks.

The manipulation of entomopathogens for insect control is not new; its roots can be traced back to early pioneers such as Agostino Bassi, Louis Pasteur, and Elie Metschnikoff (Steinhaus, 1956). In the interim, a significant amount of research has evaluated the use and development of pathogens to control various agricultural pests. However, the commercialization of microbial control agents has been limited; a 1997 estimate put the market for all biopesticides at less than 1 percent of the total crop protection market (Lisansky, 1997). This lack of adoption can likely be traced to several disadvantages that entomopathogens have compared with chemical pesticides. These disadvantages include more specific application requirements and a shorter persistence of activity, slower speed of kill, restricted host range (necessitating smaller markets), and high cost of production

(Benz, 1987; Fuxa and Tanada, 1987; Kaya and Lacey, 2000). Nevertheless, there are also several advantages, and interest in the development and deployment of microbial control agents against insect and mite agricultural pests has increased, especially in the past decade. This chapter will summarize the research in this area and highlight cases where entomopathogenic organisms are being used or show promise as biopesticides for IPM strategies in tree fruit. The challenges for the wider adoption of microbial control by the tree fruit industry will also be considered.

OVERVIEW OF MICROBIAL CONTROL IN TREE FRUIT ORCHARDS

Pests of Tree Fruit

Arthropod pests of economic importance or potential exotic pest problems in tree fruit include weevils, fruit flies, leafminers, aphids, psylla, whiteflies, thrips, stinkbugs, mites, and a range of Lepidoptera (Table 1.1). Between them, these pests attack all parts of the fruit tree, including fruits, leaves, flowers, main stems, branches, and roots. Furthermore, a number of them help spread disease by vectoring and facilitating infection by various plant pathogens. Although the number of arthropods that may constitute minor or regional pests of tree fruit is large, several important pests or pest complexes are common among particular cropping systems. For example, root weevils (Coleoptera: Curculionidae), leafminers (Lepidoptera: Gracillariidae), eriophyid mites (Acari: Eriophyidae), and armored scales (Homoptera: Diaspididae) are common citrus pests (Smith, Beattie and Broadley, 1997; McCoy, 1999). In pome fruit, codling moth (*Cydia pomonella* L.) and leafrollers (Lepidoptera: Tortricidae) are often the most important pests encountered (Barnes, 1991; Beers et al., 1993; Cross et al., 1999). Significant pests of stone fruit (peach, plum, cherry, and apricot) include the Oriental fruit moth (*Grapholita* sp.), peach tree borers (*Synanthedon* spp.), plum curculio (*Conotrachelus* sp.), cherry fruit fly (*Rhagoletis* sp.), and green peach aphid (*Myzus* sp.) (Cossentine, Banham and Jensen, 1990; Beers et al., 1993; Liu et al., 1999; Vincent, Chouinard and Hill, 1999; Yee and Lacey, 2003). The pecan weevil

TABLE 1.1. Arthropod pests of tree fruit and nuts that have been controlled by entomopathogens under natural, experimental, or operational conditions.

Crop	Target pest(s)	Entomopathogen(s)	Reference(s)
Citrus	Root weevils, *Diaprepes abbreviatus* and *Pachnaeus* spp.	*Steinernema riobrave*	Duncan et al. (1999), McCoy et al. (2000), Shapiro et al. (2000)
	Brown citrus aphid, *Toxoptera citricida*	*Beauveria bassiana* GHA strain	Poprawski, Parker and Tsai (1999)
	Whiteflies, *Dialeurodes* spp.	*Aschersonia* spp.	Fawcett (1944), Ponomarenko et al. (1975), Gao et al. (1985)
	Eriophyid mites	*Hirsutella thompsonii*	McCoy (1996), McCoy and Couch (1982)
Pome and temperate stone fruit	Codling moth, *Cydia pomonella*	Granulovirus	Lacey, Knight and Huber (2000), Arthurs and Lacey (2004)
		Steinernema and *Heterorhabditis* spp.	Lacey and Unruh (1998), Unruh and Lacey (2001), Cossentine, Jensen and Moyls (2002)
	Leafrollers and other tortricids	*Bacillus thuringiensis*	Knight et al. (1998), Cross et al. (1999), Lacey and Shapiro-Ilan (2003)
		Adoxophyes orana granulovirus	Cross et al. (1999)
	Pear thrips, *Taeniothrips inconsequens*	Hyphomycete fungi	Brownbridge et al. (1999)
	Plum curculio, *Conotrachelus nenuphar*	*Metarhizium anisopliae* and *B. bassiana*	Tedders et al. (1982)
		Steinernema carpocapsae	Olthof and Hagley (1993)
		S. riobrave	Shapiro-Ilan et al. (2004)
	Clearwing borers, *Synanthedon* spp.	*Steinernema* spp.	Deseö and Miller (1985), Gill, Davidson and Raupp (1992), Nachtigall and Dickler (1992)
		Heterorhabditis bacteriophora	Cossentine, Banham and Jensen (1990)
	Western cherry fruit fly, *Rhagoletis indifferens*	*M. anisopliae*	Yee and Lacey (2005)

TABLE 1.1 *(continued)*

Crop	Target pest(s)	Entomopathogen(s)	Reference(s)
		Steinemematid nematodes	Yee and Lacey (2003)
Blackcurrant	Currant clearwings, *Synanthedon tipuliformis*	*Steinernema feltiae*	Bedding and Miller (1981), Miller and Bedding (1982)
Papaya	Medfly, *Ceratitis capitata*	*S. feltiae*	Lindegren, Wong and McInnis (1989)
Litchi	Cerambycid beetles	*S. carpocapsae*	Xu et al. (1995, 1997)
Banana	Banana weevil borer, *Cosmopolites*	*B. bassiana*	Carballo and Lopez (1994), Nankinga and Moore (2000)
	sordidus	*Steinernema* spp.	Figueroa (1990), Schmitt, Gowen and Hague (1992)
Pecan	Pecan weevil, *Curculio caryae*	*B. bassiana* and *M. anisopliae*	Gottwald and Tedders (1983)
		S. carpocapsae	Shapiro-Ilan (2001a), Shapiro-Ilan et al. (2003)
Coconut and oil palms	Rhinoceros beetle, *Oryctes rhinoceros*	Non-occluded virus	Jacob (1996), Gopal et al. (2001)

(*Curculio* sp.), navel orangeworm (*Amyelois* sp.), and various weevils (Coleoptera: Curculionidae) are common pests associated with nut production (Harris, 1999; Siegal et al., 2004).

Pathogens Used in Microbial Control

Bacteria

A variety of insect pathogens have been used as biological control agents against tree fruit pests (Table 1.1). The most important bacterial pathogen used as a biological control agent is *Bacillus thuringiensis* Berliner *(Bt)* (Beegle and Yamamato, 1992). *Bt* is a rod-shaped bacterium found worldwide on plants and in soil that persists in the environment as resistant spores. Parasporal bodies containing proteinaceous toxins (crystals) produced by this bacterium during its spore-forming phase have insecticidal activity against a wide range of Lepidoptera, nematocerous Diptera, and certain species of Coleoptera. The *Bt* endotoxins are stomach poisons, have no contact activity, and need to be ingested to be insecticidal. In the midgut of susceptible species, the parasporal inclusions are solubilized and cleaved by proteolytic enzymes to form toxins that bind with specific sites on the surface of epithelial cells. The toxins disrupt the osmotic balance of the cells, leading to lysis and ultimately the death of the host. Subspecies of *Bt* have different toxins (endotoxins) that act on different hosts. A nomenclature system for *Bt* endotoxins was originally proposed by Höfte and Whitely (1989) and is maintained and updated with newly discovered toxins online by Crickmore et al. (http://www.lifesci.sussex.ac.uk/Home/Neil_Crickmore/*Bt*/).

Bt isolates first used for pest control contained the Cry 1 toxin class and were active against Lepidoptera larvae, with *Bt* var. *kurstaki* the most commonly used. Since then subspecies of *Bt* producing other Cry toxin classes have been produced commercially; *Bt* var. *israelensis* is active against mosquitoes and black flies, and *Bt* var. *tenebrionis* is active against several species of beetle. *Bt*-based products are widely used as inundatively applied microbial control agents and are currently among the few insecticides allowed under organic production guidelines. Worldwide, *Bt* products are used annually on several million hectares and are commonly used for Lepidoptera control in tree fruit production (Cross et al., 1999; Lacey and Shipiro-Ilan, 2003).

Viruses

Viruses are noncellular genetic elements that can replicate only within the living cell (Evans, 1997; Federici, 1997; Miller, 1997; Hunter-Fujita et al., 1998; Lacey et al., 2001). At least 13 viral families include pathogens of invertebrates; the Baculoviridae are specific to insects and related invertebrates and have the greatest potential for insect microbial control. Baculoviruses comprise the nucleopolyhedroviruses (NPVs) and granuloviruses (GVs). The virus particles of NPVs and GVs are occluded in proteinaceous occlusion bodies (OBs) that persist in the environment and form the infectious unit. As for bacteria, the route of infection is typically *per os,* and infection occurs when OBs are consumed in contaminated plant material. Baculoviruses occur mainly in the Lepidoptera and Hymenoptera, but also in Diptera and Coleoptera. The GVs of *C. pomonella* (CpGV), and to a lesser extent of the summer fruit tortrix moth, *Adoxophyes orana* Walsingham (AoGV), have been researched extensively and developed as microbial control agents for pome fruit (Luisier and Benz, 1994; Cross et al., 1999; Lacey and Shapiro-Ilan, 2003). However, commercial development of many viruses has been limited because of the high costs of *in vivo* production, slow action, short persistence, and increased specificity compared with broad-spectrum pesticides.

Nematodes

Nematodes are colorless nonsegmented and multicellular animals that contain excretory, nervous, digestive, reproductive, and muscular systems but lack circulatory and respiratory systems (Kaya and Stock, 1997). Entomopathogenic nematodes (EPNs) in the families Steinernematidae and Heterorhabditidae attack a broad range of insect pests (Gaugler and Kaya, 1990; Klein, 1990; Kaya and Gaugler, 1993). The nonfeeding infective stage juveniles (IJs) are 1-3 mm in length and infect insects through natural body openings (spiracles, mouth, or anus) or areas of thin cuticle. Once inside, the IJ releases a bacterial symbiote (*Xenorhabdus* for steinernematids and *Photorhabdus* for heterorhabditids) that multiplies rapidly and kills the host within 24-48 hours. Steinernematid and heterorhabditid species are routinely applied using common agrochemical sprayers or irrigation equipment and have been widely commercialized and marketed for use against soil-inhabiting and otherwise cryptic insects or insect stages (Klein, 1990; Kaya and Gaugler, 1993). Although EPNs have features similar to predators/parasitoids, they are associated with pathogens (presumably owing to their symbiotic relationship with bacteria) and will be included in our review.

Fungi

Although fungi are only very rarely lethal to vertebrates, they are important pathogens in arthropods, and some 700 species of entomopathogenic fungi have been recorded (McCoy, Samson and Boucias, 1988; Samson, Evans and Latgé, 1988; Hajek and St. Leger, 1994;

Lacey et al., 2001). Unlike bacteria and viruses, entomopathogenic fungi predominantly infect through the insect cuticle and are thus often the only microbial candidates for sucking pests such as aphids. Some fungi proliferate within the insect's haemocoel as hyphae and single-celled bodies called blastospores, finally killing the host through disruption of organs and tissues. Other fungi kill the host more rapidly through the production of toxins.

The most important family for natural control is the entomophthorales, which are obligate pathogens that tend to be host specific and often cause spectacular epizootics that regulate pest populations (Latgé and Paperiok, 1988; Hajek and St. Leger, 1994; Lacey et al., 2001). Most entomophthorales produce short-lived asexual spores but have complex life cycles with higher taxonomy based on their sexual forms, which produce long-lived resting spores. Difficulties encountered in the mass production of the entomophthorales limit their application to classical biological control or inoculative release.

The most important fungi used in augmentative or inundative biological control are in the subdivision Deuteromycotina or Fungi Imperfecti, in which sexual forms are rarely found. Several species in the class Hyphomycetes are good candidates for inundative biological control as they can be mass produced on inexpensive artifical media and produce infective asexual spores (conidia) that have good shelf lives (Hajek and St. Leger, 1994; Lacey et al., 2001). Despite these attributes, a decade ago only ten had reached the commercial development stage (Hajek and St. Leger, 1994). The entomopathogenic hyphomycetes most commonly investigated for use against insect and mite pests of tree fruit are *Metarhizium anisopliae* (Metsch.) Sorokin, *Beauveria bassiana* (Balsamo) Vuillemin, and *Lecanicillium* (=*Verticillium*) *lecanii* (Zimmermann) Viegas. The requirement for high humidities and moderate temperatures for spore germination and development limits the application of fungi in hot, dry conditions. Nevertheless, opportunities exist to use fungi as microbial control agents against several pests of tree fruit.

Microsporidia

Microsporidia are obligate intracellular eukaryotic parasites that probably play an important role in regulating populations of many insects (Solter and Becnel, 2000; Lacey et al., 2001). Although traditionally included among the protozoa, recent evidence indicates species in this group are highly evolved intracellular fungi (Keeling and Fast, 2002). Microsporidia have complex life cycles, although most produce environmental spores at the end of their replication cycle that must be ingested to infect new hosts. Spores contain a long coiled polar tube that penetrates the membrane of host cells and injects the infective sporoplasm. Although microsporidia have been recovered from a wide range of insects, most are host specific, slow acting (causing chronic diseases), and difficult to mass produce. As a consequence, they are rarely investigated as biological control agents of tree fruit pests.

COLEOPTERA

Citrus Root Weevils

One success story in the development of microbial control strategies in citrus has been the use of EPNs against root weevils (Coleoptera: Curculionidae), especially the diaprepes root weevil (*Diaprepes abbreviatus* L.) and blue-green citrus root weevils (*Pachnaeus* spp.). Economic damage is primarily caused by the larvae, which feed on roots. In addition to direct damage, such feeding may lead to secondary invasion of plant pathogens such as *Phytophthora* spp. (Duncan et al., 1999; McCoy, 1999). All immature root weevil stages are targeted by EPNs, which are often directly applied at the base of the trees through existing irrigation sprinklers.

In the Florida citrus industry inundative releases of EPNs to control root weevils have been of interest to commercial producers since the late 1980s. *Heterorhabditis bacteriophora* Poinar, *Heterorhabditis indica* Poinar Karunaker and David, *Steinernema carpocapsae* Weiser, and *Steinernema riobrave* (=*riobravis*) Cabanillas, Poinar, and Raulston have been marketed for root weevil control. Although *S. carpocapsae* was the first species shown to be pathogenic to *D. abbreviatus* and was briefly marketed as a bioinsecticide for root weevil control (Schroeder, 1987), its field performance was variable, and it did not always provide effective control at economic rates (Shapiro-Ilan, Gouge and Koppenhöfer, 2002). The market potential improved when a newly identified species, *S. riobrave,* was found to give consistently better control of *D. abbreviatus* and *Pachnaeus* spp. in field trials. High levels of *D. abbreviatus* control (i.e., >85 percent) could be achieved with moderately high application rates (i.e., >100 IJ/cm^2) (Schroeder, 1994; Duncan and McCoy, 1996; Duncan, McCoy and Terranova, 1996; Bullock, Pelosi and Killer, 1999). This encouraging research led to the commercial development of *S. riobrave,* which is now routinely used by many citrus growers in Florida as a component of IPM programs for control of *D. abbreviatus*. In addition to *S. riobrave, H. bacteriophora* and *H. indica* have been commercialized for use against *D. abbreviatus* on the basis of efficacy reported in laboratory or field studies (Downing, Erickson and Kraus, 1991).

The use of EPNs in the Florida citrus industry illustrates how biological, ecological, and economic factors contribute to the commercialization of microbial control agents as bioinsecticides. First, *D. abbreviatus* is a major pest of citriculture throughout the state, and thus grower demand for control is high. Second, until 1998, no effective chemical pesticides were registered for control of weevil larvae in soil; thus EPNs faced little competition from other control agents (Duncan et al., 1999; McCoy, 1999). Third, field trials showed *D. abbreviatus* could be reliably controlled in the short term by *S. riobrave*. An important economic component was that EPNs need to be applied only under the canopy where weevils occur, thus eliminating the need to broadcast applications over the whole area, as required with crops such as turf. As most citrus groves are irrigated, EPNs can be applied via under-tree sprinklers, which also provide additional moisture important for the survival of IJ stages. Additional research that may improve the efficacy of EPNs for citrus weevil control includes study of the role of endemic species of EPNs in the citrus ecosystem, the influence of soil characteristics, and the timing, rate, and method of application (Duncan et al., 1999; McCoy, 1999; Shapiro et al., 1999; McCoy et al., 2000; Shapiro et al., 2000; McCoy et al., 2002). Newly discovered strains of *S. riobrave* and another species, *S. diaprepesi* sp. nov., recently isolated from *D. abbreviatus,* also have potential to improve biological control of root weevils (Nguyen and Duncan, 2002; Stuart et al., 2004).

Weathersbee et al. (2002) tested a commercial preparation of the *Bt* var. *tenebrionis* against *D. abbreviatus*. In trials with artificial diet and potted citrus, larvae showed delayed but rate-dependent mortality. However, LC_{50} rates were high (6.2-25.4 ppm active ingredient), indicating that controlling larvae in the soil would not be economical.

Pecan Weevil

The pecan weevil, *Curculio caryae* Horn (Coleoptera: Curculionidae), is a major pest of pecans in the southeastern United States as well as parts of Texas, Oklahoma, and Kansas (Sikorowski, 1985; Harris, 1999). Both larvae and adults feed inside the developing nuts over a two- or three-year life cycle. In most years, if adult weevils are detected in the orchard but left untreated, economic damage occurs. To determine the need for insecticide treatments, weevil emergence from the soil is monitored. As an alternative, microbial control of *C. caryae* can be directed toward the larvae after it has dropped from the nut or against emerging adults.

Among pathogens, fungi have received attention as biological control agents of *C. caryae*. Most studies have focused on strains of *B. bassiana* and *M. anisopliae* that have been isolated from coleopteran hosts. In a pecan orchard, Harrison, Gardner and Kinard (1993) assessed "barrier treatments" of *B. bassiana* (three strains) targeting larvae pupating in the soil after emerging from nuts. Mortality of larvae 30 days after placement on 1 m^2 soil plots pretreated with conidial suspensions was low (2-34 percent), although the reported dose was very low (1.2×10^8 conidia/ha or 1.2 conidia/cm^2). Gottwald and Tedders (1983) also reported low mortality rates of *C. caryae* larvae exposed to *B. bassiana* (29.6 percent) and *M. anisopliae* (6 percent) at rates of up 10^7 conidia/g soil under field conditions. Higher rates of mortality were achieved when adult weevils were exposed to *B. bassiana* and *M. anisopliae* under greenhouse conditions (72 and 50 percent mortality, respectively). Gottwald and Tedders (1984) showed that *B. bassiana* that colonized *C. caryae* larvae could spread in the surrounding soils and infect new weevils up to 14 months after primary larvae were infected. Research thus far indicates that improved strains and field persistence of fungi may be needed to achieve satisfactory levels of weevil suppression (Shapiro-Ilan et al., 2003).

Initial tests with EPNs were substantially less successful against larvae of *C. caryae* than against *D. abbreviatus* (see section "Citrus root weevils"), with less than 35 percent control in field and greenhouse studies (Nyczepir, Payne and Hamm, 1992; Smith et al., 1993; Shapiro-Ilan, 2001b). Shapiro-Ilan (2001b) concluded that without improvements, suppression of *C. caryae* with EPNs is unlikely to be cost effective. However, in further studies, it was discovered that adult weevils were more susceptible than larvae to EPNs (Shapiro-Ilan, 2001a; Shapiro-Ilan, Stuart and McCoy, 2003). In laboratory studies, *S. carpocapsae* caused an average of 99 percent mortality of adult *C. caryae* (Shapiro-Ilan, 2001a). Supporting field trials are required to confirm the effectiveness under operational conditions. A possible strategy to infect adult weevils is to apply EPNs in an absorbent band placed around infested pecan trees to treat the soil during the period that adult weevils are emerging.

Plum Curculio

Native to eastern North America, the plum curculio, *Conotrachelus nenuphar* Herbst. (Coleoptera: Curculionidae), is a problem in commercial pome and stone fruit orchards, attacking plums, apples, cherries, peaches, and other stone fruits (Racette et al., 1993; Vincent, Chouinard and Hill, 1999). The adult weevil overwinters under ground litter, in woodpiles, and in other protective sites adjacent to orchards and becomes an early-season pest by feeding on the buds, flowers, leaves, and young fruit. Female curculios make a characteristic crescent-shaped cut in fruit in which they lay a single egg. In apple orchards, up to 85 percent weevil damage on fruit at harvest has been recorded in the absence of chemical treatments (Racette et al., 1993; Vincent, Chouinard and Hill, 1999). Larvae develop in the fruit and after a ten- to sixteen-day developmental period drop to the ground, where they pupate in the soil. New adults emerge after two to three weeks and produce eggs the following season.

Several studies have examined microbial control of *C. nenuphar* under laboratory conditions. Tedders et al. (1982) tested *S. carpocapsae* and several isolates of *M. anisopliae* and *B. bassiana*. Results were highly variable depending on exposure method and the isolates of fungi tested. One isolate of *M. anisopliae* (South Carolina) caused up to 100 percent mortality of *C. nenuphar* within two weeks after larvae were exposed to agar plate cultures or conidial suspensions pipetted on soil surfaces. *S. carpocapsae* was included in one test but was ineffective under the conditions tested (i.e., Petri-dish assays; Tedders et al., 1982). However, subsequent studies showed *C. nenuphar* was highly susceptible to infection by four strains of *Steinernema* spp. under such conditions (Olthof and Hagley, 1993). In further tests, *S. carpocapsae* strain All applied to a 5-cm layer of soil at 100 IJ/cm^2 killed 88-91 percent of larvae. When the same rate of nematodes was applied to natural sods (removed from an orchard floor), larval mortality rates were slightly lower (73-76 percent).

Unfortunately, few studies have quantified the feasibility of using fungi and EPNs to control *C. nenuphar* under realistic conditions. In Canada, Bélair, Vincent and Chouinard (1998) attempted to prevent weevil damage by targeting early-season adult *C. nenuphar* with broadcast sprays of *S. carpocapsae*. Three or four seasonal applications were broadcast to both foliage and soil at a rate of 50-80 IJ/cm^2. Applications were made initially at fruit set and subsequently when a threshold of 1-2 percent curculio damage was reached. Although damage at harvest was reduced in the first year of trials, nematodes did not prevent significant weevil damage in subsequent trials. The authors suggest that the rapid desiccation of IJ stages in the foliage was a problem in most years and that soil applications timed against the initial seasonal emergence of weevils under tree canopies may be more effective (Bélair, Vincent and Chouinard, 1998). Shapiro-Ilan et al. (2004) used this latter approach (soil applications) to assess the ability of *Steinernema feltiae* and *S. riobrave* to control *C. nenuphar* larvae in soil in two peach orchards. In these trials *S. feltiae* was not effective, but *S. riobrave* applied at 100 IJ/cm^2 reduced the emergence of adult weevils (in baited cages) by at least 97 percent in two out of three trials. The authors concluded that timing applications to accord with pest phenology and environmental conditions is critical to achieving good control.

Banana Weevil

Native to southeast Asia, banana weevil borer, *Cosmopolites sordidus* Germar, has become established in banana- and plantain-producing regions throughout the tropics and subtropics (Tinzaara et al., 2002). Several research programs have tested fungi and EPNs for *C. sordidus* control. Several isolates of *B. bassiana* and *M. anisopliae* may be highly pathogenic to *C. sordidus,* killing up to 100 percent of insects exposed to fresh conidial suspensions under laboratory conditions (Kaaya et al., 1993; Schoeman and Schoeman, 1999; Khan and Gangapersad, 2001). Significant but more moderate control is reported in field tests in which fungi are applied around the base of banana pseudostems, where weevils cause most damage. For example, in Costa Rica, *B. bassiana* applied to traps in pseudostems (5.8×10^{10} conidia/trap) caused 31-63 percent *C. sordidus* mortality in ten to thirteen days (Carballo and Lopez, 1994). In Uganda, a maize-based formulation of *B. bassiana* at a high rate (2×10^7 conidia/cm^2) reduced the weevil population by 63-72 percent within eight weeks of a single application (Nankinga and Moore, 2000). Field studies also report transmission of fungi between infected and healthy *C. sordidus,* suggesting that prolonged or between-season control may be possible after a single fungal application.

EPNs have also been evaluated for efficacy against *C. sordidus*. In greenhouse tests *S. carpocapsae, Steinernema glaseri,* and *S. feltiae,* applied at a rate of 4,000 and 40,000 IJ per plantain corm, reduced the number of tunnels made by *C. sordidus* and caused 100 percent larval mortality (Figueroa, 1990). In small-scale field trials in Brazil, *S. carpocapsae* All and UK strains sprayed at 5×10^6 IJ/ml in 0.4 L water onto split pseudostems and pseudostem stumps killed 70 percent of adult *C. sordidus* that were recovered from traps seven days after treatment and up to 40 percent from fourteen to twenty-one days after treatment (Schmitt, Gowen and Hague, 1992). However, in Australia, use of EPNs for banana weevil control was not considered a successful strategy after three years of plantation trials. Suspensions of *Heterorhabditis zealandica* Poinar and *S. carpocapsae* that were applied bimonthly in 200-mm-deep incisions in the residual rhizomes of harvested plants did not maintain beetle catches and damage ratings below the action thresholds that the authors proposed for the region (Smith, 1995).

Coconut Palm Rhinoceros Beetle

The coconut rhinoceros beetle, *Oryctes rhinoceros* L. (Coleoptera: Scarabaeidae), is a major pest of coconut and oil palms in the South Pacific and southeast Asia (Zelazny, Lolong and Pattang, 1992; Jacob, 1996; Gopal et al., 2001). Adult beetles bore into the heart of palm trees, and severe infestations may kill the tree. After a number of failed attempts to control this pest, a nonoccluded virus originally from Malaysia was discovered to be a good classical biological control agent. Although previously included in the baculoviruses, on the basis of molecular analysis and its specificity for scarabs, the virus has now been transferred to its own family (Evans, 1997). Infected hosts are not rapidly killed but suffer reduced fecundity and longevity (Zelanzy et al., 1992). Once established, the virus is maintained naturally in localized populations (through transmission to adults feeding on contaminated foliage and to larvae in feeding galleries and breeding sites) and seems to prevent further outbreaks of the weevil. As the virus is less stable in the environment and because most coconuts are produced on smallholdings with limited resources, the strategy adopted has been to release infected beetles to induce virus epidemics in recently invaded areas (Jacob, 1996; Hunter-Fujita et al., 1998; Gopal et al., 2001). Results documenting the effect of beetles before and after virus introductions have been striking. For example, in the Andaman Islands (India) in four locations where *Oryctes* baculovirus (OBV-KI) was released, damage to coconut palms was reduced by approximately 90 percent within forty-three months, with 60 percent of adults showing symptoms of infection thirty months after virus release (Jacob, 1996).

Longhorn Beetles

Various beetle larvae seriously damage or kill fruit trees by boring into the trunk and main branches. In addition to causing direct injury to the tree, the boreholes produced by larval mining provide pathways for infection of tree pathogens. However, compared with bark or leaf surfaces, larval boreholes also provide a favorable environment for EPNs.

The Asian longhorn beetle, *Anoplophora glabripennis* Motchulsky (Coleoptera: Cerambycidae), is an invasive pest in the United States and Canada that attacks a variety of fruit trees in its native range

(Appleby, 1999). In the United States, research on biological control of this species has included EPNs, which have been shown to be active under laboratory conditions (Solter et al., 2001; Fallon et al., 2004). In China, the Beijing strain of *S. carpocapsae* injected into insect holes or applied with sponge plugs at 7.5×10^3 IJ/ml killed 86.4 percent of *A. glabripennis* larvae (Xia, Qing and Feng, 1998). There are additional reports from China of EPNs being used to control other cerambycid beetles. Xu et al. (1995) applied preparations of *S. carpocapsae* Agriotes strain to control *Aristobia testudo* Voet infecting litchi. Nematodes injected into the top bore holes on damaged twigs in 20 ha of litchi trees killed 73.3-100 percent of larvae. The injection method appears relatively effective for longhorn beetle control, although locating the larval boreholes requires effort. Although twenty-eight of thirty-five larvae of *Rhytidodera bowringii* White inoculated with a 2 ml suspension of *S. carpocapsae* were killed in a small-scale test in mango, this technique was not considered practical for large-scale control in the field (Shen and Han, 1985). However, in follow-up tests, spraying *S. carpocapsae* on 260 ha of litchi plants reportedly killed 85-100 percent of another species, *Arbela dea* (Xu et al., 1997).

HOMOPTERA

Aphids and Psylla

Aphids and psylla (Homoptera: Aphididae, Psyllidae) are common orchard pests that left unchecked can quickly reach economic thresholds (Beers et al., 1993). Owing in part to these pests' piercing-and-sucking feeding strategy and foliar habitats, entomopathogenic fungi offer the most promise as naturally occurring and applied pathogens for these groups (Latgé and Paperiok, 1988; Lacey, Fransen and Carruthers, 1996; Milner, 1997). However, a deoxyribonucleic acid virus, belonging to the subfamily Densovirinae, was recently reported from the green peach aphid, *Myzus persicae* Sulzer (van Munster et al., 2003).

Entomophthoralean fungi, including *Pandora (=Erynia) neoaphidis, Entomophthora planchoniana,* and *Zoophthora (=Erynia) radicans,* are the major aphid pathogens (Latgé and Paperiok, 1988;

Milner, 1997; Feng and Chen, 2002). Although such fungi frequently cause epizootics at high pest densities, attempts to develop them as biopesticides have largely failed because of problems with mass production, the fragility of the conidia, and the need for suitably moist conditions (Milner, 1997). However, several hyphomycete fungi, which are inexpensive to mass produce and can be more easily developed as mycoinsecticides, have been tested against aphids. Laboratory studies have confirmed aphid-pathogenic strains of *M. anisopliae, B. bassiana, Paecilomyces fumosoroseus,* and *L. lecanii* (Butt et al., 1995; Liu et al., 1999; Poprawski, Parker and Tsai, 1999; El Salam, 2001; Xu et al., 2003; Yeo et al., 2003). A formulation of *L. lecanii,* Vertalec, was marketed by Koppert in Europe for aphids in greenhouse crops.

Unfortunately, few studies have assessed microbial control of aphids in tree fruit. One encouraging example concerns the brown citrus aphid, *Toxoptera citricida* Kirkaldy, a worldwide pest of citrus. Poprawski, Parker and Tsai (1999) tested the mycoinsecticide Mycotrol ES, containing *B. bassiana* GHA strain, against *T. citricida* in Florida citrus groves. Replicated field trials showed foliar applications at 2.5×10^5 and 5×10^5 conidia/cm^2 provided comparatively rapid mortality and caused 80 and 94 percent nymphal mortality, respectively. Two to three spray applications at three-day intervals early in the flush cycle was recommended as a strategy that could be combined with other aphid control measures (Poprawski, Parker and Tsai, 1999). As a different approach, it may be possible to inoculate fungi into aphid populations using "lure and infect"-type autodissemination traps (Hartfield et al., 2001).

The pear psylla, *Cacopsylla pyricola* Foerster, is a significant pest in the Pacific Northwest (Beers et al., 1993). Puterka, Humber and Poprawski (1994) reported good activity of several hyphomycete fungi against *C. pyricola* under laboratory conditions. In follow-up tests, Puterka (1999) did not achieve high levels of control in orchard trials. Formulated and unformulated *B. bassiana* and *P. fumosoroseus* applied at 5.4×10^5 conidia/cm^2 resulted in 18-37 percent nymphal mortality, which would be inadequate as a standalone strategy and expensive for such a low level of control.

The requirement for surface moisture for survival and movement limits the prospects for EPNs as biological control agents for foliar pests such as aphids. In apple orchards Brown et al. (1992) tested *S. carpocapsae* on soil-dwelling stages of the woolly apple aphid, *Eriosoma lanigerum* Hausmann. Neither broadcast sprays nor top-dressing applications at a fairly low dosage (38 IJ/cm^2) significantly reduced aphid populations compared with leaving trees untreated or using systemic aphicide.

Whiteflies

Entomopathogenic fungi have received considerable attention as biological control agents of whiteflies (Homoptera: Aleyrodidae). Much research, including significant foreign exploration for strains of fungi that have evolved with their hosts, stemmed from international programs set up in response to the global outbreaks of highly polyphagous biotypes of *Bemisia* spp. in the 1990s (Lacey, Fransen and Carruthers, 1996). Earlier research had been conducted into the potential of fungi to control the greenhouse whitefly, *Trialeurodes vaporariorum* Westwood, and to a lesser extent other whitefly species (Fransen, 1990; Hajek, Wraight and Vandenberg, 2001). As a result of this research, several Deuteromycete fungi (initially *L. lecanii* and more recently *P. fumosoroseus* and *B. bassiana*) received attention for commercial development against whiteflies in crops, including cucurbits, tomatoes, and cotton. Reviews on the utilization of fungi for control of *Bemisia* spp. and *T. vaporariorum* are provided elsewhere (Lacey, Fransen and Carruthers, 1996; Fransen, 1990; Faria and Wraight, 2001).

Less attention has been given to fungal pathogens of whitefly that are more commonly associated with tree fruit. However, there are reports of species of *Aschersonia* naturally controlling or being applied against outbreaks of *Dialeurodes* spp. in citrus groves in Florida, the Black Sea area, and China (Fawcett, 1944; Ponomarenko et al., 1975; Gao et al., 1985). The high ambient relative humidity found in citrus groves may promote infection and sporulation on the host, and rains may disperse the conidia. In some cases, natural epizootics of *Aschersonia* spp., which may be encouraged by the judicious use of fungicides, have been linked to reduced use of pesticides for whitefly control. Furthermore, foreign isolates of *Aschersonia* were successfully introduced into Georgia and Azerbaijan to help control *D. citri* outbreaks after it became an invasive pest (Lacey, Fransen and Carruthers, 1996). Although *Aschersonia* spp. grow on solid media, they do not sporulate in liquid culture, which may have limited mass production and commercialization of the fungus.

LEPIDOPTERA

Codling Moth

The codling moth, *C. pomonella* (Tortricidae), is a significant pest of apples and, to a lesser extent, pears and walnuts worldwide (Barnes, 1991; Beers et al., 1993; Cross et al., 1999). For reasons including the development of larvae inside the fruit and insecticide resistance, effective control of this pest is challenging. Stages that are susceptible to microbial control are the neonate larvae before they enter the fruit, the full-grown larvae after they exit the fruit, and the cocooned prepupae and pupae. Although bacteria, fungi, and microsporida have been recovered from codling moth (Falcon and Huber, 1991; Siegel, Lacey and Vossbrinck, 2001), the most promising agents for microbial control in orchards appear to be its GV and EPNs.

The GV of *C. pomonella* (CpGV) has generated extensive interest as an inundative microbial control agent. The virus is applied by spraying orchards with approximately 10^{13} granules/ha at seven- to fourteen-day intervals (Lacey, Knight and Huber, 2000). Neonate larvae ingest virus granules before or during initial entry into fruit and die before penetrating deeply into fruit. Over the past thirty years, numerous field trials have demonstrated good activity of CpGV in a variety of settings across Europe, South Africa, Australia, New Zealand, South America, and North America (Huber, 1986; Falcon and Huber, 1991; Vail et al., 1991; Jaques et al., 1994; Guillon and Biache, 1995; Cross et al., 1999; Arthurs and Lacey, 2004; Lacey et al., 2004). CpGV's specificity for codling moth and some closely related species and its safety with respect to nontarget organisms have been thoroughly documented (Gröner, 1986). The use of CpGV may therefore contribute significantly to the conservation of nontarget predacious and parasitic arthropods that suppress secondary pests in the orchard. Despite several limitations, notably a short residual activity and slow speed of kill, commercial CpGV products have been routinely used for control of codling moth in European orchards since 1988 (Cross et al., 1999; Lacey and Shapiro-Ilan, 2003), but have only recently become commercially available in the United States. However, in 2003 several organic orchardists in the Pacific Northwest of the United States used CpGV with good success (Arthurs and Lacey, 2004).

Research also indicates good control potential for EPNs (*Steinernema* spp. and *Heterorhabditis* spp.) for codling moth when adequate moisture is maintained and temperatures are above 10-15°C (Kaya et al., 1984; Nachtigall and Dickler, 1992; Lacey and Unruh, 1998; Lacey, Knight and Huber, 2000; Unruh and Lacey, 2001). One strategy involves targeting the synchronous overwintering larvae that locate hibernacula on bark surfaces. In California, Kaya et al. (1984) tested the field effectiveness of *S. carpocapsae* All strain against diapausing (winter) and nondiapausing (summer) larvae in cardboard bands placed around apple tree trunks. Control was good in February, with 95 percent mortality (58-73 percent infected with nematodes) when bands remained damp from rainfall and daytime temperatures were above the threshold for IJ activity, but decreased markedly in July applications (0-23 percent infection), when conditions were hot and dry. In Germany, Nachtigall and Dickler (1992) adopted a similar approach but used an adsorbent sponge-rubber material as "tree collars" that maintained moisture for a longer period. In orchard tests during dry conditions (50 percent relative humidity) codling moth mortality rates were higher in collars (82-89 percent) than in the tree bark (52-58 percent) when the whole trunk was sprayed with IJ. Applying a collar impregnated with EPNs as a pupation lure should be commercially feasible, but fixing and removing collars may be time consuming. It also remains unclear what proportion of larvae migrate to artificial pupation sites. However, more recent research indicates that supplemental wetting may enable full bark sprays of *S. carpocapsae* to be effective for codling moth control under various environmental conditions. Unruh and Lacey (2001) demonstrated that lightly irrigating treated areas both 0.5 hours before and up to 24 hours following treatment allowed nearly 100 percent infection of codling moth sentinel larvae or pupae in cardboard strips or logs attached to tree surfaces, compared with approximately 80 percent and 50 percent when irrigating only once or not at all. In recent trials with *S. carpo-*

capsae and *S. feltiae,* good rates of control (>90 percent) were achieved when IJs were applied to control codling moth overwintering in rough bark surfaces in pear orchards and also within various mulches around the base of trees (Lacey et al., unpublished data). In both cases, plots were irrigated before and for several hours after application.

In addition to orchard treatments, EPNs can be used to control codling moths that infest the wooden fruit bins during harvest and comprise a source of reinfestation of orchards the next year. To do this, IJs can be added to the water in the drenchers used to apply fungicides on the fruit bins or the dump tanks used for removal of fruit in packing houses. Using such approaches, more than 80 percent mortality of sentinel late-instar codling moth larvae was achieved with *S. carpocapsae* at rates of 5-50 IJ/ml water, provided high humidity and moderate temperature were maintained for several hours after treatments (Lacey and Chauvin, 1999; Cossentine, Jensen and Moyls, 2002).

Leafrollers and Other Tortricids

Several species of tortricids construct concealed habitats within leaves and other structures and are collectively referred to as leafrollers (Barnes, 1991). In the United States, some leafrollers have become more serious pests or pest complexes in pome fruit orchards since the adoption of mating disruption for codling moth (Gut and Brunner, 1998; Walker and Welter, 2001). Historically, *Bt* has been widely used against leafrollers and other tortricids in pome and other tree fruit (Knight et al., 1998; Cross et al., 1999; Lacey, Knight and Huber, 2000; Lacey and Shapiro-Ilan, 2003). Various factors, including application timing, coverage, dose, and strain used influence the efficacy of *Bt* for control of tortricids. For example, studies conducted in Washington state have shown that *Bt* applications are best applied at high rates early in the season and repeated as necessary to compensate for short residual activity and rapid foliar growth (Brunner, 1994; Knight, 1997). Knight et al. (1998) reported different susceptibilities among common leafroller species to nine formulated preparations of *Bt* Cry1 endotoxins. In Canada, a higher tolerance to *Bt* was documented among leafrollers in organic than in conventionally managed orchards, possibly reflecting a higher rate of use in organic systems (Smirle et al., 2003).

A number of baculoviruses, including several GVs and NPVs, have been reported from leafrollers in recent years (Goyer et al., 2001; Markwick et al., 2002; Pronier et al., 2002; Pfannenstiel et al., 2004). Natural virus outbreaks have been documented. For example, in the Pacific Northwest, collections of the leafroller *Pandemis pyrusana* Kearfott over three years in an unsprayed apple orchard revealed seasonal peaks of infections (up to 67 percent) by a GV (PpGV) (Pfannenstiel et al., 2004). Nevertheless, the need for multiple virus strains against pest complexes and the cost of developing viruses within living hosts (*in vivo* production) have limited the commercial adoption of insecticidal viruses for leafroller control (Cross et al., 1999; Lacey et al., 2001). However, the GV of *A. orana* (AoGV) has been researched extensively. AoGV was marketed by Andermatt Biocontrol in Europe under the tradename Capex (Luisier and Benz, 1994).

The NPV of the celery looper, *Anagrapha falcifera* Kirby (AnafaMNPV), has a broad host range within the Lepidoptera and was registered in the United States in 1996, although it was not commercially distributed. In the laboratory, Lacey, Vail and Hoffmann (2002) tested the activity of AnafaMNPV against *C. pomonella, P. pyrusana,* and two other tortricids, the oblique-banded leafroller (*Choristoneura rosaceana* Harris) and oriental fruit moth (*Grapholita molesta* Busck), pests of peaches, cherries, and other fruit (Beers et al., 1993). *C. pomonella* and *G. molesta* were significantly more susceptible to AnafaMNPV than the two leafrollers, with LC_{50} of 1.8 and 2.1×10^3 OB/mm^2 on artificial diet compared with 4.3 and 7.6×10^3 OB/mm^2 for *P. pyrusana* and *C. rosaceana,* respectively.

Several species of mermithid and steinernematid nematodes have been recovered from tortricids (Poinar, 1991). The main limitation of nematodes for their control is that the infective stages rapidly desiccate and die in foliar habitats, where the larvae of most species develop. For example, after encouraging laboratory assays, Bélair et al. (1999) attempted to control *C. rosaceana* in a Canadian apple orchard using foliar applications of *S. carpocapsae* All strain. Despite overcast conditions after the first trial, few larvae feeding within the canopy became infected (13-37 percent). However, as described earlier for codling moth, the prospects may be better for controlling species of tortricids that have feeding or diapausing stages in cryptic habitats close to the ground, since this provides more favorable conditions for EPNs. For example, overwintering *G. molesta* construct similar hibernacula to codling moth. Preliminary laboratory bioassays indicate *G. molesta* is susceptible to *S. carpocapsae* (60-80 percent mortality after exposure to 10 IJ/cm^2) and possibly *S. riobrave* and *Heterorhabditis marelatus,* provided high moisture and temperatures appropriate for activity can be maintained for several hours after application (Riga and Lacey, unpublished data). The cherry bark tortrix, *Enarmonia formosana* Scopoli, is a European species that was introduced into the Pacific Northwest within the past fifteen years and poses a considerable threat to the sweet and ornamental cherry industry. Larvae feed on phloem tissue in galleries beneath the bark and may kill trees in a relatively short period through feeding and facilitating plant pathogens such as bacterial canker. Nematodes sprayed around graft joints enter the external opening of feeding galleries and infect *E. formosana* larvae found inside (Lacey and Murray, unpublished data). Further studies on the feasibility of EPNs for control of overwintering stages of various tortricids species are warranted.

Navel Orangeworm

Navel orangeworm, *Amyelois transitella* Walker (Pyralidae), is a primary pest of almonds in California, although it also attacks figs, pistachios, and walnuts (Bentley et al., 2000). Larvae infest mature nuts on the tree and nut mummies on the tree and ground and may produce three to four adult flight periods per year. Conventionally, *A. transitella* is controlled through chemical insecticides, combined with orchard sanitation. Sanitation involves the mechanical destruction of infested nut mummies after harvest. Environmental concerns over such approaches include pesticide residues and air pollution resulting from disking, blowing, and flail mowing (J. Siegel and L. Lacey, personal observation).

One strategy for microbial control of *A. transitella* is to target infested nut mummies using high-volume sprays of EPNs. In a Californian almond orchard, Agudelo-Silva et al. (1995) sprayed overwintering larval populations of *A. transitella* in nuts left on trees after harvest with *S. carpocapsae* All strain and a cold-tolerant *Heterorhabditis* species. Although the population of *A. transitella* was reduced significantly (11.8 percent mortality), the authors did not conclude that this approach was economically viable. A better strategy might be to apply EPNs against larvae in fallen nuts, where IJs are less susceptible to desiccation and may be more persistent and efficacious. Using this latter approach in a pistachio orchard, Siegel et al. (2004) demonstrated *A. transitella* larvae were highly susceptible to *S. carpocapsae* and moderately susceptible to *S. feltiae* applied in well-irrigated plots. Spraying *S. carpocapsae* at 10^5 and 10^4 IJ/m^2 in 1 m^2 sandy loam plots resulted in more than 94 percent and 72 percent mortality of larvae in pistachios and almonds, respectively, when nighttime temperatures were above freezing.

Among viruses, AnafaMNPV is pathogenic to *A. transitella* neonates under laboratory conditions (Vail et al., 1993; Cardenas et al., 1997). Further research is needed into the production characteristics of the AnafaMNPV virus and its potential against *A. transitella* under field conditions.

Wood-Boring Lepidoptera

Wood-boring larvae are good candidates for EPNs (see "Longhorn Beetles"). Application strategies for controlling wood borers are bark surface treatments and direct gallery injections. In bark surface treatments, IJs are sprayed over the entire trunk and may be concentrated around heavily infested areas. Direct gallery injections apply IJ suspensions to gallery openings with a stream nozzle or squirt bottle.

Several studies report encouraging results from using EPNs against the clearwing moths, *Synanthedon* spp. (Sesiidae), infesting pome and stone fruits. Cossentine, Banham and Jensen (1990) demonstrated that spraying the base of trunks with *H. bacteriophora* in peach orchards in Canada reduced the emergence of adult clearwing peachtree borers, *Synanthedon exitiosa* Say, by 80 percent up to 90 days after treatment. In similar studies, moderate control of *S. exitiosa* larvae (66 percent) was achieved after spraying cherry laurel with *S. carpocapsae* (Gill, Davidson and Raupp, 1992). In Europe, damaging stages of the apple clearwing, *Synanthedon myopaeformis* Brkh., have been targeted with trunk sprays of EPNs. Fall and spring applications of *S. carpocapsae* (three strains) and *S. feltiae* (two strains) were evaluated in north Italian (Deseö and Miller, 1985) and German orchards (Nachtigall and Dickler, 1992). Results obtained during cool, humid conditions showed *S. myopaeformis* removed from treated tree trunks and main branches had infection levels of 74-94 and 64-87 percent, respectively. Both studies reported infection rates of less than 20 percent when IJs were applied at low rates or, during hot, dry conditions, or when *H. bacteriophora* were applied.

In Australia, a locally adapted strain of *S. feltiae* was successfully used to control currant clearwings, *Synanthedon tipuliformis* Clerk, in cane cuttings and established field plantings of blackcurrants, *Ribes nigrum* L. Since many blackcurrant infestations are thought to originate from planting stock, Bedding and Miller (1981) investigated the feasibility of disinfesting the 0.5-1 m lengths of year-old woodcuttings before planting. Spraying canes and enclosing them with damp paper in polyethylene bags at 22°C for two days to maximize penetration resulted in up to 99 percent mortality of diapausing and nondiapausing *S. tipuliformis*. Miller and Bedding (1982) further showed that 68-90 percent of *S. tipuliformis* in established blackcurrant bushes could be controlled with field sprays at the early grape stage. It was concluded that using EPNs to control *S. tipuliformis* in blackcurrants is a feasible strategy.

In contrast to the above examples, attempts to disinfest stone fruit trees (tart cherry and peach) of the American plum borer, *Euzophera semifuneralis* Walker (Pyralidae), using commercial formulations of *S. feltiae* and *H. bacteriophora* were unsuccessful (Kain and Agnello, 1999). In two years of trials, both coarse sprays to trunks and lower scaffold branches and direct gallery injection were ineffective compared with systemic insecticides. It was suggested that tree gum secreted in feeding galleries may have resulted in inadequate contact of the nematodes with the target host. Other studies have demonstrated the efficacy of *Bt* var. *kurstaki* against the peach twig borer, *Anarsia lineatella* Zeller (Barnett et al., 1993), and peach fruit borer, *Carposina niponensis* Wals. (Lu et al., 1993).

Leafminers

Some dipteran and lepidopteran larvae mine within the leaf tissue. Because of their cryptic habitats, these leafminers are often protected from nonsystemic insecticides and pathogens on the leaf surfaces.

Theoretically, the mine protects against desiccation and solar radiation, representing a suitable habitat for EPNs. Foliar sprays of EPNs may allow the IJs to enter leaf tissue via oviposition holes, feeding scars, or tears in the leaf surface (Harris, Begley and Warkentin, 1990; Lebeck et al., 1993) and locate and infect larvae or early-stage pupae by following chemical trails (Schmidt and All, 1979; Gaugler et al., 1980). This approach to controlling leafmining flies (Diptera: Agromyzidae) has generated interest in covered ornamental and vegetable crops. Good rates of control (>80 percent larval mortality) have been achieved in some trials under conditions of high (>80 percent) humidity and moderate temperature on lettuce, cabbage, and tomato (Williams and MacDonald, 1995; Williams and Walters, 2000).

However, few studies have evaluated EPNs against leafminers in tree fruit. In Australia, Beattie et al. (1995) tested *S. carpocapsae* against the citrus leafminer, *Phyllocnistis citrella* Stainton (Lepidoptera: Gracillariidae). In two field studies, the effects of nematode treatments on mortality were variable, but damage (percentage of mined leaves) was not reduced by treatments and nematodes were considered ineffective compared with several alternatives tested.

OTHER ARTHROPODS

Mites

The increasing economic importance of mites in a range of agricultural crops has stimulated attention in acaropathogens. Fungal pathogens, which have been recovered from at least five of the seven orders of Acari, appear to be the most promising microbial control agents (McCoy, 1996; Chandler et al., 2000; van der Geest et al., 2000). Most applied research on mite pathogens concerns fungal pathogens of phytophagous eriophyoids and spider mites. Because fungi may decimate mite populations, research has also considered factors leading to fungi epizootics of mites and ways to enhance natural pest control.

The most common phytophagous mites associated with tree fruit are rust mites, notably the citrus rust mite, *Phyllocoptruta oleivora* Ashmead (Acari: Eriophyidae) (Browning, 1999). A strain of *Hirsutella thompsonii* Fisher (class Hyphomycetes) has been shown to cause natural epizootics on *P. oleivora* and other eriophyids (McCoy, 1996). The commercial mycoacaricide Mycar (based on *H. thompsonii* mass produced in artificial media) was marketed in Florida by Abbot Laboratories for the control of eriophyoid mites in citrus. However, control was highly variable and dependent on environmental conditions, especially high humidity, and Mycar was withdrawn from sale in the 1980s (McCoy and Couch, 1982). As an alternative to direct application of the fungus, the isolation and application of toxins from *in vitro* grown *H. thompsonii* was found to be a possible alternative for *P. oleivora* control (Omoto and McCoy, 1998). In addition to applying *H. thompsonii* in response to a pest outbreak, the augmentation or inoculation of fungal pathogens is an option to be considered within mite IPM. In Mexico, Rosas-Acevedo and Sampedro-Rosas (2000) showed that *H. thompsonii* reduced populations of *Brevipalpus* spp. (Acari: Tenuipalpidae) and persisted in infested lemon groves for up to seven months after foliar application. In aerial monitoring of *P. oleivora* dispersal from Florida citrus groves, mites infected with *H. thompsonii* were captured in two out of three sites (Bergh and McCoy, 1997). In these studies, the highest capture of infected mites was followed by a decline in the number of mites caught in traps, suggesting that a fungal epizootic may have been responsible for the decline. There are also reports of entomopathogenic fungi attacking spider mites (Acari: Tetranychidae). Strains of *B. bassiana, M. anisopliae, L. lecanii,* and *P. fumosoroseus* (Wize) Brown and Smith have activity against *Tetranychus cinnabarinus* Boisd. and *Eutetranychus orientalis* Klein under laboratory conditions (Batta, 2003; Shi et al., 2004). Further work on the potential of Hyphomycetes for mite control is warranted.

There is evidence that fungi may be suitable for development as mycoacaricides for a range of pests, either as standalone treatments or for use in mite IPM. Further research is required to evaluate such fungi in field trials and to develop improved mass-production techniques (Chandler et al., 2000). Occasionally viruses, bacteria, and microsporidia are reported from mites, although van der Geest et al. (2000) suggest that manipulating these as biological control agents is impractical.

Fruit Flies

Fruit flies, particularly the Mediterranean fruit fly, *Ceratitis capitata* Wiedemann (Diptera: Tephritidae), are common pests of citrus and other fruit and often necessitate chemical interventions to protect crops (Smith, Beattie and Broadley, 1997). In recent years, several researchers have documented attempts at microbial control of tephritid pests using fungi and EPNs. Fungi in the class Hyphomycetes, particularly *M. anisopliae* and *B. bassiana*, have reportedly caused high mortality rates (>80 percent) of *C. capitata* and other tephritids under laboratory conditions (Castillo et al., 2000; Lezama-Gutierrez et al., 2000; Ekesi, Maniania and Lux, 2002, 2003; Dimbi et al., 2004). *M. anisopliae* caused lower mortality (15-68 percent) in the western cherry fruit fly, *Rhagoletis indifferens* Curran, exposed as teneral adults in soil (Yee and Lacey, 2005). Several authors suggest that inoculating soil under infested fruit trees with conidial suspensions against pupating larvae or emerging adults provides a novel alternative to chemical control for tephritids. However, the variable rates of control reported among the range of laboratory assays suggest that environmental conditions, a correct match of strain and host, timing applications to the most susceptible stage, and the use of additional control tactics will be important to the success of this approach under operational conditions. In a field-cage experiment *M. anisopliae* applied at 2.5×10^6 conidia/ml to sandy loam and loam soil reduced emergence of the adult Mexican fruit fly, *Anastrepha ludens* Loew, only moderately, by 22 and 43 percent, respectively (Lezama-Gutierrez et al., 2000).

Among EPNs, *Steinernema* spp., and to a lesser extent *Heterorhabditis* spp., have shown good efficacy against tephritid larvae in a range of laboratory tests. The best rates of control have been reported against larvae under moist conditions at rates between 50 and 5,000 IJ/cm^2. Under such conditions more than 80 percent larval and prepupal mortality has been obtained for *R. indifferens* (Yee and Lacey, 2003), Carribean fruit fly (*Anastrepha suspensa* Loew; Beavers and Calkins, 1984), *C. capitata* (Lindegren and Vail, 1986; Gazit, Rossler and Glazer, 2000; Laborda et al., 2003), and peach fruit fly (*Bactrocera zonata* Saunders; Attalla, Fatma and Eweis, 2002). Unfortunately, few studies have assessed EPNs against fruit flies under operational conditions. However, in two years of field trials in Hawaii, Lindegren, Wong and McInnis (1989) demonstrated 52-97 and 92-100 percent reductions in *C. capitata* adult emergence in soil treated with a high or very high rate (500 and 5,000 IJ/cm^2, respectively) of *S. feltiae* (=*bibionis*) Mexican strain. In Spain, Laborda et al. (2003) recorded 70 percent overall *C. capitata* larval mortality in field plots after treatment with a lower rate (50 IJ/cm^2) of *Steinernema* spp. The highest mortalities were observed in soil plots that were shaded, wet, or covered with mulches compared with those exposed to the sun, weed-covered plots, or bare soil. In Pacific Northwest cherry orchards, soil treatment of *R. indifferens* larvae with *S. feltiae* at 50 or 100 IJ/cm^2 produced mortalities of 77 and 86 percent (Yee and Lacey, 2003). The high pathogenicity of *S. carpocapsae, S. feltiae,* and *S. riobrave* against tephritid larvae and emerging adults and the abilities of IJs to persist in soil indicate that EPNs offer a nontoxic alternative to chemical soil treatments for fly control. However, these data show that nematode species and concentration combined with soil type and other environmental conditions influence the success of this approach. EPNs do not readily infect fruit fly puparia (Beavers and Calkins, 1984; Lindegren and Vail, 1986; Laborda et al., 2003; Yee and Lacey, 2003). Thus applications should be timed to coincide with larvae emerging from fruit and entering the soil before pupation or against emerging adults.

Thrips

Thrips are common pests on a variety of crops, including tree fruits (Lewis, 1997). Depending on the season, thrips may cause economic damage in leaves, fruit, and blossoms (Parker, Skinner and Lewis, 1995). The western flower thrips (WFT), *Frankliniella occidentalis* Pergande, develops on many weeds and low-growing vegetation inside and outside orchards and often becomes a pest of blossoms during fruit set. Female WFTs also cause oviposition scars on developing fruit. On apples, this damage is called "pansy spot."

The WFT spends about one-third of its developmental time as a prepupa or pupa in the ground, providing a window of opportunity for microbial control of soil stages. Several species of EPNs and fungi have activity against WFTs in a range of laboratory and greenhouse trials. For example, *S. feltiae, S. carpocapsae,* and *H. bacteriophora* controlled up to 80 percent of WFTs when applied at a concentration of 200-1,000 IJ/cm^2 against late-second instar larvae, pepupae, and pupae on the soil surface (Helyer et al., 1995; Ebssa et al., 2001a,b). In other studies, all WFT stages were susceptible to strains of *L. lecanii, M. anisopliae,* and *B. bassiana* (Schreiter et al., 1994; Helyer et al., 1995; Vestergaard et al., 1995; Butt and Brownbridge, 1997; Jacobson et al., 2001). In Europe, fungi among the entomophthorales have also induced epizootics among WFTs in greenhouses (Vacante, Cacciola and Pennisi, 1994; Montserrat, Castane and Santamaria, 1998), though these fungi cannot currently be manipulated as microbial pesticides. Reports on effective microbial control of WFTs in tree fruit are lacking, and chemical control remains the tactic of choice for WFT management by many growers.

In the Pacific Northwest, the pear thrips, *Taeniothrips inconsequens* Uzel, is an introduced species that has also become a common pest in cherry, pear, and apple. In a field survey in Vermont, Brownbridge et al. (1999) recorded fungal disease of *T. inconsequens* and found up to 12 percent infection in larvae that were recovered from the soil. The predominant fungi were *Lecanicillium* spp., *P. farinosus,* and *B. bassiana.* Further studies to characterize and test strains of fungi and EPNs and assess their potential in tree fruit IPM programs are warranted.

FUTURE PROSPECTS

Microbial Control and IPM

The commercial development and adoption of microbial control agents is most common within IPM programs where pesticides are not effective or cannot be used, such as with organic production. The advent of IPM has also produced a niche for microbial pesticides, which are often not as effective as chemical pesticides. This is because IPM frequently relies on the cumulative effect of several control strategies, each of which might not provide effective control independently (Dent, 1995).

There is potential for integrating microbial control with other soft technologies, including mating disruption with synthetic pheromones, cultural practices, and other reduced-risk pesticides, including insect growth regulators (IGRs), neonicotinals, botanicals, and fermentation byproducts such as spinosad and abamectin. In a few cases, microbial pesticides may work as standalone treatments. The judicious use of microbial control agents that are relatively specific to a key pest may preserve native beneficial species that can regulate secondary pests that otherwise might require treatment. Thus the adoption of microbial pesticides relies on cost-effectiveness and compatibility with other biological and chemical control agents, as well as overcoming specific limitations such as sensitivity to environmental conditions. The wider adoption of microbial pathogens to control tree fruit pests will require an increased awareness of these considerations by growers and the supporting industries. These subjects are discussed in more detail by several authors (Milner, 1997; Waage, 1997; Cross et al., 1999; Lacey et al., 2001).

Commercial Development

Many biopesticides are developed following the same procedures of design, testing, and registration as their chemical counterparts (Waage, 1997). There are problems with this approach. The high cost of pesticide registration places microbial pesticides, which often target small niche markets, at a considerable disadvantage. It has been argued that without some regulatory changes, biopesticides will not realize their full potential (Goettel and Jaronski, 1997; Waage, 1997; Lacey et al., 2001). Similarly, the evaluation of microbial pesticides is often solely from the perspective of their short-term efficacy and cost compared with alternatives, without allowing for the various advantages associated with their use (Lacey et al., 2001). To expect a slow-acting biological control agent to compete with a powerful, fast-acting synthetic chemical on these terms alone is unrealistic. Agroecosystems are variable and unpredictable, and resident pests or biological control agents may respond differently over space or time. However, in practice, studies with microbial agents are often conducted on spatial and temporal scales that do not allow for much of the inherent variability or longer-term population control to be observed. For example, sublethal effects on pest populations and beneficial effects on natural enemies may not be recognized. Many microbial pesticides are being used with little understanding of economic thresholds, optimal application strategies, or realistic expectations (Goettel and Jaronski, 1997; Waage, 1997; Lacey et al., 2001). Given that the efficacy of microbial pesticides depends more closely on the host/pathogen/environment relationship than is the case with chemical pesticides, benefits may accrue from testing agents at an operational scale over different regions and longer time periods to minimize misleading results and possible failure.

Research Priorities

The efficacy of most microbial control products is limited by their short activity, narrow host range, storage potential, and application challenges. Research to improve the efficacy of these products through strain selection, mass production, formulation technology, and development of optimal application strategies is underway at research institutions around the world. Additional strategies to increase the efficacy of entomopathogens include low level inoculation, synergism with pesticides, induction of epizootics, and environmental manipulation.

An interesting area that warrants further attention concerns novel attempts to disperse entomopathogens into populations of tree fruit pests by using pests as vectors, so-called autodissemination. Such techniques often rely on host-specific pheromones to lure vectors into traps, where they become contaminated with a pathogen and then disperse to spread infection to new individuals (i.e., horizontal or secondary transmission). Several attract-and-infect traps for dissemination of a variety of pathogens into insect pest populations are reviewed by Vega et al. (2000). Newly patented electrostatic dusts or powders may enhance the effectiveness of such techniques. Currently, traps utilizing specific pheromones mixed with electrostatic powders are marketed in Europe as a novel mating confusion technique for a number of species (Ian Baxter, personal communication). The inclusion of pathogens into the mix may enhance the effectiveness of such techniques for pest control.

REFERENCES

Agudelo-Silva, F., F.G. Zalom, A. Hom and L. Hendricks. (1995). Dormant season application of *Steinernema carpocapsae* (Rhabditida, Steinernematidae) and Heterorhabditis sp. (Rhabditida, Heterorhabditidae) on almond for control of overwintering *Amyelois transitella* and *Anarsia lineatella* (Lepidoptera, Gelechiidae). *Florida Entomologist* 78: 516-523.

Appleby, J.E. (1999). The pine shoot beetle and the Asian longhorned beetle, two new exotic pests. *Phytoprotection* 80: 97-101.

Arthurs, S.P. and L.A. Lacey. (2004). Field evaluation of commercial formulations of the codling moth granulovirus: persistence of activity and success of seasonal applications against natural infestations of codling moth in Pacific Northwest apple orchards. *Biological Control* 31: 388-397.

Attalla, A., A. Fatma and M.A. Eweis. (2002). Preliminary investigation on the utilization of entomopathogenic nematodes as biological control agents against the peach fruit fly, *Bactrocera zonata* (Saunders) [Diptera: Tephritidae]. *Egyptian Journal of Agricultural Research* 80: 1045-1053.

Barnes, M.M. (1991). Tortricids in pome and stone fruits, codling moth occurrence, host race formation and damage. In *Tortricid Pests, Their Biology, Natural Enemies and Control,* L.P.S. van der Geest and H.H. Evenhuis, eds. Elsevier Science Publishers, Amsterdam, The Netherlands, pp. 313-327.

Barnett, W.W., J.P. Edstrom, R.L. Coviello and F.P. Zalom. (1993). Insect pathogen "*Bt*" controls peach twig borer on fruits and almonds. *California Agriculture* 47: 4-6.

Batta, Y.A. (2003). Symptomatology of tobacco whitefly and red spidermite infection with the entomopathogenic fungus *Metarhizium anisopliae* (Metsch.) Sorokin. *Dirasat Agricultural Sciences* 30: 294-303.

Beattie, G.A.C., V. Somsook, D.M. Watson, A.D. Clift and L. Jiang. (1995). Field evaluation of *Steinernema carpocapsae* (Weiser) (Rhabditida: Steinernematidae) and selected pesticides and enhancers for control of *Phyllocnistis citrella* Stainton (Lepidoptera: Gracillariidae). *Journal of the Australian Entomological Society* 34: 335-342.

Beavers, J.B. and C.O. Calkins. (1984). Susceptibility of *Anastrepha suspensa* (Diptera: Tephritidae) to steinernematid and heterorhabditid nematodes in laboratory studies. *Environmental Entomology* 13: 137-139.

Bedding, R.A. and L.A. Miller. (1981). Disinfesting blackcurrant cuttings of *Synanthedon tipuliformis*, (Lepidoptera, Sesiidae) using the insect parasitic nematode, *Neoaplectana bibionis* (Nematoda, Steinernematidae). *Environmental Entomology* 10: 449-453.

Beegle, C.C. and T. Yamamato. (1992). History of *Bacillus thuringensis* Berliner research and development. *Canadian Entomologist* 124: 587-616.

Beers, E.H., J.F. Brunner, M.J. Willett and G.M. Warner. (1993). *Orchard Pest Management: A Resource Book for the Pacific Northwest,* Good Fruit Grower, Yakima, WA.

Bélair, G., C. Vincent and G. Chouinard. (1998). Foliar sprays with *Steinernema carpocapsae* against early season apple pests. *Journal of Nematology* 30: 599-606.

Bélair, G., C. Vincent, S. Lemire and D. Coderre. (1999). Laboratory and field assays with entomopathogenic nematodes for the management of oblique-banded leafroller *Choristoneura rosaceana* (Harris) (Tortricidae). *Journal of Nematology* 31: 684-689.

Bentley, W.J., R.E. Rice, R.H. Beede and K. Daane. (2000). *UC IPM Pest Management Guidelines: Pistachio Insects and Mites,* University of California ANR Publication 3461.

Benz, G. (1987). Environment. In *Epizootiology of Insect Diseases,* J.R. Fuxa and Y. Tanada, eds. John Wiley and Sons, New York, pp. 177-215.

Bergh, J.C. and C.W. McCoy. (1997). Aerial dispersal of citrus rust mite (Acari: Eriophyidae) from Florida citrus groves. *Environmental Entomology* 26: 256-264.

Brown, M.W., J.J. Jaeger, A.E. Pye and J.J. Schmitt. (1992). Control of edaphic populations of woolly apple aphid using entomopathogenic nematodes and a systemic aphicide. *Journal of Entomological Science* 27: 224-232.

Brownbridge, M., A. Adamowicz, M. Skinner and B.L. Parker. (1999). Prevalence of fungal entomopathogens in the life cycle of pear thrips, *Taeniothrips inconsequens* (Thysanoptera: Thripidae), in Vermont sugar maple forests. *Biological Control* 16: 54-59.

Browning, H. (1999). Arthropod pests of fruit and foliage. In *Citrus Health Management,* L.W. Timmer and L.W. Duncan, eds. American Phytopathological Society, St. Paul, MN, pp. 116-123.

Brunner, J.F. (1994). Using *Bt* products as tools in pest control. *Good Fruit Grower* 45: 34-38.

Bullock, R.C., R.R. Pelosi and E.E. Killer. (1999). Management of citrus root weevils (Coleoptera: Curculionidae) on Florida citrus with soil-applied entomopathogenic nematodes (Nematoda: Rhabditida). *Florida Entomologist* 82: 1-7.

Butt, T.M. and M. Brownbridge. (1997). Fungal pathogens of thrips. In *Thrips as Crop Pests,* T. Lewis, ed. CAB International, Wallingford, Oxon, pp. 399-434.

Butt, T.M., L. Ibrahim, S.J. Clark and A. Beckett. (1995). The germination behaviour of *Metarhizium anisopliae* on the surface of aphid and flea beetle cuticles. *Mycological Research* 99: 945-950.

Carballo, V.M. and M. Lopez. (1994). Evaluation of *Beauveria bassiana* for the control of *Cosmopolites sordidus* and *Metamasius hemipterus* (Coleoptera: Curculionidae) under field conditions. *Manejo Integrado de Plagas* 31: 22-24 (in Spanish).

Cardenas, F.A., P.V. Vail, D.F. Hoffman, J.S. Tebbetts and F.E. Schreiber. (1997). Infectivity of celery looper (Lepidoptera: Noctuidae) multiple nucleopcapsid polyhedrosis virus to navel orangeworm (Lepidoptera: Pyralidae). *Environmental Entomology* 26: 131-134.

Castillo, M.A., P. Moya, E. Hernandez and E. Primo-Yufera. (2000). Susceptibility of *Ceratitis capitata* Wiedemann (Diptera: Tephritidae) to entomopathogenic fungi and their extracts. *Biological Control* 19: 274-282.

Chandler, D., G. Davidson, J.K. Pell, B.V. Ball, K. Shaw and K.D. Sunderland. (2000). Fungal biocontrol of Acari. *Biocontrol Science and Technology* 10: 357-384.

Cossentine, J.E., F.L. Banham and L.B. Jensen. (1990). Efficacy of the nematode *Heterorhabditis bacteriophora* (Rhabditida: Heterorhabditidae) against the peachtree borer, *Synanthedon exitosa* (Lepidoptera: Sesiidae) in peach trees. *Journal of the Entomological Society of British Columbia* 87: 82-84.

Cossentine, J.E., L.B. Jensen and L. Moyls. (2002). Fruit bins washed with *Steinernema carpocapsae* (Rhabditida: Steinernematidae) to control *Cydia pomonella* (Lepidoptera: Tortricidae). *Biocontrol Science and Technology* 12: 251-258.

Cross, J.V., M.G. Solomon, D. Chandler, P. Jarrett, P.N. Richardson, D. Winstanley, H. Bathon, et al. (1999). Biocontrol of pests of apples and pears in northern and central Europe: 1. Microbial agents and nematodes. *Biocontrol Science and Technology* 9: 125-149.

Dent, D. (1995). *Integrated Pest Management,* Chapman & Hall, London.

Deseö, K.V. and L.A. Miller. (1985). Efficacy of entomogenous nematodes, *Steinernema* spp, against clearwing moths, *Synanthedon* spp., in north Italian apple orchards. *Nematologica* 31: 100-108.

Dimbi, S., N.K. Maniania, S.A. Lux, S. Ekesi and J.K. Mueke. (2004). Pathogenicity of *Metarhizium anisopliae* (Metsch.) Sorokin and *Beauveria bassiana* (Balsamo) Vuillemin, to three adult fruit fly species: *Ceratitis capitata* (Weidemann), *C. rosa* var. *fasciventris* Karsch and *C. cosyra* (Walker) (Diptera: Tephritidae). *Mycopathologia* 157: 375-382.

Downing, S.A., C.G. Erickson and M.J. Kraus. (1991). Field evaluations of entomopathogenic nematodes against citrus root weevils (Coleoptera: Curculionidae) in Florida citrus. *Florida Entomologist* 74: 584-586.

Duncan, L.W. and C.W. McCoy. (1996). Vertical distribution in soil, persistence and efficacy against citrus root weevil (Coleoptera: Curculionidae) of two species of entomogenous nematodes (Rhabditida: Steinernematidae: Heterorhabditidae). *Environmental Entomology* 25: 174-178.

Duncan, L.W., C.W. McCoy and A.C. Terranova. (1996). Estimating sample size and persistence of entomogenous nematodes in sandy soils and their efficacy against the larvae of *Diaprepes abbreviatus* in Florida. *Journal of Nematology* 28: 56-67.

Duncan, L.W., D.I. Shapiro, C.W. McCoy and J. Graham. (1999). Entomopathogenic nematodes as a component of citrus root weevil IPM. In *Optimal Use of Insecticidal Nematodes in Pest Management. Proceedings of Workshop, 28-30 August,* S. Polavarapu, ed. Rutgers University, New Brunswick, NJ, pp. 69-78.

Ebssa, L., C. Borgemeister, O. Berndt and H.M. Poehling. (2001a). Efficacy of entomopathogenic nematodes against soil-dwelling life stages of western flower thrips, *Frankliniella occidentalis* (Thysanoptera: Thripidae). *Journal of Invertebrate Pathology* 78: 119-127.

Ebssa, L., C. Borgemeister, O. Berndt and H.M. Poehling. (2001b). Impact of entomopathogenic nematodes on different soil-dwelling stages of western flower thrips, *Frankliniella occidentalis* (Thysanoptera: Thripidae), in the laboratory and under semi-field conditions. *Biocontrol Science and Technology* 11: 515-525.

Ekesi, S., N.K. Maniania and S.A. Lux. (2002). Mortality in three African tephritid fruit fly puparia and adults caused by the entomopathogenic fungi, *Metarhizium anisopliae* and *Beauveria bassiana*. *Biocontrol Science and Technology* 12: 7-17.

Ekesi, S., N.K. Maniania and S.A. Lux. (2003). Effect of soil temperature and moisture on survival and infectivity of *Metarhizium anisopliae* to four tephritid fruit fly puparia. *Journal of Invertebrate Pathology* 83: 157-167.

El Salam, S.A.A. (2001). Toxicity and biochemical effects of some safe alternative materials against *Pterochloroides persicae* Chlod. individuals (Order Homoptera: Fam. Aphididae) on peach trees. *Egyptian Journal of Agricultural Research* 79: 1387-1397.

Evans, H. and M. Shapiro. (1997). Viruses. In *Manual of Techniques in Insect Pathology,* L.A. Lacey, ed. Academic Press, London, pp. 17-53.

Falcon, L.A. and J. Huber. (1991). Biological control of the codling moth. In *Tortricid Pests, their Biology, Natural Enemies and Control,* L.P.S. van der Geest and H.H. Evenhuis, eds. Elsevier Science Publishers, Amsterdam, The Netherlands, pp. 355-369.

Fallon, D.J., L.F. Solter, M. Keena, M. McManus, J.R. Cate and L.M. Hanks. (2004). Susceptibility of Asian longhorned beetle, *Anoplophora glabripennis* Motchulsky (Coleoptera: Cerambycidae) to entomopathogenic nematodes. *Biological Control* 30: 430-438.

Faria, M. and S. Wraight. (2001). Biological control of *Bemisia tabaci* with fungi. *Crop Protection* 20: 767-778.

Fawcett, H.S. (1944). Fungus and bacterial diseases of insects as factors in biological control. *Botanical Review* 10: 327-348.

Federici, B.A. (1997). Baculovirus pathogenesis. In *The Baculoviruses,* L.K. Miller, ed. Plenum Press, New York, pp. 33-59.

Feng, M.G. and C. Chen. (2002). Incidence of infected *Myzus persicae* alatae trapped in flight imply place-to-place dissemination of entomophthoralean fungi in aphid populations through migration. *Journal of Invertebrate Pathology* 81: 53-56.

Figueroa, W. (1990). Biocontrol of the banana root borer weevil, *Cosmopolites sordidus* (Germar), with steinernematid nematodes. *Journal of Agriculture of the University of Puerto Rico* 74: 15-19.

Fransen, J.J. (1990). Natural enemies of whiteflies: Fungi. In *Whiteflies: Their Bionomics, Pest Status and Management,* D. Gerling, ed. Intercept, Andover, UK, pp. 187-210.

Fuxa, J.R. and Y. Tanada. (1987). Epidemiological concepts applied to insect epizootiology. In *Epizootiology of Insect Diseases,* J.R. Fuxa and Y. Tanada, eds. Wiley, New York, pp. 3-21.

Gao, R., Z. Ouyang, Z. Gao and J. Zheng. (1985). A preliminary report on the application of *Aschersonia aleyrodis* for the control of citrus whitefly. *Chinese Journal of Biological Control* 1: 45-46 (in Chinese; English summary).

Gaugler, R. and H.K. Kaya. (1990). *Entomopathogenic Nematodes in Biological Control,* CRC Press, Boca Raton, FL.

Gaugler, R., L. Lebeck, B. Nakagaki and G.M. Boush. (1980). Orientation of the entomogenous nematode *Neoaplectana carpocapsae* to carbon-dioxide. *Environmental Entomology* 9: 649-652.

Gazit, Y., Y. Rossler and I. Glazer. (2000). Evaluation of entomopathogenic nematodes for the control of Mediterranean fruit fly (Diptera: Tephritidae). *Biocontrol Science and Technology* 10: 157-164.

Gill, S., J.A. Davidson and M.J. Raupp. (1992). Control of peachtree borer using entomopathogenic nematodes. *Journal of Arborculture* 18: 184-187.

Goettel, M.S. and S.T. Jaronski. (1997). Safety and registration of microbial control agents of grasshoppers and locusts. *Memoirs of the Entomological Society of Canada* 171: 83-99.

Gopal , M., A. Gupta , B. Sathiamma and C.P. Nair. (2001). Control of the coconut pest *Oryctes rhinoceros* L. using the *Oryctes* virus. *Insect Science And Its Application* 21: 93-101.

Gottwald, T.R. and W.L. Tedders. (1983). Suppression of pecan weevil (Coleoptera: Curculionidae) populations with entomopathogenic fungi. *Environmental Entomology* 12: 471-474.

Gottwald, T.R. and W.L. Tedders. (1984). Colonization, transmission, and longevity of *Beauveria bassiana* and *Metarhizium anisopliae* (Deuteromycotina: Hypomycetes) on pecan weevil larvae (Coleoptera: Curculionidae) in the soil. *Environmental Entomology* 13: 557-560.

Goyer, R.A., H. Wei, J.R. Fuxa and H.X. Wei. (2001). Prevalence of viral diseases of the fruittree leafroller, *Archips argyrospila* (Walker) (Lepidoptera: Tortricidae), in Louisiana. *Journal of Entomological Science* 36: 17-22.

Gröner, A. (1986). Specificity and safety of baculoviruses. In *The Biology of Baculoviruses, Vol. I, Biological Properties and Molecular Biology,* R.R. Granados and B.A. Federici, eds. CRC Press, Boca Raton, FL, pp. 177-202.

Guillon, M. and G. Biache. (1995). IPM strategies for control of codling moth (*Cydia pomonella* L.) (Lepidoptera Olethreutidae) interest of CmGV for long term biological control of pest complex in orchards. *Mededelingen van de Faculteit Landbouwwetenschappen Rijksuniversiteit Gent* 60: 695-705.

Gut, L.J. and J.F. Brunner. (1998). Pheromone-based management of codling moth (Lepidoptera: Tortricidae) in Washington apple orchards. *Journal of Agricultural Entomology* 15: 387-405.

Hajek, A.E. and R.J. St. Leger. (1994). Interactions between fungal pathogens and insect hosts. *Annual Review of Entomology* 39: 293-321.

Hajek, A.E., S.P. Wraight and J.D. Vandenberg. (2001). Control of arthropods using pathogenic fungi. In *Bio-Exploitation of Filamentous Fungi,* S.B. Pointing and K.D. Hyde, eds. Fungal Diversity Press, Hong Kong, pp. 309-347.

Harris, M.A., J.W. Begley and D.L. Warkentin. (1990). *Liriomyza trifolii* (Diptera, Agromyzidae) suppression with foliar applications of *Steinernema carpocapsae* (Rhabditida, Steinernematidae) and Abamectin. *Journal of Economic Entomology* 83: 2380-2384.

Harris, M.K. (1999). Pecan weevil management considerations. In *Pecan Industry: Current Situation and Future Challenges. Third National Pecan Workshop*

Proceedings, Las Cruces New Mexico, June 20-23, 1998, B. McGraw, E.H. Dean and B.W. Wood, eds. pp. 66-73.

Harrison, R.D., W.A. Gardner and D.J. Kinard. (1993). Relative susceptibility of pecan weevil fourth instars and adults to selected isolates of *Beauveria bassiana. Biological Control* 3: 34-38.

Hartfield, C.M., C.A.M. Campbell, J. Hardie, J.A. Pickett and L.J. Wadhams. (2001). Pheromone traps for the dissemination of an entomopathogen by the damson-hop aphid *Phorodon humuli. Biocontrol Science and Technology* 11: 401-410.

Helyer, N.L., P.J. Brobyn, P.N. Richardson and R.N. Edmondson. (1995). Control of western flower thrips (*Frankliniella occidentalis* Pergande) pupae in compost. *Annals of Applied Biology* 127: 405-412.

Höfte, H. and H.R. Whitely. (1989). Insecticidal crystal proteins of *Bacillus thuringiensis. Microbiological Review* 53: 242-255.

Hokkanen, H.M.T. and A.E. Hajek. (2004). *Environmental Impacts of Microbial Pesticides,* Springer, Dordrecht, The Netherlands, 272 p.

Huber, J. (1986). Use of Baculoviruses in pest management programs. In *The Biology of Baculoviruses. Vol. II, Practical Application for Insect Control,* R.R. Granados and B.A. Federici, eds. CRC Press, Boca Raton, FL, pp. 181-202.

Hunter-Fujita, F.R., P.F. Entwistle, H.F. Evans and N.E. Crook. (1998). *Insect Viruses and Pest Management*, Wiley, Chichester, UK.

Jacob, T.K . (1996). Introduction and establishment of baculovirus for the control of rhinoceros beetle *Oryctes rhinoceros* (Col: Scarabaeidae) in the Andaman Islands (India). *Bulletin of Entomological Research* 86: 257-262.

Jacobson, R.J., D. Chandler, J. Fenlon and K.M. Russell. (2001). Compatibility of *Beauveria bassiana* (Balsamo) Vuillemin with *Amblyseius cucumeris* Oudemans (Acarina: Phytoseiidae) to control *Frankliniella occidentalis* Pergande (Thysanoptera: Thripidae) on cucumber plants. *Biocontrol Science and Technology* 11: 391-400.

Jaques, R., J. Hardman, J. Laing and R. Smith. (1994). Orchard trials in Canada on control of *Cydia pomonella* (LEP: Tortricidae) by granulosis virus. *Entomophaga* 39: 281-292.

Kaaya, G.P., K.V. Seshu Reddy, E.D. Kokwaro and D.M. Munyinyi. (1993). Pathogenicity of *Beauveria bassiana, Metarhizium anisopliae* and *Serratia marcescens* to the banana weevil *Cosmopolites sordidus. Biocontrol Science and Technology* 3: 177-187.

Kain, D.P. and A.M. Agnello. (1999). Pest status of American plum borer (Lepidoptera: Pyralidae) and fruit tree borer control with synthetic insecticides and entomopathogenic nematodes in New York State. *Journal of Economic Entomology* 92: 193-200.

Kaya, H.K. and R. Gaugler. (1993). Entomopathogenic Nematodes. *Annual Review of Entomology* 38: 181-206.

Kaya, H.K., J.L. Joos, L.A. Falcon and A. Berlowitz. (1984). Suppression of the codling moth (Lepidoptera, Olethreutidae) with the entomogenous nematode, *Steinernema feltiae* (Rhabditida, Steinernematidae). *Journal of Economic Entomology* 77: 1240-1244.

Kaya, H.K. and L.A. Lacey. (2000). Introduction to microbial control. In *Field Manual of Techniques in Invertebrate Pathology: Application and Evaluation of Pathogens for Control of Insects and Other Invertebrate Pests,* L.A. Lacey and H.K. Kaya, eds. Kluwer Academic Publishers, Dordrecht, pp. 1-4.

Kaya, H.K. and S.P. Stock. (1997). Techniques in insect nematology. In *Manual of Techniques in Insect Pathology,* L.A. Lacey, ed. Academic Press, London, pp. 281-324.

Keeling, P.J. and N.M. Fast. (2002). Microsporedia: biology and evolution of highly reduced intracellular parasites. *Annual Review of Microbiology* 56: 93-116.

Khan, A. and G. Gangapersad. (2001). Comparison of the effectiveness of three entomopathogenic fungi in the management of the banana borer weevil, *Cosmopolites sordidus* (Germar) (Coleoptera: Curculionidae). *International Pest Control* 43: 208-213.

Klein, M.G. (1990). Efficacy against soil-inhabiting insect pests. In *Entomopathogenic Nematodes in Biological Control,* R. Gaugler and H.K. Kaya, eds. CRC Press, Boca Raton, FL, pp. 195-214.

Knight, A. (1997). Optimizing the use of *Bts* for leafroller control. *Good Fruit Grower* 48: 47-49.

Knight, A.L., L.A. Lacey, B. Stockoff and R. Warner. (1998). Activity of Cry 1 endotoxins of *Bacillus thuringiensis* for four tree fruit leafroller pest species (Lepidoptera: Tortricidae). *Journal of Agricultural Entomology* 15: 93-103.

Laborda, R., L. Bargues, C. Navarro and O. Barajas. (2003). Susceptibility of the Mediterranean fruit fly (*Ceratitis capitata*) to *Steinernema* spp. *Bulletin OILB/SROP* 26: 95-97.

Lacey, L.A., S. Arthurs, A. Knight, K. Becker and H. Headrick. (2004). Efficacy of codling moth granulovirus: effect of adjuvants on persistence of activity and comparison with other larvicides in a Pacific Northwest apple orchard. *Journal of Entomological Science* 39: 500-513.

Lacey, L.A. and R.L. Chauvin. (1999). Entomopathogenic nematodes for control of diapausing codling moth (Lepidoptera: Tortricidae) in fruit bins. *Journal of Economic Entomology* 92: 104-109.

Lacey, L.A., J.J. Fransen and R. Carruthers. (1996). Global distribution of naturally occurring fungi of *Bemisia*, their biologies and use as biological control agents. In *Bemisia 1995: Taxonomy, Biology, Damage, and Management,* D. Gerling and R. Mayer, eds. Intercept, Andover, UK, pp. 401-433.

Lacey, L.A., R. Frutos, H.K. Kaya and P. Vail. (2001). Insect pathogens as biocontrol agents: do they have a future? *Biological Control* 21: 230-248.

Lacey, L.A., A. Knight and J. Huber. (2000). Microbial control of lepidopteran pests of apple orchards. In *Field Manual of Techniques in Invertebrate Pathology: Application and Evaluation of Pathogens for Control of Insects and Other Invertebrate pests,* L.A. Lacey and H.K. Kaya, eds. Kluwer Academic Publishers, Dordrecht, pp. 557-576.

Lacey, L.A. and D.L. Shapiro-Ilan. (2003). The potential role for microbial control of orchard insect pests in sustinable agriculture. *Journal of Food, Agriculture and Environment* 1: 326-331.

Lacey, L.A. and T.R. Unruh. (1998). Entomopathogenic nematodes for control of codling moth, *Cydia pomonella* (Lepidoptera: Tortricidae): effect of nematode species, concentration, temperature, and humidity. *Biological Control* 13: 190-197.

Lacey, L.A., P.V. Vail and D.F. Hoffmann. (2002). Comparative activity of baculoviruses against the codling moth *Cydia pomonella* and three other totricid pests of tree fruit. *Journal of Invertebrate Pathology* 80: 64-68.

Laird, M., L.A. Lacey and E.W. Davidson. (eds.) (1990). *Safety of Microbial Insecticides*. CRC Press, Boca Raton, FL.

Latgé, J.P. and B. Paperiok. (1988). Aphid pathogens. In *World Crop Pests, 2B-Aphids: Their Biology, Natural Enemies and Control,* A.K. Minks and P. Harrewijn, eds. Elsevier, Amsterdam, pp. 323-335.

Lebeck, L.M., R. Gaugler, H.K. Kaya, A.H. Hara and M.W. Johnson. (1993). Host stage suitability of the leafminer *Liriomyza trifolii* (Diptera, Agromyzidae) to the entomopathogenic nematode *Steinernema carpocapsae* (Rhabditida, Steinernematidae). *Journal of Invertebrate Pathology* 62: 58-63.

Lewis, T. (1997). Pest thrips in perspective. In *Thrips as Crop Pests,* T. Lewis, ed. CAB International, Wallingford, UK, pp. 1-13.

Lezama-Gutierrez, R., A. Trujillo-De La Luz, J. Molina-Ochoa, O. Rebolledo-Dominguez, A.R. Pescador, M. Lopez-Edwards and M. Aluja. (2000). Virulence of *Metarhizium anisopliae* (Deuteromycotina: Hyphomycetes) on *Anastrepha ludens* (Diptera: Tephritidae): laboratory and field trials. *Journal of Economic Entomology* 93: 1080-1084.

Lindegren, J.E. and P.V. Vail. (1986). Susceptibility of Mediterranean fruit fly, melon fly, and oriental fruit fly (Diptera: Tephritidae) to the entomogenous nematode *Steinernema feltiae* in laboratory tests. *Environmental Entomology* 15: 465-468.

Lindegren, J.E., T.T. Wong and D.O. McInnis. (1989). Response of Mediterranean fruit fly (Diptera: Tephritidae) to the entomogenous nematode *Steinernema feltiae* in field tests in Hawaii. *Environmental Entomology* 19: 383-386.

Lisansky, S. (1997). *Microbial Biopesticides,* British Crop Protection Council, Farnham, UK. pp. 3-10.

Liu, Y., M. Feng, S. Liu, Y.Q. Liu, M.G. Feng and S.S. Liu. (1999). Virulence of *Beauveria bassiana* against the green peach aphid, *Myzus persicae. Acta Phytophylacica Sinica* 26: 347-352 (in Chinese; English summary).

Lu, Q.G., N.X. Zhang, Y.Z. Jiang and S.F. Zhang. (1993). Laboratory and field tests on the efficacy of a Chinese produced *Bacillus thuringiensis* wettable powder against peach fruit borer, *Carposina niponensis* (Lep.: Carposinidae). *Chinese Journal of Biological Control* 9: 156-159.

Luisier, N. and G. Benz. (1994). Improving the efficiency of the granulosis virus of *Adoxophyes orana* F. v. R. for microbiological control. *Bulletin IOBC/SROP* 17: 274.

Markwick, N., S. Graves, F.M. Fairbairn, L.C. Docherty, J. Poulton and V.K. Ward. (2002). Infectivity of *Epiphyas postvittana* nucleopolyhedrovirus for New Zealand leafrollers. *Biological Control* 25: 207-214.

McCoy, C.W. (1996). Pathogens of Eriophyid mites. In *Eriophyid Mites. Their Biology, Natural Enemies and Control Vol. 6,* E.E. Lindquist, M.W. Sabelis and J. Bruim, eds. Elsevier, Dordrecht, pp. 481-490.

McCoy, C.W. (1999). Arthropod pests of citrus roots. In *Citrus Health Management,* L.W. Timmer and L.W. Duncan, eds. American Phytopathological Society, St. Paul, pp. 149-156.

McCoy, C.W. and T.L. Couch. (1982). Microbial control of the citrus rust mite with the Mycoacaricide, Mycar. *Florida Entomologist* 65: 116-126.

McCoy, C.W., R.A. Samson and D.R. Boucias. (1988). Entomogenous fungi. In *Handbook of Natural Pesticides Vol. V, Microbial Insecticides Part A, Entomogenous Fungi and Protozoa,* C.M. Ignoffo and N. Bushan, eds. CRC Press, Boca Raton, Florida, pp. 151-236.

McCoy, C.W., D.I. Shapiro, L.W. Duncan and K. Nguyen. (2000). Entomopathogenic nematodes and other natural enemies as mortality factors for larvae of *Diaprepes abbreviatus* (Coleoptera: Curculionidae). *Biological Control* 19: 182-190.

McCoy, C.W., R.J. Stuart, L.W. Duncan and K. Nguyen. (2002). Field efficacy of two commercial preparations of entomopathogenic nematodes against larvae of *Diaprepes abbreviatus* (Coleoptera: Curculionidae) in alfisol type soil. *Florida Entomologist* 85: 537-544.

Miller, L.A. and R.A. Bedding. (1982). Field testing of the insect parasitic nematode, *Neoaplectana bibionis* [Nematoda, Steinernematidae] against currant borer moth, *Synanthedon tipuliformis* [Lep, Sesiidae] in blackcurrants. *Entomophaga* 27: 109-114.

Miller, L.K. (ed.). (1997). *The Baculoviruses,* Plenum Press, New York.

Milner, R.J. (1997). Prospects for biopesticides for aphid control. *Entomophaga* 42: 227-239.

Montserrat, M., C. Castane and S. Santamaria. (1998). *Neozygites parvispora* (Zygomycotina: Entomophthorales) causing an epizootic in *Frankliniella occidentalis* (Thysanoptera: Thripidae) on cucumber in Spain. *Journal of Invertebrate Pathology* 71: 165-168.

Nachtigall, S. and E. Dickler. (1992). Experiences with field applications of entomoparasitic nematodes for biological control of cryptic living insects in orchards. *Acta Phytopathologica et Entomologica Hungarica* 27: 485-490.

Nankinga, C.M. and D. Moore. (2000). Reduction of banana weevil populations using different formulations of the entomopathogenic fungus *Beauveria bassiana. Biocontrol Science and Technology* 10: 645-657.

Nguyen, K.B. and L.W. Duncan. (2002). *Steinernema diaprepesi* n. sp. (Rhabditida: Steinernematidae), a parasite of the citrus root weevil *Diaprepes abbreviatus* (L) (Coleoptera: Curculionidae). *Journal of Nematology* 34: 159-170.

Nyczepir, A.P., J.A. Payne and J.J. Hamm. (1992). *Heterorhabditis bacteriophora*: a new parasite of pecan weevil *Curculio caryae. Journal of Invertebrate Pathology* 60: 104-106.

Olthof, T.H.A. and E.A.C. Hagley. (1993). Laboratory studies of the efficacy of steinernematid nematodes against the plum curculio (Coleoptera, Curculionidae). *Journal of Economic Entomology* 86: 1078-1082.

Omoto, C. and C.W. McCoy. (1998). Toxicity of purified fungal toxin Hirsutellin A to the citrus rust mite *Phyllocoptruta oleivora* (Ash.). *Journal of Invertebrate Pathology* 72: 319-322.

Parker, B.L., M. Skinner and T. Lewis. (1995). *Thrips Biology and Management,* NATO ASI Series. Series A: Life Sciences, New York.

Pfannenstiel, R.S., M. Szymanski, L.A. Lacey, J.F. Brunner and K. Spence. (2004). Discovery of a granulovirus of *Pandemis pyrusana* (Lepidoptera: Tortricidae), a leafroller pest of apples in Washington. *Journal of Invertebrate Pathology* 86: 124-127.

Poinar, G.O. (1991). Nematode parasites. In *Tortricid Pests: Their Biology, Natural Enemies and Control,* L.P.S. van de Geest and H.H. Evenhuis, eds. Elsevier, Amsterdam, The Netherlands, pp. 273-281.

Ponomarenko, N.G., N.A. Prilepskaja, M. Murvanidze and L.A. Stoljarova. (1975). *Aschersonia* against whitefly. *Zashchita Rastenii* 6: 44-45.

Poprawski, T.J., P.E. Parker and J.H. Tsai. (1999). Laboratory and field evaluation of hyphomycete insect pathogenic fungi for control of brown citrus aphid (Homoptera: Aphididae). *Environmental Entomology* 28: 315-321.

Pronier, I., J. Pare, J.C. Wissocq and C. Vincent. (2002). Nucleopolyhedrovirus infection in obliquebanded leafroller (Lepidoptera: Tortricidae). *Canadian Entomologist* 134: 303-309.

Puterka, G.J. (1999). Fungal pathogens for arthropod pest control in orchard systems: mycoinsecticidal approach for pear psylla control. *BioControl* 44: 183-210.

Puterka, G.J., R.A. Humber and T.J. Poprawski. (1994). Virulence of fungal pathogens (imperfect fungi: Hyphomycetes) to pear psylla (Homoptera: Psyllidae). *Environmental Entomology* 23: 514-520.

Racette, G., G. Chouinard, C. Vincent and S.B. Hill. (1993). Ecology and management of plum curculio, *Conotrachelus nenuphar* (Coleoptera: Curculionidae). *Phytoprotection* 75: 85-100.

Rosas-Acevedo, J.L. and L. Sampedro-Rosas. (2000). Biological control of *Brevipalpus* spp. on Citrus aurantifolia in Guerrero, Mexico. *Manejo Integrado de Plagas* 55: 56-59.

Samson, R.A., H.C. Evans and J.P. Latgé. (1988). *Atlas of Entomopathogenic Fungi,* Springer Verlag, New York, 187 pp.

Schmidt, J. and J.N. All. (1979). Attraction of *Neoaplectana carpocapsae* (Nematoda, Steinernematidae) to common excretory products of insects. *Environmental Entomology* 8: 55-61.

Schmitt, A.T., S.R. Gowen and N.G.M. Hague. (1992). Baiting techniques for the control of *Cosmopolites sordidus* Germar (Coleoptera, Curculionidae) by *Steinernema carpocapsae* (Nematoda, Steinernematidae). *Nematropica* 22: 159-163.

Schoeman, P.S. and M.H. Schoeman. (1999). Transmission of *Beauveria bassiana* from infected to uninfected adults of the banana weevil *Cosmopolites sordidus* (Coleoptera: Curculionidae). *African Plant Protection* 5: 53-54.

Schreiter, G., T.M. Butt, A. Beckett, S. Vestergaard and G. Moritz. (1994). Invasion and development of *Verticillium lecanii* in the western flower thrips, *Frankliniella occidentalis*. *Mycological Research* 98: 1025-1034.

Schroeder, W.J. (1987). Laboratory bioassays and field trials of entomogenous nematodes for control of *Diaprepes abbreviatus*. *Environmental Entomology* 16: 987-989.

Schroeder, W.J. (1994). Comparison of two Steinernematid species for control of the root weevil *Diaprepes abbreviatus*. *Journal of Nematology* 26: 360-362.

Shapiro, D.I., J.R. Cate, J. Pena, A. Hunsberger and C.W. McCoy. (1999). Effects of temperature and host range on suppression of *Diaprepes abbreviatus* (Coleoptera: Curculionidae) by entomopathogenic nematodes. *Journal of Economic Entomology* 92: 1086-1092.

Shapiro, D.I., C.W. McCoy, A. Fares, T. Obreza and H. Dou. (2000). Effects of soil type on virulence and persistence of entomopathogenic nematodes in relation to control of *Diaprepes abbreviatus* (Coleoptera: Curculionidae). *Environmental Entomology* 29: 1083-1087.

Shapiro-Ilan, D.I. (2001a). Virulence of entomopathogenic nematodes to pecan weevil (Coleoptera: Curculionidae) adults. *Journal of Entomological Science* 36: 325-328.

Shapiro-Ilan, D.I. (2001b). Virulence of entomopathogenic nematodes to pecan weevil larvae *Curculio caryae* (Coleoptera: Curculionidae) in the laboratory. *Journal of Economic Entomology* 94: 7-13.

Shapiro-Ilan, D.I., W.A. Gardner, J.R. Fuxa, B.W. Wood, K.B. Nguyen, B.J. Adams, R.A. Humber and M.J. Hall. (2003). Survey of entomopathogenic nematodes and fungi endemic to pecan orchards of the Southeastern United States and their virulence to the pecan weevil (Coleoptera: Curculionidae). *Environmental Entomology* 32: 187-195.

Shapiro-Ilan, D.I., D.H. Gouge and A.M. Koppenhöfer. (2002). Factors affecting commercial success: case studies in cotton, turf and citrus. In *Entomopathogenic Nematology*, R. Gaugler, ed. CABI, New York, pp. 333-356.

Shapiro-Ilan, D.I., R.F. Mizell, T.E. Cottrell and D.L. Horton. (2004). Measuring field efficacy of *Steinernema feltiae* and *Steinernema riobrave* for suppression of plum curculio, *Conotrachelus nenuphar*, larvae. *Biological Control* 30: 496-503.

Shapiro-Ilan, D.I., R. Stuart and C.W. McCoy. (2003). Comparison of beneficial traits among strains of the entomopathogenic nematode, *Steinernema carpocapsae*, for control of *Curculio caryae* (Coleoptera: Curculionidae). *Biological Control* 28: 129-136.

Shen, J.D. and Q.Y. Han. (1985). A preliminary study of controlling *Rhytidodera bowrigii* with DD-136 nematode. *Natural Enemies of Insects* 7: 28-29.

Shi, W., M. Feng, W.B. Shi and M.G. Feng. (2004). Lethal effect of *Beauveria bassiana, Metarhizium anisopliae,* and *Paecilomyces fumosoroseus* on the eggs of *Tetranychus cinnabarinus* (Acari: Tetranychidae) with a description of a mite egg bioassay system. *Biological Control* 30: 165-173.

Siegel, J.P., L.A. Lacey and C.R. Vossbrinck. (2001). Impact of a North American isolate of the microsporidiurn *Nosema carpocapsae* on a laboratory population of the codling moth, *Cydia pomonella*. *Journal of Invertebrate Pathology* 78: 244-250.

Siegel, J., L.A. Lacey, R.J. Fritts, B.S. Higbee and P. Noble. (2004). Use of steinernematid nematodes for post harvest control of navel orangeworm (Lepidoptera: Pyralidae, *Amyelois transitella*) in fallen pistachios. *Biological Control* 30: 410-417.

Sikorowski, P.P. (1985). Pecan weevil pathology. In *Pecan Weevil: Research Perspective* W.W. Neel, ed. Quail Ridge Press, Brandon, pp. 87-101.

Smirle, M.J., D.T. Lowery and C.L. Zurowski. (2003). Susceptibility of leafrollers (Lepidoptera: Tortricidae) from organic and conventional orchards to azinphosmethyl, spinosad, and *Bacillus thuringiensis*. *Journal of Economic Entomology* 96: 879-884.

Smith, D. (1995). Banana weevil borer control in south-eastern Queensland. *Australian Journal of Experimental Agriculture* 35: 1165-1172.

Smith, D., G.A.C. Beattie and R. Broadley. (1997). *Citrus Pests and Their Natural Enemies: Integrated Pest Management in Australia,* Department of Primary Industries, Brisbane, Queensland, Australia.

Smith, M.T., R. Georgis, A.P. Nyczepir and R.W. Miller. (1993). Biological control of the pecan weevil, *Curculio caryae* (Coleoptera: Curculionidae), with entomopathogenic nematodes. *Journal of Nematology* 25: 78-82.

Solter, L.F. and J.J. Becnel. (2000). Entomopathogenic microsporedia. In *Field Manual of Techniques in Invertebrate Pathology: Application and Evaluation of Pathogens for Control of Insects and Other Invertebrate Pests,* L.A. Lacey and H.K. Kaya, eds. Kluwer Academic, Dordrecht, pp. 231-254.

Solter, L.F., M. Keena, J.R. Cate, M.L. McManus and L.M. Hanks. (2001). Infectivity of four species of nematodes (Rhabditoidea: Steinernematidae, Heterorhabditidae) to the Asian longhorn beetle, *Anoplophora glabripennis* Motchulsky (Coleoptera: Cerambycidae). *Biocontrol Science and Technology* 11: 547-552.

Steinhaus, E.A. (1956). Microbial control: the emergence of an idea. *Hilgardia* 26: 107-160.

Stuart, R.J., D.I. Shapiro-Ilan, R.R. James, K.B. Nguyen and C.W. McCoy. (2004). Virulence of new and mixed strains of the entomopathogenic nematode *Steinernema riobrave* to larvae of the citrus root weevil *Diaprepes abbreviatus*. *Biological Control* 30: 439-445.

Tanada, Y. and H.K. Kaya. (1993). *Insect Pathology,* Academic Press, San Diego, CA.

Tedders, W.L., D.J. Weaver, E.J. Wehunt and C.R. Gentry. (1982). Bioassay of *Metarhizium anisopliae, Beauveria bassiana,* and *Neoaplectana carpocapsae* against larvae of the plum curculio *Conotrachelus nenuphar* (Herbst) (Coleoptera: Curculionidae). *Environmental Entomology* 11: 901-904.

Tinzaara, W., M. Dicke, A. van Huis and C.S. Gold. (2002). Use of infochemicals in pest management with special reference to the banana weevil, *Cosmopolites sordidus* (Germar) (Coleoptera: Curculionidae). *Insect Science and Its Application* 22: 241-261.

Unruh, T.R. and L.A. Lacey. (2001). Control of codling moth, *Cydia pomonella* (Lepidoptera: Tortricidae), with *Steinernema carpocapsae*: effects of supplemental wetting and pupation site on infection rate. *Biological Control* 20: 48-56.

Vacante, V., S.O. Cacciola and A.M. Pennisi. (1994). Epizootiological study of *Neozygites parvispora* (Zygomycota, Entomophthoraceae) in a population of *Frankliniella occidentalis* (Thysanoptera, Thripidae) on pepper in Sicily. *Entomophaga* 39: 123-130.

Vail, P.V., W. Barnett, D.C. Cowan, S. Sibbett, R. Beede and J.S. Tebbets. (1991). Codling moth (Lepidoptera: Tortricidae) control on commercial walnuts with a granulosis virus. *Journal of Economic Entomology* 84: 1448-1453.

Vail, P.V., D.F. Hoffman, D.A. Streett, J.S. Manning and J.S. Tebbets. (1993). Infectivity of a nuclear polyhedrosis virus isolated from *Anagrapha falcifera* (Lepidoptera: Noctuidae) against production and postharvest pests and homologous lines. *Environmental Entomology* 22: 1140-1145.

van der Geest, L.P.S., S.L. Elliot, J.A.J. Breeuwer and E.A.M. Beerling. (2000). Diseases of mites. *Experimental and Applied Acarology* 24: 497-560.

van Munster, M., A.M. Dullemans, M. Verbeek, J.F. van den Heuvel, C. Reinbold, V. Brault, A. Clérivet and F. van der Wilk. (2003). Characterization of a new densovirus infecting the green peach aphid *Myzus persicae. Journal of Invertebrate Pathology* 84: 6-14.

Vega, F.E., P.F. Dowd, L.A. Lacey, J.K. Pell, D.M. Jackson and M.G. Klein. (2000). Dissemination of beneficial microbial agents by insects. In *Field Manual of Techniques in Invertebrate Pathology: Application and Evaluation of Pathogens for Control of Insects and Other Invertebrate Pests,* L.A. Lacey and H.K. Kaya, eds. Kluwer Academic Publishers, Dordrecht, pp. 152-177.

Vestergaard, S., A.T. Gillespie, T.M. Butt, G. Schreiter and J. Eilenberg. (1995). Pathogenicity of the hyphomycete fungi *Verticillium lecanii* and *Metarhizium anisopliae* to the western flower thrips, *Frankliniella occidentalis. Biocontrol Science and Technology* 5: 185-192.

Vincent, C., G. Chouinard and S.B. Hill. (1999). Progress in plum curculio management: a review. *Agriculture Ecosystems and Environment* 73: 167-175.

Waage, J.K. (1997). *Biopesticides at the Crossroads: IPM Products or Chemical Clones?* British Crop Protection Council; Farnham; UK. pp. 11-19.

Walker, K.R. and S.C. Welter. (2001). Potential for outbreaks of leafrollers (Lepidoptera: Tortricidae) in California apple orchards using mating disruption for codling moth suppression. *Journal of Economic Entomology* 94: 373-380.

Weathersbee, A.A., Y.Q. Tang, H. Doostdar and R.T. Mayer. (2002). Susceptibility of *Diaprepes abbreviatus* (Coleoptera: Curculionidae) to a commercial preparation of *Bacillus thuringiensis* subsp. *tenebrionis. Florida Entomologist* 85: 330-335.

Williams, E.C. and O.C. MacDonald. (1995). Critical factors required by the nematode *Steinernema feltiae* for the control of the leafminers *Liriomyza huidobrensis, Liriomyza bryoniae* and *Chromatomyia syngenesiae. Annals of Applied Biology* 127: 329-341.

Williams, E.C. and K.F.A. Walters. (2000). Foliar application of the entomopathogenic nematode *Steinernema feltiae* against leafminers on vegetables. *Biocontrol Science and Technology* 10: 61-70.

Xia, L.H., C. Qing and M. Feng. (1998). Field experiment of *Anoplophora glabripennis* and *Melanophila decastigma* control with pathogenic nematode *Steinernema carpocapsae. Ningxia Journal of Agricultural and Forestry Science and Technology,* 5: 15-17 (in Chinese; English summary).

Xu, J., M. Feng, X. Tong, J.H. Xu, M.G. Feng and X.M. Tong. (2003). Screening of conidia germination stimulus as an additive to *Beauveria bassiana* formulation to enhance its efficacy for aphid control. *Acta Phytophylacica Sinica* 30: 45-50 (in Chinese; English summary).

Xu, J.L., R.H. Han, X.L. Liu, L. Cao and P. Yang. (1995). The application of codling moth nematode against the larvae of the litchi longicorn beetle. *Acta Phytophylacica Sinica* 22: 12-16 (in Chinese; English summary).

Xu, J., R. Xie, J.L. Xu and R.C. Xie. (1997). Mass application of entomogenous nematode against litchi stem borer. *Acta Phytophylacica Sinica* 24: 150-154.

Yee, W.L. and L.A. Lacey. (2003). Stage-specific mortality of *Rhagoletis indifferens* (Diptera: Tephritidae) exposed to three species of Steinernema nematodes. *Biological Control* 27: 349-356.

Yee, W.L. and L.A. Lacey. (2005). Mortality of different life stages of *Rhagoletis indifferens* (Diptera: Tephritidae) exposed to the entomopathogenic fungus *Metarhizium anisopliae*. *Journal of Entomological Science* 40: 167-177.

Yeo, H., J.K. Pell, P.G. Alderson, S.J. Clark and B.J. Pye. (2003). Laboratory evaluation of temperature effects on the germination and growth of entomopathogenic fungi and on their pathogenicity to two aphid species. *Pest Management Science* 59: 156-165.

Zelazny, B, A. Lolong and B. Pattang. (1992). *Oryctes rhinoceros* (Col: Scarabaeidae) populations suppressed by a baculovirus. *Journal of Invertebrate Pathology* 59: 61-68.

Chapter 2

Biology and Control of Vectors of *Xylella fastidiosa* in Fruit and Nut Crops

J. D. Dutcher
G. W. Krewer

INTRODUCTION

The bacterium *Xylella fastidiosa* Wells causes disease in a range of host plants, including many important fruit and nut crops (Wells, Raju and Nyland, 1983, 1987; Hopkins, 1989). *X. fastidiosa* exclusively colonizes the degenerated plant cells that make up the xylem elements of the plant and is transmitted by insects that ingest fluids from the plant through their piercing-sucking mouthparts, which they insert into the xylem (Meadows, 2001). Soon after transmission, the bacterial cells colonize the xylem and impede the flow of water and nutrients within the plant (Fletcher and Wayadande, 2002). Symptoms in the host plant range from low plant vigor to leaf scorch and on to plant death. Plants with infected xylem elements may also be predisposed to other plant diseases. *X. fastidiosa* is currently classified as one species with many distinct strains (Henderson et al., 2001). Climate is a major factor controlling the incidence of diseased plants, and abrupt geographic changes in the incidence of disease are common. Disease development is curtailed by prolonged freezing each winter season that kills the bacteria in the stem. Disease incidence increases after a mild winter, and after a cool winter in areas with variable winter conditions (Hoddle, 2004; Redak et al., 2004). In the moderate Gulf

Management of Nematode and Insect-Borne Plant Diseases

doi:10.1300/5754_02

Coast region of the southeastern United States, Pierce's disease, phony peach disease, plum leaf scald, and pecan bacterial leaf scorch, caused by *X. fastidiosa* and transmitted by sharpshooters (Auchenorrhyncha: Cicadellini), are limiting factors in fruit and nut production. The disease is pernicious in bunch grape vineyards and peach, nectarine, and plum orchards in areas where the bacteria survive the winter in the stem (Wells, Raju and Nyland, 1983). Pecan bacterial leaf scorch results in early defoliation but not tree death in certain pecan varieties (Sanderlin, 1998; Sanderlin and Heyderich-Alger, 2000). Pecan nut production, leaf size, and nut quality in 'Cape Fear', a highly susceptible pecan variety, are reduced by the disease (Sanderlin and Heyderich-Alger, 2003). Pierce's disease prevents the development of bunch grape culture based on *Vitis labrusca* and *Vitis vinifera* in the southernmost areas of the southeastern United States, where grapes are produced on resistant hybrid and muscadine, *Vitis rotundifolia,* grapevines (Dutcher, McGiffin and All, 1988). Entire peach and nectarine orchards have to be replanted every seven to ten years owing to rapid decline and death of the trees. Japanese plum trees cannot survive more than five years. The more common plants that are reservoirs of the bacterium are muscadine grapevines, wild plum, and sycamore trees. The more important vectors, such as *Homalodisca coagulata* (Say), *Homalodisca insolita* (Walker), *Graphocephala versuata* (Say), and *Oncometopia orbona* (F.), feed on a variety of wild host plants and fly into vineyards and orchards, where they transmit the disease. Controls for the vectors and cures for diseased plants are very limited. Control of phony peach disease has been achieved in middle and north Georgia by early detection, removal of trees with symptoms from the orchard, and removal of wild plums adjacent to the orchard. Rouging is less effective in southern Georgia, where milder winters lead to higher disease incidence. Detection is based on sighting of symptoms in trees, which may occur several years after the initial infection. In this region, the disease has reduced commercial orchard life to ten years or less (Dutcher, Krewer and Mullinix, 2005). Peach tree life in central Texas varies from one to fifteen years and is usually six to ten years (McEachern, Stein and Kamas, 2004). Plum leaf scald limits the life of Japanese plum orchards to less than five years in this region. Only immediate rouging of infected trees has been successful (Hutchins, 1993). It may be possible to control these diseases and extend orchard life by combining treatments, even though no single treatment effectively eliminates disease. The glassy-winged sharpshooter, *H. coagu-*

lata, has caused significant economic losses to the bunch grape vineyards of southern California since its introduction in 1994 and 1996, presumably from the southeastern United States. This catastrophe and the possibility of citrus variegated chlorosis being introduced from South and Central America to the citrus orchards of California and Florida have spurred a significant research effort to develop controls for the vectors and cures for diseased plants across all host plants (Blua et al., 2001; Purcell and Feil, 2001). This chapter reviews the most recent ideas on management of the vectors and disease. Practical applications of the research are rapidly developing, particularly in citrus, grape, and coffee.

RESOURCES

The references cited in this chapter serve to lead the reader to major sources; they do not comprise a comprehensive bibliography (refer to Purcell, 2005). Research into the basic biology and control of diseases caused by *X. fastidiosa* has been reviewed (Hopkins, 1989; Purcell and Hopkins, 1996; Hopkins and Purcell, 2002; Mizell et al., 2003; Redak et al., 2004). Control recommendations, including detection of bacteria in vectors and host plants, suppression of vectors with insecticide, removal of alternate host plants, rouging diseased plants, and antibiotic injections, have been outlined for citrus (Grafton-Cardwell, Batkin and Gorden, 2003), grapes in the eastern United States (Hartman et al., 2001; Mizell et al., 2003), grapes in the western United States (Varela, Smith and Phillips, 2001; McEachern, Stein and Kamas, 2004; Gubler et al., 2005), peach and nectarine (Simeone, 2000), almond (Teviotdale, 2003), and pecan (Sanderlin and Heyderich-Alger, 2003). The complete genome of the 9a5c citrus strain of *X. fastidiosa,* which causes citrus variegated chlorosis, has been characterized (Simpson et al., 2000), and the phylogenetic relationship between strains is an area of intensive research (Hendson et al., 2001). This basic genetic research is used to determine the types of proteins produced by the bacteria and their functions. The research is also leading to better methods of detecting the bacteria in host plants and vectors (Minsavage et al., 1994) that will eventually lead to better control techniques (Rodrigues et al., 2003).

BIOLOGY

Host Plants

Hosts of *X. fastidiosa* include woody and herbaceous plants (Hoddle, Triapitsyn and Morgan, 2003). In fact, *X. fastidiosa* has been detected in the xylem of seventy-five host plants, including woody and herbaceous species. *X. fastidiosa* usually causes disease in the plant, though the severity of disease varies considerably among plant species, among plant varieties within species, and according to climate, and some host plants do not display symptoms (Fletcher and Wayadande, 2002). The range of host plants affected includes important fruit and nut crops in the Americas. The more economically important diseases of fruit and nut crops caused by *X. fastidiosa* are Pierce's disease in grapes (*Vitis* L.), phony peach disease in peach and nectarine (*Prunus persicae* [L.] Batsch.), almond leaf scald in almond, citrus variegated chlorosis in citrus (*Citrus* [L.]), plum leaf scald in Japanese plum (*Prunus saliçina* Lindl.), and pecan bacterial leaf scorch in pecan (*Carya illinoinensis* [Wangenheim] K. Koch).

Bacteria

The bacteria colonizing the xylem of plants are transmitted to the plant by xylem-feeding insects (Raju and Wells, 1986). Vectors are able to acquire the *X. fastidiosa* from infected plants after an incubation time in the plant ranging from four to twenty-nine days in different host plants (Hill and Purcell, 1997). Vectors can transmit the disease within one hour and for up to sixty days after acquiring the bacterium. The bacterial cells reproduce in a foregut of the sharpshooter, increasing to a dense colony on the inner surface over a seven-day period (Hill and Purcell, 1995a). The number of bacterial cells in the foregut is apparently not directly related to the success of disease transmission, and insects with as few as 200 bacterial cells in the foregut can transmit the disease (Redak et al., 2004). Once transmitted to a plant, the bacterial cells move systemically through the plant at rates that vary depending on plant species (Hill and Purcell, 1995b). Microscopic examination of *X. fastidiosa* colonies in the xylem elements in citrus trees with citrus variegated chlorosis and coffee trees with coffee leaf scorch (Lima et al., 1997; Lima, 2005) indicates that the xylem elements in the stems and roots are infested. Xylem elements within the same infested plant can be clear or filled with bacterial colonies. Clear elements can be adjacent or interspersed with colonized elements. The naturally occurring pores between elements are larger than the bacterial cells. The pores are covered by a membrane that often prevents the transfer of bacterial cells between elements. The bacteria readily ascend and descend in the xylem elements but have limited lateral movement in citrus and coffee. A comparison of infected xylem elements using scanning and transmission electron microscopy in plum, coffee, citrus, tobacco, sycamore, grape, and pecan revealed that tyloses were formed in grape and pecan xylem elements after infection; bacteria moved laterally between the xylem elements of the leaf petiole where pit membranes between xylem elements were found to be altered after infection (Alves, 2004).

Vectors

The most important insect vectors are sharpshooters (Auchenorrhyncha: Cicadellini) (Redak et al., 2004). Spittlebugs (Auchenorrhyncha: Cercopoidea) and cicadas (Auchenorrhyncha: Cicadoidea) are also vectors (Severin, 1950; Paiao et al., 2002). These insects transmit and acquire the bacterium when they feed on fluids extracted from the plant xylem through a piercing-sucking mouthpart and a muscular pump, the cibarium, in their head. Vectors have to ingest copious amounts of the xylem fluid to extract sufficient nutrients for life. The bacteria are transmitted during feeding from the insect foregut to the plant xylem. The inner lining of the foregut is ectodermal in origin and is cast off with each molt, so vectors lose the ability to transmit the disease and have to reacquire bacteria after each molt (Hill and Purcell, 1995a, 1997; Almeida and Purcell, 2003; Redak et al., 2004).

Environment

Suitable environments for *X. fastidiosa* disease and its vectors exist around the world (Hoddle, Triapitsyn and Morgan, 2003). However, the major infestations of *X. fastidiosa* are found in the Americas (Purcell, 1997), though pear leaf scorch has been identified in Taiwan (Leu and Su, 1993). Hartman et al. (2001) state, "The severity of PD appears to depend on climate. The bacterium appears to be sensitive to cold winter temperatures and produces milder symptoms at higher altitudes, farther inland from ocean influences and at more northern latitudes, even where vectors are plentiful." Containment of the disease in the Americas requires the export of disease-free plants from nurseries in the region and the eradication of vectors if they are introduced into exotic areas. Sensitive bioassays using polymerase chain reaction techniques for the detection of *X. fastidiosa* in plants and insects are available, but a nested procedure is often required to find bacteria in the vectors (Pooler, Myung and Bentz, 1997).

CULTURAL CONTROL

Rouging infected trees is an effective method to control phony peach disease in areas of moderate disease intensity. This technique becomes less practical as disease intensity increases, and in regions of intense disease the entire orchard must be removed, a new orchard planted on a second plot of land, and the old site left fallow or planted with cover crops for two years before being replanted. Replanting a field of peaches to peaches is not recommended as this encourages high populations of damaging ring nematodes, *Criconemella xenoplax* Raski, in the soil that lead to peach trees having a short life (Nyczepar et al., 1983). Pruning is effective in the control of citrus variegated chlorosis (Amaral, Paiva and Souza, 1994). Removal of wild plum trees and plants from woodlots adjacent to orchards and vineyards is recommended. Before the glassy-winged sharpshooter was introduced into California, planting vineyards away from riparian areas was an effective way to control Pierce's disease. Indigenous sharpshooters in California are more abundant in the vegetated areas near streams and rivers. In contrast, the glassy-winged sharpshooter disperses over a large area and feeds and reproduces on many types of plants, many of them commonly found outside riparian areas.

CHEMICAL CONTROL

Chemical insecticides are used to control vectors of *X. fastidiosa* in commercial vineyards and orchards but have limited effectiveness in controlling disease development. A rating of the efficacy, residual action, and safety to beneficial insects of insecticides against glassy-winged sharpshooters on citrus (Grafton-Cardwell, Batkin and Gorden, 2003) indicates that attempts to control leafhopper and phony peach disease with dichloro-diphenyl-trichloroethane and encapsulated methyl parathion have not been successful (Evert, 1989). Imidacloprid is a systemic insecticide that has efficacy against homopterans in many crops (Mullins, 1993) and slows the development of Pierce's disease in grapes (Krewer, Dutcher and Chang, 2002) and of phony peach disease in peach and nectarine for the first five years after planting (Figures 2.1-2.5). It is also somewhat effective in reducing the rate of plum leaf scald development in plum (Dutcher, Krewer and Mullinix, 2005). The use of imidacloprid to reduce the rate of development of phony peach disease is the first time since rouging was introduced (Hutchins, 1933) that a treatment has shown any indication of reducing the severity of these diseases.

Vector control techniques that have shown promise against Pierce's disease include particle film sprays (Puterka et al., 2003). A thin layer

FIGURE 2.1. A vineyard in which susceptible bunch grapes have been planted among resistant muscadine grapes in southern Georgia offers a test area for control strategies for Pierce's disease. Muscadines offer a constant source of inoculum, and the climate supports a diverse array of vectors. Untreated bunch grapes typically die within two to five seasons after planting. Note the yellow sticky boards used for monitoring vectors (Krewer, Dutcher and Chang, 2002).

FIGURE 2.2. A thriving 4.5-year-old bunch grape vine in a vineyard treated each year with imidacloprid in southern Georgia (Krewer, Dutcher and Chang, 2002).

FIGURE 2.3. A dying 4.5-year-old bunch grape vine in the untreated control vineyard in southern Georgia (Krewer, Dutcher and Chang, 2002).

FIGURE 2.4. Scorch along the leaf margin is a common symptom on bunch grape vine infected with *X. fastidiosa* (Krewer, Dutcher and Chang, 2002).

FIGURE 2.5. Defoliation at the distal end of the leaf petiole is a common symptom on bunch grape vine infected with *X. fastidiosa* (Krewer, Dutcher and Chang, 2002).

of particle film formulated as Surround WP (Engelhard Corp., Iselin, NJ) in water and applied to the plant surface with an airblast sprayer effectively repels the glassywinged sharpshooter from host plants, leading to significant reductions in sharpshooter abundance and oviposition. The research, conducted in large replicated plots and supplemented with detailed insect behavior studies (Puterka et al., 2003), indicated that the particle film-treated plants were not preferred as host plants over untreated control plants.

BIOLOGICAL CONTROL

Host Plant Resistance

X. fastidiosa and other bacteria are commonly found in the xylem of plants, yet only certain plants have disease symptoms. Among host plants that have disease, there is often considerable variation in the severity of the symptoms among varieties. Plant breeders, especially in grape, have used this variation to find varieties and hybrids of high crop value that are also resistant to disease. In areas where *X. fastidiosa* is endemic, wild plants are generally resistant to disease, whereas disease occurs in introduced plants (Redak et al., 2004). Grape varieties and hybrids have a wide range of susceptibility to Pierce's disease (Loomis, 1958; Lu, 2000). Most bunch grape varieties die when planted in areas where Pierce's disease is endemic. European bunch varieties *(V. vinifera)* die within two to five years. American bunch varieties *(V. labrusca)* often live longer than five years. French-American hybrids are intermediate in susceptibility (Hartman et al., 2001). Southeastern U.S. grape production in areas prone to Pierce's disease is primarily based on muscadine grapes. Florida hybrid bunch grapes, a distinct race of grape developed by the Florida Agricultural Experiment Station, have high fruit quality and are resistant to Pierce's disease. The hybrids are produced by crossing *V. vinifera* with native grape species. One variety, 'Southern Home', was hybridized from muscadine (40 chromosomes) and bunch grape (38 chromosomes). 'Mortenson Hardy' is a bunch grape variety that has resistance to Pierce's disease. Resistance to Pierce's disease is also found in *Vitis caribea* × *V. vinifera* hybrids, but resistance breaks down when the vines reach eight to thirteen years of age.

These hybrids express low vigor but not marginal leaf scorch and have a life span of fifteen years or more in Central America (Ingalls and Jimenez, 2003). Susceptibility to infection varies among *V. vinifera* varieties. 'Chenin Blanc', 'Sylvaner', 'Ruby Cabernet', and 'White Riesling' are somewhat tolerant of the disease and recovery is faster than in less tolerant varieties after the infection. 'Thompson Seedless', 'Cabernet Sauvignon', 'Gray Riesling', 'Merlot', 'Napa Gamay', 'Petite Sirah', and 'Sauvignon Blanc' have intermediate recovery rates. 'Barbera', 'Chardonnay', 'Mission', 'Fiesta', and 'Pinot Noir' have lower recovery rates (Gubler et al., 2005).

Pecan in the southern United States is susceptible to pecan bacterial leaf scorch caused by *X. fastidiosa* (Sanderlin and Heyderich-Alger, 2000). The symptoms on pecan are necrosis of the leaf tissue beginning at the tip of the leaflet and progressing to its base. The leaflet usually abscises before the necrosis extends to its base (Sanderlin, 1998). 'Cape Fear' and 'Kiowa' varieties are more likely to show leaf scorch symptoms than other varieties. Disease in 'Cape Fear' pecan trees causes prolonged defoliation from mid-summer until the first freezes, and this reduces nut quality (Sanderlin and Heyderich-Alger, 2003). There is currently no control recommendation.

Plum leaf scald-tolerant Japanese plum varieties developed by plant breeders (Tangsukkasemsan, Norton and Boyhan, 1995) at Auburn University (AU-Cherry, AU-Roadside, AU-Rosa, and AU-Rubrum; Norton, Boyhan and Tangsukkasemsan, 1995) and the University of Florida (Gulfbeauty, Gulfruby, Gulfblaze, and Gulfrose; Gray, 2003) can be grown in the southeastern United States, though AU-Rosa has been found to develop symptoms and die within four years after planting in one highly infested area (Dutcher, Krewer and Mullinix, 2005; Figure 2.6).

Peach and nectarine varieties have no appreciable tolerance to phony peach disease, and no tolerant varieties are recommended (Ferree and Krewer, 1996). Producers in areas with significant incidence of phony peach disease have to start new orchards every seven to ten years. Orchards in these areas in southern Georgia typically have a very early harvesting period, and producers are able to sell the fresh peaches at a premium market price to make up for additional costs caused by phony peach disease. These early-season

FIGURE 2.6. Comparison of imidacloprid-treated (a) and untreated (b) 'AU-Rosa' Japanese plum trees 3.5 years after planting indicates that the treated tree has better foliage retention than the untreated tree. Treatment was applied in May and August each season during the first 3 years and once in May of the fourth year (Dutcher, Krewer and Mullinix, 2005). Pictures were taken in August of the fourth year.

peaches are quickly harvested, shipped, and sold across the eastern United States, often as individual fruit.

Natural Enemies

Biological control of glassy-winged sharpshooters in California is possible with mymarid egg parasites (Triaptisyn et al., 1998; Lopez et al., 2004). Egg-laying activity peaks in spring (April) and late summer (August). Parasitic hymenopterans attack the vector population at both peaks, though the incidence is higher in the late summer eggs. Mymarids (Hymenoptera: Mymaridae) that attack the eggs in the spring are imported to California from northern Mexico. Entomologists are searching the native range of glassy-winged sharpshooters in the southeastern United States, northern Mexico, and South America for additional natural enemy species. The parasites can be reared and imported into different areas as single species or combinations of species to complement the naturally occurring predators and parasites. Fungal entomopathogens are natural enemies of glassy-winged sharpshooters that are promising as a biological control (Kanga et al., 2004). In the southeastern United States, *H. coagulata* is one of several vagile, polyphagous sharpshooter species. *H. insolita, O. orbona,* and *G. versuata* are equally abundant (Krewer, Dutcher and Chang, 2002) and are vectors of *X. fastidiosa*. Sharpshooters are present in the orchards continuously from April to November in all fruit and nut crops. Peak activity varies considerably from year to year and from crop to crop within a given year. *O. orbona* typically has an early spring period of high abundance and is more abundant in muscadine grape and peach than in bunch grape or plum (Krewer, Dutcher and Chang, 2002). In pecan orchards, where trees are 20 m or more in height, these same sharpshooters are collected all season from the understory plants in the orchard and rarely from the tree crown. Spittlebugs are potential vectors, however, and are very common in pecan (Dutcher, unpublished material). These factors indicate that a generalist predator, parasite, or entomopathogen with a wide host range may be required for biological control of sharpshooters in the southeastern United States.

Symbiote Control

A new type of biological control is symbiote control, where bacteria that are found in the host or vector are genetically transformed to produce gene products that are lethal to the disease-causing agent. Innocuous bacteria in the system can be genetically transformed to produce proteins or other compounds that kill or neutralize the disease-causing bacteria. The genetically engineered microbes (GEMs) are ingested by the vectors and destroy the disease within the vector, thus breaking the disease cycle. The vector introduces the GEM to the host, where it kills the disease-causing agent. Paratransgenesis is a technique used in insect-borne diseases of man, such as Chagas disease, that cannot be practically controlled with conventional vaccines or insecticides (Beard, Durvasula and Richards, 1998, Beard et al., 2001; Durvasula et al., 2003). In grapevines, the xylem harbors many species of endophytic bacteria that have a limited effect on vine health. One research team working on Pierce's disease (Miller, 2003) has found and transformed a suitable bacterium and inserted a gene that forms a product that is lethal to *X. fastidiosa*. Limited field trials have shown that the GEM does not move to the fruit or stems after it is introduced into the xylem, is rapidly killed in the soil, and does not survive well among normal bacteria in the wild. The GEM is viewed as a microbial pesticide by the U.S. Environmental Protection Agency and may be considered for registration in the future. Considerable public opposition to the release of GEMs into the environment may slow the process even after an effective technique is found.

INTEGRATED MANAGEMENT

Two general types of control strategies are emerging for diseases caused by *X. fastidiosa*. The first is to break a weak link in the chain of events leading to disease using deactivating bacteria in the vector. The second is to apply controls wherever possible to stall the development of disease effectively. Control of the disease involves a cumulative effect through control of alternative host plants, control of vectors, and control of the bacteria within the plant and within the vector. No single tactic can control all these components. An integrated management strategy is required, and often used, by producers to keep disease incidence low in orchards and vineyards. Vector control techniques vary in cost of application and efficacy. Contact insecticides kill vectors quickly after application and can be reapplied when the residual activity expires on the plant. This can range from a few days in the case of natural pyrethrins and piperonyl butoxide, to several weeks for neonicotinoids applied to the foliage and to several months for imidacloprid applied to the soil as a systemic insecticide (Krewer, Dutcher and Chang, 2002; Grafton-Cardwell, Batkin and Gorden, 2003; Dutcher, Krewer and Mullinix, 2005). Injection of the stem base with an antibiotic solution (Hopkins and Mortenson, 1971) cures diseased grapevines that have symptoms but is costly (Peterson, 1998). Pressurized injection methods, originally developed to apply pesticides, silvicides, and nutrients in urban shade trees (Dutcher, Worley and Littrell, 1983), are more efficient, and new pressurized injectors can be used to treat sixty trees per hour (Doccola et al., 2003). Research is needed to test the efficacy of pressurized injection of antibiotics for control of bacteria in the xylem and insecticides for control of vectors.

REFERENCES

Almeida, R.P.P. and A.H. Purcell. (2003). Transmission of *Xylella fastidiosa* to grapevines by *Homalodisca coagulata* (Hemiptera: Cicadellidae). *Journal of Economic Entomology* 96: 265-271.

Alves, E. (2004). Application of electron microscopy to the study of host-pathogen interactions in plants infected by the bacterium *Xylella fastidiosa. Microscopy and Microanalysis* 10: 1438-1439.

Amaral, A.M.D., L.V. Paiva and M.D. Souza. (1994). Effect of pruning in Valencia and Pera Rio orange trees (*Citrus sinensis* L. Osbeck) with symptoms of citrus variegated chlorosis (CVC). *Ciencia e Pratica* 18: 306-307.

Beard, C.B., E.M. Dotson, P.M. Pennington, S. Eichler, C. Cordon-Rosales and R.V. Durvasula. (2001). Bacterial symbiosis and paratransgenic control of vector-borne Chagas disease. *International Journal for Parasitology* 31: 621-627.

Beard, C.B., R.V. Durvasula and F.F. Richards. (1998). Bacterial symbiosis in arthropods and the control of disease transmission. *Emerging Infectious Diseases* 4: 581-591.

Blua, M.J., R.A. Redak, D.J.W. Morgan and H.S. Costa. (2001). Seasonal flight activity of two *Homalodisca* species (Homoptera: Cicadellidae) that spread *Xylella fastidiosa* in southern California. *Journal of Economic Entomology* 94: 1506-1510.

Doccola, J.J., P.M. Wild, I. Ramasamy, P. Castillo and C. Taylor. (2003). Efficacy of arbojet viper microinjections in the management of hemlock woolly adelgid. *Journal of Arboriculture* 29: 327-330.

Durvasula, R.V., R.K. Sundaram, C. Cordon-Rosales, P. Pennington and C.B. Beard. (2003). *Rhodnius prolixus* and its symbiont, *Rhodococcus rhodnii*: A model for paratransgenic control of disease transmission. In *Insect Symbiosis*, K. Bourtzis and T.A. Miller, eds. CRC Press, Florida, pp. 83-95.

Dutcher, J.D., G.W. Krewer and B. Mullinix. (2005). Imidacloprid insecticide slows development of phony peach and plum leaf scald. *HortTechnology* 15: 642-645.

Dutcher, J.D., K.C. McGiffin and J.N. All. (1988). Entomology of muscadine grapes. In *The Entomology of Indigenous and Naturalized Systems in Agriculture,* M.K. Harris and C.R. Rogers, eds. Westview Studies in Insect Biology, Westview Press, Boulder, Colorado, 238p.

Dutcher, J.D., R.E. Worley and R.H. Littrell. (1983). Trunk injection for pecan tree health. *University of Georgia Experiment Station Bulletin* 296, 12p.

Evert, D.R. (1989). Phony peach. In *Peach Production Handbook,* S.C. Myers, ed. Univesity of Georgia Cooperative Extension Service Handbook, Vol. 1, pp. 33-37.

Ferree, M.E. and G.W. Krewer. (1996). Peaches and nectarines. *University of Georgia Cooperative Extension Service Circular* 741(revised), 2p.

Fletcher, J. and A. Wayadande. (2002). Fastidious vascular-colonizing bacteria. *The Plant Health Instructor.* American Phytopathological Society APSnet Education Center. DOI: 10.1094/PHI-I-2002-1218-02.

Grafton-Cardwell, B., T. Batkin and J. Gorden. (2003). *Glassy-Winged Sharpshooter and Related Bacterial Disease, 2nd edn.* Citrus Research Board, Visalia,

California, 15p. Website: http://citrusent.uckac.edu/sharpshooter%202003.pdf (last visted July 11, 2005).

Gray, D. (2003). New plants in Florida: Grape. *University of Florida, Fla. Agric. Exp. Stn. Florida Cooperative Extension Service Circular* 1440 p.

Gubler, W.D., J.J. Stapleton, G.M. Leavitt, A.H. Purcell, L.G. Varela and R.J.Smith. (2005). Grape pierce's disease pathogen: *Xylella fastidiosa.* In *University of California. IPM Pest Management Guidelines: Grape.* UC ANR Publ. 3448.

Hartman, J., D. Saffray, D. Perkins, J. Strang, R. Bessin and J. Beale. (2001). Pierce's disease: A new disease of grapes in Kentucky. University of Kartucky Co-operative Extension Service. Headsup Pest Alert. Website: http://www.uky.edu/Agriculture/IPM/headsup/grapes.htm (last visited March 27, 2005).

Henderson, M., A.H. Purcell, D. Chen, C. Smart, M. Guilhabert, and B. Kirkpatrick. (2001). Genetic diversity of pierce's disease strains and other pathotypes of *Xylella fastidiosa. Applied and Environmental Microbiology* 67: 895-903.

Hill, B.L. and A.H. Purcell. (1995a). Acquisition and retention of *Xylella fastidiosa* by an efficient vector. *Phytopathology* 85: 209-212.

Hill, B.L. and A.H. Purcell. (1995b). Multiplication and movement of *Xylella fastidiosa* within grape and four other plants. *Phytopathology* 85: 1368-1372.

Hill, B.L. and A.H. Purcell. (1997). Populations of *Xylella fastidiosa* in plants required for transmission by an efficient vector. *Phytopathology* 87: 1197-1201.

Hoddle, M.S. (2004). The potential adventive geographic range of glassy-winged sharpshooter, *Homalodisca coagulata* and the grape pathogen, *Xylella fastidiosa*: Implications for California and other grape growing regions of the world. *Crop Protection* 23: 691-699.

Hoddle, M.S., S.V. Triapitsyn and D.J.W. Morgan. (2003). Distribution and plant association records for *Homalodisca coagulata* (Hemiptera: Cicadellidae) in Florida. *Florida Entomologist* 86: 89-91.

Hopkins, D.L. (1989). *Xylella fastidiosa*: Xylem-limited bacterial pathogen of plants. *Annual Review Phytopathology* 27: 271-290.

Hopkins, D.L. and J.A. Mortensen. (1971). Suppression of Pierce's disease symptoms by tetracycline antibiotics. *Plant Disease Reporter* 55: 610-612.

Hopkins, D.L. and A.H. Purcell. (2002). *Xylella fastidiosa*: Cause of Pierce's disease of grapevine and other emergent diseases. *Plant Disease* 86:1056-1066.

Hutchins, L.M. (1993). Identification and control of the phony disease of peach. *Georgia Official State Entomology Bulletin* 78: 1-55.

Ingalls, A. and A.L.G. Jimenez. (2003). Twenty years of grape breeding in the tropics: The stress pathogen *Botryodiplodia* confused the ability to select for resistance to Pierce's disease. *International Society for Horticultural Science Acta Horticulturae* 528 p.

Kanga, L.H.B., W.A. Jones, R.A. Humber and D.W. Boyd Jr. (2004). Fungal pathogens of the glassy-winged sharpshooter *Homalodisca coagulata* (Homoptera: Cicadellidae). *Florida Entomologist* 87: 225-228.

Krewer, G., J.D. Dutcher and C.J. Chang. (2002). Imidacloprid insecticide slows development of Pierce's disease in bunch grapes. *Journal of Entomological Science* 37: 101-112.

Leu, L.S. and C.C. Su. (1993). Isolation, cultivation, and pathogenicity of *Xylella fastidiosa,* the causal bacterium of pear leaf scald in Taiwan. *Plant Disease* 77: 642-646.

Lima, J. (2005). Photo gallery, *Xyella fastidiosa*—Amarelinho (citrus variegated chlorosis, CVC) bacterium. Agrcitrus Mudas e Productos Agricolus Ltda. Casa Branca, Brazil. Website: http://www.citrolima.com.br/enindicexylella.htm (last visited March 28, 2005).

Lima, J.E.O., V.S. de Miranda, S.R. Roberto, A. Coutinho, R. Palma and A.C. Pizzolitto. (1997). Diagnose da clorose variegada dos citros por microscopia ótica. *Fitopatol Bras* 22: 370-374.

Loomis, N.H. (1958). Performance of *Vitis* species in the south as an indication of their relative resistance to Pierce's disease. *Plant Disease Reporter* 42: 833-836.

Lopez, R., R.F. Mizell, P.C. Andersen and B.V. Brodbeck. (2004). Overwintering biology, food supplementation and parasitism of eggs of *Homalodisca coagul*ata (Say) (Homoptera: Cicadellidae) by *Gonatocerus ashmeadi* Girault and *Gonatocerus morrilli* (Howard) (Hymenoptera: Mymaridae). *Journal of Entomological Science* 39(2): 214-222.

Lu, J. (2000). The Pierce's disease resistant grapes in the southeast United States. *American Journal Enology and Viticulture* 51: 285-286.

McEachern, G.R., L. Stein and J. Kamas. (2004). Growing Pierce's disease resistant grapes in central, south and east Texas. Texas Cooperative Extension Service, Ext. Info. Resources. Website: http://aggie-horticulture.tamu.edu/extension/publications.html (last visited March 23, 2005).

Meadows, R. (2001). Scientists, state aggressively pursue Pierce's disease. *California Agriculture* 55: 4.

Miller, T.A. (2003). Designing Insects. Issues in Biotechnology. ActionBioscience.org. American Institute of Biological Science. Online article. Website: http://www.actionbioscience.org/biotech/miller.html (last visited March 31, 2005).

Minsavage, G.V., C.M. Thompson, D.L. Hopkins, L.C. Leite and R.E. Leite. (1994). Development of a polymerase chain reaction protocol for detection of *Xylella fastidiosa* in plant tissue. *Phytopathology* 84: 456-461.

Mizell, R.F., P.C. Anderson, C. Tipping and B. Brodbeck. (2003). *Xylella fastidiosa* diseases and their leafhopper vectors. *University of Florida Extension IFAS Bulletin.* ENY-683: 1-7.

Mullins, J.W. (1993). Imidacloprid, a new nitroguanidine insecticide. ACS Symposium Series. *American Chemical Society* 524: 183-198.

Norton, J.D., G.E. Boyhan and B. Tangsukkasemsan. (1995). Breeding for resistance to plum leaf scald. *Hortscience* 30: 431.

Nyczepar, A.P., E.I. Zehr, A.C. Lewis and D.C. Harshman. (1983). Short life of peach induced by *Criconemella xenoplax. Plant Disease* 67: 507-508.

Paiao, F.G., A.M. Meneguim, E.C. Casagrande and R.P. Leite. (2002). Envolvimento de cigarras (Homoptera, Cicadidae) na tansmisao de *Xylella fastidiosa* em cafeeiro. *Fitopatol Bras* 27: 67.

Peterson, R. (1998). Pierce's disease cure?—Injection of antibiotics into living grapevines. *Wines and Vines* 79: 66-68.

Pooler, M.R., I.S. Myung and J. Bentz. (1997). Detection of *Xylella fastidiosa* in potential insect vectors by immunomagnetic separation and nested polymerase chain reaction. *Letters in Applied Microbiolilogy* 25: 123-126.

Purcell, A.H. (1997). *Xylella fastidiosa,* a regional problem or global threat? *Journal of Plant Pathology* 79: 99-105.

Purcell, A.H. (2005). *Xylella fastidiosa*—A scientific and community internet resource on diseases caused by the bacterium *Xylella fastidiosa.* University of California, Berkley, California. Website: http://nature.berkeley.edu/xylella/index.html (last visited March 29, 2005).

Purcell, A.H. and H. Feil. (2001). Glassy-winged sharpshooter. *Pesticide Outlook* 11: 199-203.

Purcell, A.H. and D.L. Hopkins. (1996). Fastidious xylem-limited bacterial plant pathogens. *Annual Review of Phytopathology* 34: 131-151.

Puterka, G.J., M. Reinke, D. Luvisi, M.A. Ciomperik, D. Bartels, L. Wendel and D.M. Glenn. (2003). Particle film, surround WP, effects on glassy-winged sharpshooter behavior and its utility as a barrier to sharpshooter infestations in grape. Online Publication. Plant Health Progress. DOI:10.1094/PHP-2003-0321-01-RS.

Raju, B.C. and J.M. Wells. (1986). Diseases caused by fastidious xylem-limited bacteria and strategies for management. *Plant Disease* 70: 182-186.

Redak, R.A., A.H. Purcell, J.R.S. Lopes, M.J. Blua, R.F. Mizell and P.C. Anderson. (2004). The biology of xylem fluid-feeding insect vectors of *Xylella fastidiosa* and their relation to disease epidemiology. *Annual Review of Entomology* 49: 243-270.

Rodrigues, J.L.M., M.E. Silva-Stenico, J.E. Gomes, J.R.S. Lopes and S.M. Tsai. (2003). Detection and diversity assessment of *Xylella fastidiosa* in field-collected plant and insect samples by using 16S rRNA and gyrB sequences. *Applied Environmental Microbiology* 69: 4249-4255.

Sanderlin, R.S. (1998). Evidence that *Xylella fastidiosa* is associated with pecan fungal leaf scorch. *Plant Disease* 82: 264.

Sanderlin, R.S. and K.I. Heyderich-Alger. (2000). Evidence that *Xylella fastidiosa* can cause leaf scorch disease of pecan. *Plant Disease* 84(12): 1282-1286.

Sanderlin, R.S. and K.I. Heyderich-Alger. (2003). Effects of pecan bacterial leaf scorch on growth and yield components of cultivar Cape Fear. *Plant Disease* 87(3): 259-262.

Severin, H.H.P. (1950). Spittle-insect vectors of Pierce's disease virus. II. Life history and virus transmission. *Hilgardia* 19: 357-382.

Simeone, G.W. (2000). Disease control in peach (*Prunus persicae*). *University of Florida, Institute of Food and Agricultural Sciences.* PDMG-V3-26: 1-20.

Simpson, A.J., F.C. Reinach, P. Arruda, F.A. Abreu, M. Acencio, R. Alvarenga, L.M. Alves, et al. (2000). The genome sequence of the plant pathogen *Xylella fastidiosa. Nature* 406: 151-157.

Tangsukkasemsan, B., J.D. Norton and G.E. Boyhan. (1995). The occurrence of plum leaf scald on plum cultivars in Alabama. *Hortscience* 30: 437.

Teviotdale, B.L. (2003). Almond leaf scorch. *University of California Division of Agriculture and Natural Resources Publication* 8106: 1-4.

Triaptisyn, S.V., R.F. Mizell, J.L. Bosart and C.E. Carlton. (1998). Egg parasitoids of *Homalodisca coagulata* (Homoptera: Cicadellidae). *Florida Entomologist* 8: 241-243.

Varela, L.G., R.J. Smith and P.A. Phillips. (2001). Pierce's disease. *University of California Division of Agriculture and Natural Resources Publication* 21600: 1-20.

Wells, J.M., B.C. Raju and G. Nyland. (1983). Isolation, culture and pathogenicity of the bacterium causing phony disease of peach. *Phytopathology* 73: 859-862.

Wells, J.M., B.C. Raju, H.Y. Hung, W.G. Weisburg, P.L. Mandelco and D.J. Brenner. (1987). *Xylella fastidiosa* gen. nov., sp. nov.: Gram-negative, xylem-limited, fastidious plant bacteria related to *Xanthomonas* spp. *International Journal of Systematic Bacteriology* 37: 136-143.

Chapter 3

Synthetic and Living Mulches for Control of Homopteran Pests and Diseases in Vegetables

Oscar E. Liburd
Daniel L. Frank

INTRODUCTION

The management of plant viruses is often a complex and challenging task because they depend almost entirely on insect vectors for their transmission. This dependence can be complicated by a number of relationships that occur between virus, vector, and host. It is because of this that crop-plant physiological disorders and insect-transmitted diseases have become serious problems for many growers around the world. It has been estimated that homopteran pests spread more than 90 percent of insect-borne plant diseases (Eastop, 1977). Among the most important vectors of plant viruses are aphids, which have been estimated to transmit 275 different viral disorders (Nault, 1997). Whiteflies, too, are considered important vectors of plant viruses but can also be problematic because of their ability to induce plant physiological disorders such as squash silverleaf disorder (SSL).

In addition to disease transmission, heavy infestations of aphids and whiteflies generally cause a reduction in plant vigor (Barlow, Randolph and Randolph, 1977; Buntin, Gilbertz and Oetting, 1993; Frank, 2004). The excretion of honeydew by these insects can serve as an important medium for promoting growth of sooty mold fungi

Management of Nematode and Insect-Borne Plant Diseases

doi:10.1300/5754_03

(*Capnodium* spp.), which can further reduce plant vigor and yield (Byrne and Miller, 1990; Palumbo et al., 2000). The unpredictability and severity of these pests and associated diseases in conjunction with injury from secondary pests make efficient management strategies necessary in any production system.

Numerous management practices have been evaluated for control of these pest species. However, only sound production practices will incorporate management that not only reduces aphid and whitefly numbers but also reduces vector efficiencies and disease spread. Living and synthetic mulches have been used as an effective means for accomplishing these goals (Figures 3.1 and 3.2). Mulches can reduce aphid and whitefly numbers while also delaying the onset and spread of associated insect-borne diseases. Several studies have shown that synthetic mulch offers significant protection from these pests compared with bare-ground plantings (Brown et al., 1993; Summers et al., 1995; Summers and Stapleton, 2002; Frank, 2004). In addition, research conducted by Hooks, Valenzuela and Defrank (1998) and Frank (2004) has shown that living mulches are an effective means for reducing multiple pest complexes and the incidence of associated diseases compared with bare ground (Figure 3.3).

FIGURE 3.1. Squash grown with buckwheat (living) mulch (Photograph courtesy of Daniel L. Frank, University of Florida).

FIGURE 3.2. Squash grown with reflective (synthetic) mulch (Photograph courtesy of Daniel L. Frank, University of Florida).

FIGURE 3.3. Squash grown with white (synthetic) mulch (Photograph courtesy of Daniel L. Frank, University of Florida).

To understand why mulches are effective at reducing these vector populations, we must first understand the biology and behavior of these insects. We must then consider other management practices that are associated with controlling these pests. Efficient control of aphids and whiteflies as well as any associated diseases requires a broad knowledge of current management tactics and the limitations involved with each.

APHIDS

Aphids can be found on a variety of host plants throughout the world. Their ability to transmit numerous plant viruses coupled with their high reproductive rates make them an important pest in agricultural systems. Although feeding by aphids can reduce plant vigor and yield (Barlow, Randolph and Randolph, 1977; Breen and Teetes, 1986), transmission of viruses is a more significant problem to growers worldwide. In addition, viruses are typically transmitted by transient alate species, creating further management difficulties (Broadbent, Chaudhuri and Kapica, 1950; Swenson, 1968). Other factors, such as their small size and highly polymorphic nature, can make identification of problematic aphid species difficult.

Biology and Behavior

The life cycle of many aphid species can become rather complex. Variations in hosts, reproduction, biology, and activity can occur within the annual cycle of a single species (Kring, 1959). Aphids can be grouped as monoecious (non-host alternating) or heteroecious (host alternating). Monoecious species require only one or a few related host plant species, whereas heteroecious species generally require two or more different species of plants in separate families to complete their life cycle.

Reproduction is accomplished through sexual (holocyclic) and/or parthenogenetic (anholocyclic) forms. In warm climates and areas where continual production of crops occurs, sexual forms tend not to be as important (Miyazaki, 1987). Females that reproduce parthenogenetically are viviparous and can produce a new generation in as little time as a week. With a short reproductive and developmental time, numerous generations are possible within one production season.

Dispersal is accomplished by alate, or winged morphs, and apterous, or wingless aphids, typically remain restricted to the plant surface. During flight, alate aphids respond strongly to visual stimuli and locate potential host plants by contrasting the soil background with plant foliage (Kring, 1972; Liburd, Casagrande and Alm, 1998). Aphids determine host suitability after landing and making brief, shallow test probes with their stylets on potential host plants. If the plant is not suitable, the aphid will move to another potential host, repeating the process. This behavior is especially important as transmission of stylet-borne viruses can occur from these test probes before aphid management is implemented (Nault, 1997). Moreover, these viruses can reach epidemic proportions as a result of secondary spread.

Monitoring

Aphids generally respond to two different wavelengths of light. Once alate aphids have completed their development of wings, they become strongly attracted to shortwave or ultraviolet light. This attraction induces the aphid to fly toward the sky in either a high-level migratory flight, which can last from one to several hours, or a low-level nonmigratory flight over short distances (Kring, 1972). High wavelengths of light, such as yellow and green, tend to stimulate alighting and settling behavior (Kring, 1967; Webb, Kok-Yokomi and Voegtlin, 1994; Liburd, Casagrande and Alm, 1998). In general, plants that show obvious yellowing symptoms will have greater aphid populations (Zitter and Simons, 1980). As a correlation can generally be seen between the number of alate aphids trapped and the percentage of virus infection for a particular crop (Zitter and Simons, 1980), water pan-traps have been used as an effective monitoring tool (Heathcote, 1957; Adlerz, 1987; Frank 2004). Pan-traps are an efficient means to track aphid activity within a particular field.

SILVERLEAF WHITEFLIES

The silverleaf whitefly, *Bemisia argentifolii* (Bellows and Perring), occurs in many tropical and subtropical habitats as well as in greenhouses. Research has shown that direct feeding pressure and deposition of honeydew by adult whitefly populations can cause significant damage and economic loss to a variety of cultivars (Van Lenteren and Noldus, 1990; Perring et al., 1993). However, the potential for silverleaf whitefly to act as a vector for virus and its ability to induce plant impairment have recently garnered particular attention (Costa and Brown, 1990; Yokomi, Hoelmer and Osborne, 1990; Schuster, Kring and Price, 1991; Polston and Anderson, 1997). One important disorder, SSL, is characterized by the silvering of the adaxial leaf surface of cucurbit plants. This disorder is typically seen in late summer and fall crops and can cause substantial economic losses to growers.

Biology and Behavior

The silverleaf whitefly is primarily polyphagous and colonizes predominantly annual, herbaceous plants (Brown, Frohlich and Rosell, 1995). Four nymphal instars characterize whitefly development from egg to pupa. The first nymphal instars are usually referred to as crawlers and are capable of limited movement, whereas the second, third, and fourth instars are immobile. Completion of all four instars takes approximately seventeen to twenty-one days under optimum conditions. Females are generally larger than males throughout all instars (Tsai and Wang, 1996). The nymphal stage of the whitefly life cycle can induce a variety of phytotoxic disorders.

After development through the nymphal and pupal stages, adult whiteflies emerge. Movement of adults for feeding and oviposition occurs on the shaded abaxial surface of suitable host leaves (Simmons, 1994). Females live for ten to twenty-four days and can lay anywhere from 66 to 300 eggs (Tsai and Wang, 1996). They prefer to oviposit on young leaves, which creates a stratification of different life stages as the host plant grows (Gould and Naranjo, 1999). Similarly, factors such as leaf shape, color, and nitrogen content can affect ovipositional preference for whiteflies (Mound, 1962; Butler, Henneberry and Wilson, 1986; Bentz et al., 1995). Wind is the main mechanism for dispersal over long and short distances. Unintentional transport by people is also responsible for dispersal of whiteflies (Blackmer and Byrne, 1993; Brown, Frohlich and Rosell, 1995).

Monitoring

Immature whiteflies spend a large amount of their time in an immobile feeding stage. Monitoring for immatures has involved removal of disks from selected leaves. These leaf disks are then examined under a microscope for identification of nymphal stages and/or species (Hook et al., 1998; Gould and Naranjo, 1999; Frank, 2004).

Whitefly adults respond strongly to visual cues in their environment. For instance, adult whiteflies have a strong attraction to wavelengths of light falling in the yellow spectrum of the visible color range (Mound, 1962). Yellow sticky traps have been used effectively in many crop systems for monitoring whiteflies and their distribution among selected plots (Frank, 2004; Frank and Liburd, 2005).

APHID AND WHITEFLY DISEASES

Aphids and whiteflies are excellent vectors for a number of plant viruses. Their piercing-sucking mouthparts cause minimal damage to plant cells when they feed, which is optimal for viruses because they require living cells to reproduce. In addition, aphids and whiteflies are tissue-specific feeders that require phloem, which is where many plant viruses reside.

Aphids transmit more viruses than any other homopteran insect (Nault, 1997). These viruses can be transmitted in several different ways and can have several different modes of action. The most common aphid-transmitted viruses are the potyviruses, which are transmitted nonpersistently. It is believed that the main attachment of these viruses occurs within the maxillary food canal in aphids. These viruses can be particularly problematic because acquisition and inoculation can occur in a matter of seconds. Moreover, they have no latent period within the vector and have an infectivity of a few hours or days. Efforts to control these viruses can be made more difficult because transient alate aphid species are generally the cause of infection in plants.

Whiteflies are the second-most important vectors of plant viruses within Homoptera and are known to transmit at least 41 different viral disorders (Nault, 1997). The most important whitefly-transmitted viruses are the geminiviruses, which are transmitted persistently. These circulative viruses are acquired and inoculated after long feeding probes and are generally transmitted for weeks or the life of the insect. Efforts to control these viruses are usually not as complex as with aphids because of the sedentary nature of immatures (Perring,

Gruenhagen and Farrar, 1999). However, the silverleaf whitefly's ability to induce SSL can be of particular concern. SSL (Figure 3.4) is associated with the feeding of immature whiteflies. Variations in the feeding densities of immatures have been shown to affect the severity of SSL symptoms, which can develop in as little as 14 days (Yokomi, Hoelmer and Osborne, 1990; Schuster, Kring and Price, 1991; Costa et al., 1993). In squash, feeding by as few as two to three nymphs per plant can induce SSL (Costa et al., 1993). The silvering of leaves is believed to be a plant response to feeding, and its severity varies directly with the number of feeding immatures (Yokomi, Hoelmer and Osborne, 1990; Schuster, Kring and Price, 1991; Costa et al., 1993; Jiménez, Shapiro and Yokomi, 1993). It is hypothesized that air spaces between the epidermis and palisade cells of leaves cause the characteristic silver color. This change in color, chlorophyll content, and light reflectance reduces the photosynthetic ability of the plant species (Burger, Schwartz and Paris, 1988; Jimenez et al., 1995). It has been

FIGURE 3.4. Squash silver leaf disorder in zucchini (Photograph courtesy of Daniel L. Frank, University of Florida).

estimated that photosynthesis can be reduced by up to 30 percent in severe cases (Burger, Schwartz and Paris, 1988; Cardoza, McAuslane and Webb, 2000).

MANAGEMENT OF APHIDS AND WHITEFLIES

Broad-Spectrum and Reduced-Risk Pesticides

Traditionally, broad-spectrum pesticides play a major role in the management of the most important agricultural insect pests. However, this control strategy is problematic and can lead to increased production costs as well as an increased potential for resistance. The problems associated with the use of broad-spectrum pesticides and their effects on nontarget organisms and the environment have spurred the search for alternative methods of pest control. Oils have been regarded as an alternative and have been used effectively to inhibit the transmission of aphid and whitefly viruses (Simons and Zitter, 1980; Butler, Coudriet and Henneberry, 1989; Puri et al., 1994). Oils are thought to affect the behavior of vectors or impede the infection process even though the virus is still transmitted (Perring, Gruenhagen and Farrar, 1999). Other alternatives to broad-spectrum insecticides include reduced-risk insecticides, which are more selective and less toxic. Commonly used reduced-risk insecticides for aphid and whitefly management include the neonicotinoids imidacloprid (Admire) and thiamethoxam (Actara and Platinum). These reduced-risk insecticides can be used effectively to suppress homopteran pests on commercial farms.

Host-Plant Resistance

Economic losses via virus or plant disorders have forced some growers to use seedlings crossprotected with mild virus strains for resistance (Cho et al., 1992). However, the highly variable nature of these virus strains (Lisa and Lecoq, 1984) and the fact that homopteran pests can injure plants in several ways (Jackson et al., 2000) make the success of using crossprotection uncertain in infected areas. Furthermore, crop plantings may be prone to attack by several viral diseases within the same planting period, making it difficult to develop varieties that are resistant to all of these strains (Purcifull et al., 1988).

In addition, developing plants to be physically resistant to a specific insect pest may also have a detrimental impact on their associated parasitoids by disrupting their searching behavior or entrapping them (Gruenhagen and Perring, 1999), subsequently reducing natural control of insect pests.

Row Covers

Row covers have been used successfully to reduce the incidence of aphids and whiteflies and also delay the spread of associated virus symptoms (Perring, Royalty and Farrar, 1989; Webb and Linda, 1992; Costa, Johnson and Ullman, 1994). In addition, they are effective at excluding a number of other pest species. Despite these benefits, row covers are often difficult and time consuming to use. They can reduce the level of light to plants, reducing photosynthesis and yields. In addition, weed management under covers can be difficult, and the amount of effective pollination can be reduced.

Synthetic Mulches

Synthetic mulches (polyethylene films) have been commercially available to vegetable growers for several decades. They have been used for a variety of vegetable crops, including tomatoes, peppers, corn, potatoes, and various cucurbits (Greer and Dole, 2003). Synthetic mulches have been used as an efficient means to reduce many crop pests (Costa, Johnson and Ullman, 1994; Csizinszky, Schuster and Kring, 1997; Smith et al., 2000, Frank and Liburd, 2004, 2005). In general, they regulate soil temperature, can increase the speed of above-ground plant growth (Schalk et al., 1979; Decoteau, Kasperbauer and Hunt, 1989), and also affect the flight pattern and behavior of insect pest species (Zitter and Simons, 1980; Costa and Robb, 1999). This ability to attract or repel insects has been important in reducing the incidence and severity of pest species and associated diseases. The type and color of these mulches have changed over the years. In Florida, where a large percentage of the winter vegetables for North America are produced, vegetable growers have used white or white-on-black mulches in the fall, and black in the winter and spring. Black plastic mulches are often the least expensive choice because of their wide availability, and they are used more often than any other plastic mulch (Lamont, 1993). Black plastic mulches are ideally suited for growing winter and spring vegetable crops because they can warm the soil 4-5°F and eliminate most weed problems that occur on vegetable beds (Hopen and Oebker, 1976; Greer and Dole, 2003). White and white-on-black plastic mulches have been used extensively for summer and fall crops in Florida because they keep soil temperatures cooler than other plastic mulch types.

During the 1950s, research began focusing on the use of reflective or aluminum mulches as an alternative to the traditional standards. As the name implies, a reflective mulch reflects light, which repels aphids, whiteflies, and other pest species (Adlerz and Everett, 1968; Greer and Dole, 2003; Summers, Mitchell and Stapleton, 2004). Studies involving reflective mulches have shown mostly positive benefits such as reduced pest numbers, higher yields, delayed or reduced viral onset, and increased plant size and growth. In addition, reflective mulches as well as white and black synthetic mulches have been used as an alternative to conventional pesticides (Frank and Liburd, 2004, 2005).

Within the past decade a variety of other color mulches have been introduced. These include blue, red, yellow, orange, green, pink, and even clear mulch (Greer and Dole, 2003; Csizinszky, Schuster and Kring, 1995, 1997). Studies involving colored mulches have produced mixed results. A variety of factors, including climate, seasonal condition, and plant taxa, have been postulated as potential causes for the varying results in terms of crop growth and productivity (Mahmoudpour and Stapleton, 1997).

Despite the many benefits that synthetic mulches offer, disposal following crop termination can be problematic and may interfere with routine farming practices such as cultivation. Moreover, a reduction in the effectiveness of synthetic mulches can occur as the area of foliage increases throughout the growing season. One alternative to synthetic mulch is water-soluble biodegradable silver spray mulch (Liburd, Casagrande and Alm, 1998). However, weathering and dust and soil accumulation can decrease the effectiveness and make biodegradable mulches ineffective (Summers et al., 1995).

Living Mulches

Living mulches (diversified crops) can be thought of as "protection crops" if they are planted within or around a crop field. Living mulches increase diversity and the overall sustainability of many cropping systems. They are essentially cover crops grown within a marketable cash crop to reduce erosion, enhance fertility, improve soil quality, and suppress weeds while also reducing the incidence and severity of pest insects. Root (1973) suggested two hypotheses as to why herbivore loads are reduced in these diverse crop habitats. The resource concentration hypothesis predicts that herbivores will more readily find, reproduce in, and remain in monoculture environments of their host plants because resources are more localized and habitable within these areas. His other hypothesis theorized that predators and parasitoids are more effective in floristically diverse ecosystems because of the availability of greater and more diverse food resources and refuge.

Living mulches can provide additional feeding sites for potential infectious insects, reducing the spread of nonpersistent viruses and disorders throughout the cash crop. Toba et al. (1977) demonstrated that wheat, *Triticum aestivum* L., delayed the frequency and severity of aphid transmitted nonpersistent viruses in cantaloupe. In a similar study, Bernays (1999) showed that the whitefly, *Bemisia tabaci* Gennadius, spent less time on plants exposed to environments containing multiple plant species. These whiteflies also demonstrated greater levels of restlessness than whiteflies in monoculture environments, which led Bernays to suggest that whitefly movement within field situations could be limited in homogeneous environments.

Additional studies have shown that fewer whiteflies (Hooks, Valenzuela and Defrank, 1998), aphids (Smith, 1969; Kloen and Altieri, 1990; Costello and Altieri, 1995; Hooks, Valenzuela and Defrank, 1998), and insect-transmitted diseases (Power,1987; Hussein and Samad, 1993; Hooks, Valenzuela and Defrank, 1998) occur in habitats with diversified crops than in habitats with monoculture crops. For instance, Hooks, Valenzuela and Defrank (1998) planted zucchini (*Cucurbita pepo* L.) between rows containing buckwheat (*Fagopyrum esculentum* Moench) and yellow mustard (*Sinapis alba* L.) and found that densities of *Aphis gossypii* Glover and *B. argentifolii* (Figures 3.5 and 3.6) and the conveyance of aphid-transmitted diseases and SSL on zucchini were reduced in these living mulch treatments when compared with monoculture plantings. Additional research by Frank (2004) and Frank and Liburd (2005) has shown that living mulch

FIGURE 3.5. Melon aphids, *Aphis gossypii* (Photograph courtesy of James L. Castner, University of Florida).

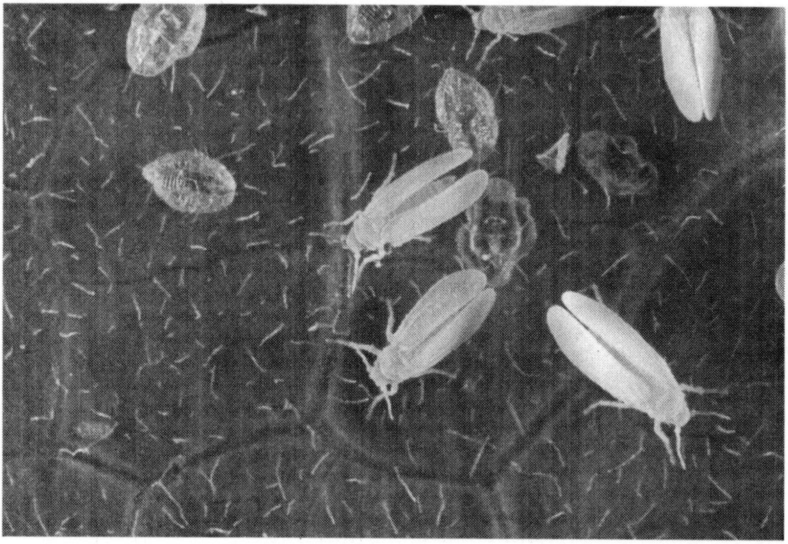

FIGURE 3.6. Silverleaf whiteflies, *Bemesia argentifolii* (Photograph courtesy of Lyle J. Buss, University of Florida).

plots containing buckwheat and white clover have higher numbers of natural enemies than synthetic mulch or bare-ground plots.

Despite these benefits, it is often difficult to determine the best type of living mulch to use in a cropping system containing multiple pest species. Special care must be taken to find mulches that will not act as an additional host for disease transmission. Additional factors such as climate, soil type, timing, and establishment play an important role in determining the effectiveness of living mulches. Other factors such as competition between living mulch and crops can significantly reduce yields (Andow et al., 1986; Andow, 1991; Frank, 2004; Frank and Liburd, 2005).

ECONOMICS

Economic gains and ease of incorporation play central roles in whether a particular tactic is incorporated into an IPM program. Although synthetic mulches are likely to provide higher yields, the cost of establishment is higher than with living mulch. The cost of seeds for living mulch for one productive season is significantly less than that of synthetic mulches. Also, the cost of seeds can be further reduced if proper management is initiated. For instance, if living mulch is managed so that it produces seeds throughout the year, it will not require subsequent planting later in the season (Platt, Caldwell and Kok, 1999). This can reduce the overall cost of implementing living mulch. It should be noted that although the costs of living mulches are less than those of synthetic mulches, they require additional upkeep and management. For instance, living mulches require additional water and must be maintained so that they do not become a weed problem later in the year.

Synthetic mulch may be easier to maintain than living mulch. It requires no establishment period before vegetable crops are planted and is not weather dependent. In addition, no special watering or additional drip lines are needed in plots treated with synthetic mulch. Harvesting fruit on synthetic mulches can also be easier, because individual fruits are not concealed by extra vegetation.

Current management of aphids and whiteflies within vegetable crops requires the use of multiple IPM tactics. The use of pesticides, oils, row covers, and cross-protection in conjunction with mulches may help to limit pest outbreaks and the occurrence of transmitted diseases. Understanding how whiteflies and aphids respond to different mulching systems allows development of new and effective approaches for integration with current management practices.

REFERENCES

Adlerz, W.C. (1987). Cucurbit potyvirus transmission by alate aphids (Homoptera: Aphididae) trapped alive. *Journal of Economic Entomology* 80: 87-92.

Adlerz, W.C. and P.H. Everett. (1968). Aluminum foil and white polyethylene mulches to repel aphids and control watermelon mosaic. *Journal of Economic Entomology* 61: 1276-1279.

Andow, D.A. (1991). Yield loss to arthropods in vegetationally diverse agroecosystems. *Environmental Entomolology* 20: 1228-1235.

Andow, D.A., A.G. Nicholson, H.C. Wien and H.R. Wilson. (1986). Insect population on cabbage grown with living mulches. *Environmental Entomolology* 15: 293-299.

Barlow, C.A., P.A. Randolph and J.C. Randolph. (1977). Effects of pea aphids (Homoptera: Aphididae) on growth and productivity of pea plants. *Canadian Entomologist* 109: 1491-1502.

Bentz, J.A., J. Reeves, P. Barbosa and B. Francis. (1995). Within-plant variation in nitrogen and sugar content of poinsettia and its effects on the oviposition pattern, survival and development of *Bemisia argentifolii* (Homoptera: Aleyrodidae). *Environmental Entomolology* 24: 271-277.

Bernays, E.A. (1999). When host choice is a problem for a generalist herbivore: Experiments with the whitefly, *Bemisia tabaci*. *Ecological Entomology* 24: 260-267.

Blackmer, J.L. and D.N. Byrne. (1993). Flight behavior of *Bemisia tabaci* in a vertical flight chamber: Effect of time of day, sex, age and host quality. *Physiological Entomology* 18: 223-232.

Breen, J.P. and G.L. Teetes. (1986). Relationships of yellow sugarcane aphid (Homoptera: Aphididae) density to sorghum damage. *Journal of Economic Entomology* 79: 1106-1110.

Broadbent, L., R.P. Chaudhuri and L. Kapica. (1950). The spread of virus diseases to single potato plants by winged aphids. *Annals of Applied Biology* 37: 355-362.

Brown, J.E., J.M. Dangler, F.M. Woods, M.C. Henshaw and W.A. Griffy. (1993). Delay in mosaic virus onset and aphid vector reduction in summer squash grown on reflective mulches. *HortScience* 28: 895-896.

Brown, J.K., D.R. Frohlich and R.C. Rosell. (1995). The sweetpotato or silverleaf whiteflies: Biotypes of *Bemisia tabaci* or a species complex? *Annual Review of Entomology* 40: 511-534.

Buntin, G.D., D.A. Gilbertz and R.D. Oetting. (1993). Chlorophyll loss and gas exchange in tomato leaves after feeding injury by *Bemisia tabaci* (Homoptera: Aleyrodidae). *Journal of Economic Entomology* 86: 517-522.

Burger, Y., A. Schwartz and H.S. Paris. (1988). Physiological and anatomical features of the silvering disorder of *Cucurbita*. *Journal of Horticultural Science* 63: 635-640.

Butler, G.D. Jr., D.L. Coudriet and T.J. Henneberry. (1989). Sweetpotato whitefly: Host plant preference and repellent effect of plant-derived oils on cotton, squash, lettuce and cantaloupe. *The Southwestern Entomologist* 13: 81-86.

Butler, G.D. Jr., T.J. Henneberry and F.D. Wilson. (1986). *Bemisia tabaci* (Homoptera: Aleyrodidae) on cotton: Adult activity and cultivar oviposition preference. *Journal of Economic Entomology* 79: 350-354.

Byrne, D.N. and W.B. Miller. (1990). Carbohydrate and amino acid composition of phloem sap and honeydew produced by *Bemisia tabaci*. *Journal of Insect Physiology* 36: 433-439.

Cardoza, Y.J., H.J. McAuslane and S.E. Webb. (2000). Effect of leaf age and silverleaf symptoms on oviposition site selection and development of *Bemisia argentifolii* (Homoptera: Aleyrodidae) on zucchini. *Environmental Entomology* 29: 220-225.

Cho, J.J., D.E. Ullman, E. Wheatley, J. Holly and D. Gonsalves. (1992). Commercialization of ZYMV cross protection for zucchini production in Hawaii. *Phytopathology* 82: 1073.

Costa, H.S. and J.K. Brown. (1990). Variability in biological characteristics, isozyme patterns and virus transmission among populations of *Bemisia tabaci*. *Phytopathology* 80: 888.

Costa, H.S., M.W. Johnson and D.E. Ullman. (1994). Row covers effect on sweetpotato whitefly (Homoptera: Aleyrodidae) densities, incidence of silverleaf and crop yield in zucchini. *Journal of Economic Entomology* 87: 1616-1621.

Costa, H.S. and K.L. Robb. (1999). Effects of ultraviolet-absorbing greenhouse plastic films on flight behavior of *Bemesia argentifolii* (Homoptera: Aleyrodidae) and *Frankliniella occidentalis* (Thysanoptera: Thripidae). *Journal of Economic Entomology* 92: 557-562.

Costa, H.S., D.E. Ullman, M.W. Johnson and B.E. Tabashnik. (1993). Squash silverleaf symptoms, induced by immature, but not adult, *Bemisia tabaci*. *Phytopathology* 83: 763-766.

Costello, M.J. and M.A. Altieri. (1995). Abundance, growth rate and parasitism of *Brevicoryne brassicae* and *Myzus persicae* (Homoptera: Aphididae) on broccoli grown in living mulches. *Agriculture, Ecosystems and Environment* 52: 187-196.

Csizinszky, A.A., D.J. Schuster and J.B. Kring. (1995). Color mulches influence yield and insect pest populations in tomatoes. *Journal of the American Society for Horticultural Science* 120: 778-784.

Csizinszky, A.A., D.J. Schuster and J.B. Kring. (1997). Evaluation of color mulches and oil sprays for yield and for the control of silverleaf whitefly, *Bemisia argentifolii* (Bellows and Perring) on tomatoes. *Crop Protection* 16: 475-481.

Decoteau, D.R., M.J. Kasperbauer and P.G. Hunt. (1989). Mulch surface color affects yield of fresh-market tomatoes. *Journal of the American Society for Horticultural Science* 114: 216-219.

Eastop, V.F. (1977). Worldwide importance of aphids as virus vectors. In *Aphids as Virus Vectors,* K.F. Harris and K. Maramorosch, eds. Academic Press, New York, pp. 3-62.

Frank, D.L. (2004). Evaluation of living and synthetic mulches in zucchini for control of homopteran pests. Thesis submitted to Graduate School of the University of Florida in partial fulfillment of the requirements for the degree of Master of Science in Entomology subject category: Pest Management, Gainesville, Florida.

Frank, D.L. and O.E. Liburd. (2004). Comparison of living and synthetic mulches in zucchini. *Citrus and Vegetable Magazine* 69: 24-27.

Frank, D.L. and O.E. Liburd. (2005). Effects of living and synthetic mulch on the population dynamics of whiteflies and aphids, their associated natural enemies and insect-transmitted plant diseases in zucchini. *Environmental Entomology* 34: 857-865.

Gould, J.R. and S.E. Naranjo. (1999). Distribution and sampling of *Bemisia argentifolii* (Homoptera: Aleyrodidae) and *Eretmocerus eremicus* (Hymenoptera: Aphelinidae) on cantaloupe vines. *Journal of Economic Entomology* 92: 402-408.

Greer, L. and J.M. Dole. (2003). Aluminum foil, aluminum-painted plastic and degradable mulches increase yields and decrease insect-vectored viral diseases of vegetables. *HortTechnology* 13: 276-284.

Gruenhagen, N.M. and T.M. Perring. (1999). Velvetleaf: A plant with adverse impacts on insect natural enemies. *Environmental Entomology* 28: 884-889.

Heathcote, G.D. (1957). The comparison of yellow cylindrical, flat water traps and of Johnson suction traps, for sampling aphids. *Annals of Applied Biology* 45: 133-139.

Hooks, C.R.R., H.R. Valenzuela and J. Defrank. (1998). Incidence of pest and arthropod natural enemies in zucchini grown in living mulches. *Agriculture, Ecosystems and Environment* 69: 217-231.

Hopen, H.J. and N.F. Oebker. (1976). Vegetable crop responses to synthetic mulches: An annotated bibliography. Illinois University College of Agriculture Cooperative Extension Service.

Hussein, M.Y. and N.A. Samad. (1993). Intercropping chilli with maize or brinjal to suppress populations of *Aphis gossypii* Glov. and transmission of chilli viruses. *International Journal of Pest Management* 39: 216-222.

Jackson, D.M., M.W. Farnham, A.M. Simmons, W.A. Van Giessen and K.D. Elsey. (2000). Effects of planting pattern of collards on resistance to whiteflies (Homoptera: Aleyrodidae) and on parasitoid abundance. *Journal of Economic Entomology* 93: 1227-1236.

Jiménez, D.R., J.P. Shapiro and R.K. Yokomi. (1993). Biotype-specific expression of dsRNA in the sweetpotato whitefly. *Entomologia Experimentalis et Applicata* 70: 143-152.

Jiménez, D.R., R.K. Yokomi, R.T. Mayer and J.P. Shapiro. (1995). Cytology and physiology of silverleaf whitefly-induced squash silverleaf. *Physiological and Molecular Plant Pathology* 46: 227-242.

Kloen, H. and M.A. Altieri. (1990). Effect of mustard (*Brassica hirta*) as a non-crop plant on competition and insect pests in broccoli (*Brassica oleracea*). *Crop Protection* 9: 90-96.

Kring, J.B. (1959). The life cycle of the melon aphid, *Aphis gossypii,* an example of facultative migration. *Annals of the Entomological Society of America* 52: 284-286.

Kring, J.B. (1967). Alighting of aphids on colored cards in a flight chamber. *Journal of Economic Entomology* 60: 1207-1210.

Kring, J.B. (1972). Flight behavior of aphids. *Annual Review of Entomology* 17: 461-492.

Lamont, W.J. (1993). Plastic mulches for the production of vegetable crops. *Hort-Technology* 3: 35-39.

Liburd, O.E., R.A. Casagrande and S.R. Alm. (1998). Evaluation of various color hydromulches and weed fabric on broccoli insect populations. *Journal of Economic Entomology* 91: 56-262.

Lisa, V. and H. Lecoq. (1984). Zucchini yellow mosaic virus. In *Descriptions of Plant Viruses.* Commonwealth Mycological Institute/Association of Applied Plant Biology, Kew, Surrey, England, p. 282.

Mahmoudpour, M.A. and J.J. Stapleton. (1997). Influence of sprayable mulch colour on yield of eggplant (*Solanum melongena* L. cv. Millionaire). *Scientia Horticulturae* 70: 331-338.

Miyazaki, M. (1987). Forms and morphs of aphids. In *Aphids: Their Biology, Natural Enemies and Control,* Vol. A, A.K. Minks and P. Harrewijn, eds. Elsevier Science Publishing Company Inc., New York, USA, pp. 27-47.

Mound, L. (1962). Studies on the olfaction and colour sensitivity of *Bemisia tabaci* (Genn.) (Homoptera: Aleyrodidae). *Entomologia Experimentalis et Applicata* 5:99-104.

Nault, L.R. (1997). Arthropod transmission of plant viruses: A new synthesis. *Annals of the Entomological Society of America* 90: 521-541.

Palumbo, J.C., N.C. Toscano, M.J. Blua and H.A. Yoshida. (2000). Impact of *Bemisia* whiteflies (Homoptera: Aleyrodidae) on alfalfa growth, forage yield and quality. *Journal of Economic Entomology* 93: 1688-1694.

Perring, T.M., A.D. Cooper, R.J. Rodriguez, C.A. Farrar and T.S. Bellows. (1993). Identification of a whitefly species by genomic and behavioral studies. *Science* 259: 74-77.

Perring, T.M., N.M. Gruenhagen and C.A. Farrar. (1999). Management of plant viral diseases through chemical control of insect vectors. *Annual Review of Entomology* 44: 457-481.

Perring, T.M., R.N. Royalty and C.A. Farrar. (1989). Floating row covers for the exclusion of virus vectors and the effect in disease incidence and yield of cantaloupe. *Journal of Economic Entomology* 82: 1709-1715.

Platt, J.O., J.S. Caldwell and L.T. Kok. (1999). Effect of buckwheat as a flowering border on populations of cucumber beetles and their natural enemies in cucumber and squash. *Crop Protection* 18: 305-313.

Polston, J.E. and P.K. Anderson. (1997). The emergence of whitefly-transmitted Gemini viruses in tomato in the western hemisphere. *Plant Disease* 81: 1358-1369.

Power, A.G. (1987). Plant community diversity, herbivore movement and an insect-transmitted disease of maize. *Ecology* 68: 1658-1669.

Purcifull, D.E., G.W. Simone, C.A. Baker and E. Hiebert. (1988). Immunodiffusion tests for six viruses that infect cucurbits in Florida. *Proclamation of the Florida State Horticultural Society* 101: 401-403.

Puri, S.N., B.B. Bhosle, M. Ilyas, G.D. Butler and T.J. Henneberry. (1994). Detergents and plant-derived oils for control of the sweetpotato whitefly on cotton. *Crop Protection* 13: 45-48.

Root, R.B. (1973). Organization of a plant-arthropod association in simple and diverse habitats: The fauna of collards (*Brassica oleracea*). *Ecological Monographs* 43: 95-120.

Schalk, J.M., C.S. Creighton, R.L. Fery, W.R. Sitterly, B.W. Davis, T.L. McFadden and A. Day. (1979). Reflective film mulches influences insect control and yield in vegetables. *Journal of the American Society for Horticultural Science* 104: 759-762.

Schuster, D.J., J.B. Kring and J.F. Price. (1991). Association of the sweetpotato whitefly with a silverleaf disorder of squash. *HortScience* 26: 155-156.

Simmons, A.M. (1994). Ovipositon on vegetables by *Bemisia tabaci* (Homoptera: Aleyrodidae): Temporal and leaf surface factors. *Environmental Entomology* 23: 381-389.

Simons, J.N. and T.A. Zitter. (1980). Use of oils to control aphid borne viruses. *Plant Disease* 64: 542-546.

Smith, J.G. (1969). Some effects of crop background on populations of aphids and their natural enemies on brussels sprouts. *Annals of Applied Biology* 63: 326-329.

Smith, H.A., R.L. Koenig, H.J. McAuslane and R. McSorley. (2000). Effect of silver reflective mulch and a summer squash trap crop on densities of immature *Bemisia argentifolii* (Homoptera: Aleyrodidae) on organic bean. *Journal of Economic Entomology* 93: 726-731.

Summers, C.G., J.P. Mitchell and J.J. Stapleton. (2004). Management of aphid-borne viruses and *Bemisia argentifolii* (Homoptera: Aleyrodidae) in zucchini squash by using UV reflective plastic and wheat straw mulches. *Environmental Entomology* 33: 1447-1457.

Summers, C.G. and J.J. Stapleton. (2002). Reflective mulches for management of aphids and aphid-borne virus diseases in late-season cantaloupe (*Cucumis melo* L. var. *cantalupensis*). *Crop Protection* 21: 891-898.

Summers, C.G., J.J. Stapleton, A.S. Newton, R.A. Duncan and D. Hart. (1995). Comparison of sprayable and film mulches in delaying the onset of aphid-transmitted virus diseases in zucchini squash. *Plant Disease* 79: 1126-1131.

Swenson, K.G. (1968). Role of aphids in the ecology of plant viruses. *Annual Review of Phytopathology* 6: 351-374.

Toba, H.H., A.N. Kishaba, G.W. Bohn and H. Hield. (1977). Protecting muskmelon against aphid-borne viruses. *Phytopathology* 67: 1418-1423.

Tsai, J.H. and K. Wang. (1996). Development and reproduction of *Bemisia argentifolii* (Homoptera: Aleyrodidae) on five host plants. *Environmental Entomology* 25: 810-816.

Van Lenteren, J.C. and P.J. Noldus. (1990). Whitefly-plant relationships: Behavioural and ecological aspects. In *Whiteflies: Their Bionomics, Pest Status and Management,* D. Gerling, ed. Intercept, Andover, UK, pp. 47-89.

Webb, S.E., M.L. Kok-Yokomi and D.J. Voegtlin. (1994). Effect of trap color on species composition of alate aphids (Homoptera: Aphididae) caught over watermelon plants. *Florida Entomologist* 77: 146-154.

Webb, S.E. and S.B. Linda. (1992). Evaluation of spunbounded polyethylene row covers as a method of excluding insects and viruses affecting fall-grown squash in Florida. *Journal of Economic Entomology* 85: 2344-2352.

Yokomi, R.K., K.A. Hoelmer and L.S. Osborne. (1990). Relationships between the sweetpotato whitefly and the squash silverleaf disorder. *Phytopathology* 80: 895-900.

Zitter, T.A. and J.N. Simons. (1980). Management of viruses by alteration of vector efficiency and by cultural practices. *Annual Review of Phytopathology* 18: 289-310.

Chapter 4

The Potential of Genetically Enhancing the Microbial Control of Insect and Plant Disease Pests

Wayne A. Gardner
Maria Nenmaura Gomes Pessoa
Ericka Scocco

INTRODUCTION

Biological control, as a tactic in pest management systems, is based upon pest species being controlled or regulated by naturally occurring enemies (i.e., predators, parasites, pathogens, competitors, antagonists) or by manipulating and enhancing the occurrence and activity of these agents (Huffaker, Messenger and DeBach, 1971). Microbial control is only one component of biological control and includes all aspects of using microorganisms or their by-products in the control of pest species.

The microbial control of insects has been reviewed by several scholars, including Burges and Hussey (1971), Burges (1981), Ignoffo (1988), and Lacey and Goettel (1995). Insect pathogens may be used in pest management systems in the four approaches of classical introduction, augmentation, inundation, and conservation of naturally occurring pathogens. In brief, the classical method involves the introduction and establishment of exotic microorganisms from one geographical area in another. Target pests are usually nonindigenous and lack natural enemies in their expanded ranges. Augmentative approaches are used when naturally occurring pathogens require

Management of Nematode and Insect-Borne Plant Diseases

doi:10.1300/5754_04

supplementation or enhancement to reach epidemic proportions. Methods include modification of the physical environment to favor survival and infection of the target pests or periodic release of infective propagules of the pathogen. Inundative augmentation involves the mass production, standardization, formulation, and application of the microorganism to initiate epidemics of the disease (e.g., microbial insecticides).

Microbial control of insect pests is most evident in causing mortality of infected insect hosts, although insect pathogens may be more insidious by interfering with host development, altering reproduction, and lowering host resistance to other control agents. By comparison, microbial control of plant pathogens is accomplished by competitive or antagonistic actions of microorganisms that reduce the inoculum density or alter the pathogenicity or virulence of the plant pathogen, thereby reducing or preventing plant disease (Baker and Cook, 1974).

Historically, microbial control dates to Agostino Maria Bassi's work with *Beauveria bassiana* (Balsamo) Vuillemin, the causative pathogen of white muscardine disease in silkworms. This work was the foundation of the modern germ theory in that it demonstrated that the disease could be artificially passed to other silkworms or other species (Steinhaus, 1975). Large-scale applications of mass-produced pathogens for microbial control were first conducted in the late 1800s in Russia and the United States against wheat pests. These yielded mixed results in terms of control but showed feasibility and promise as possible control agents.

Little to no development of microbial control, however, occurred during much of the 1900s, primarily because of the availability of relatively inexpensive, but highly efficacious, chemical insecticides. A resurgence of interest in, and subsequent development and adoption of, microbial agents in insect pest management occurred during the 1960s and 1970s. This followed the demise of the chlorinated hydrocarbons in the United States and many world markets because of development of genetic resistance in target pest populations and widespread health and environmental concerns. However, interest in microbial control waned again with the availability of synthetic pyrethroids and other effective chemistries for insect control in the 1980s. Microbial control of plant diseases rejuvenated interest in microbial pesticides in the 1990s. The market for biopesticides grew almost 80 percent from

1991 to 1995 (Georgis, 1997) and, by the end of the century, there were 59 species of microorganisms that were registered as pesticides worldwide, including 36 for insect control, 20 for plant disease control, and 3 for weed control (Copping, 1999).

Microbial agents offer environmentally friendly alternatives to conventional chemical pesticides, which often produce harmful effects in nontargets and the environment and cause development of resistance among target pest populations. In addition, little to no tolerance is to be found on the part of consumers or regulatory entities of chemical residues on food products, especially vegetables and fruits, created by application of chemical pesticides. Yet, microbial control has been limited by numerous problems, usually inherent to the microorganisms or their microbial products. These limiting factors include, but are not restricted to, production, shelf life, field stability, host range, dependence upon abiotic conditions, efficacy, and market conditions.

Research is showing that many microbial control agents are amenable to genetic manipulation and that the resulting recombinant agents possess enhanced capabilities, causing greater levels of mortality at faster rates than experienced with wild-type nonrecombinant counterparts. Thus, genetic studies and biotechnological methodologies can improve the performance and efficacy of microbial agents in pest management systems.

The use of genetics and capitalization upon genetic modifications are not new to the agricultural sciences. Traditional Mendelian genetics have long been a foundation of plant and animal breeding and the development of improved varieties, hybrids, and lines. However, only recently have biotechnological methods been explored and utilized to enhance the activity of microorganisms used in managing insect and plant disease pests. Efforts are generally directed to improving some inherent characteristic (e.g., field stability, pathogenicity, virulence, transmission) of the microorganism that otherwise limits its natural performance and efficacy. Indeed, we must understand the factors that make the microorganism effective as a microbial control agent to improve it via genetic modification (Frey, 2001).

Our objective here is to provide an overview of some accomplishments and successes in this area and to discuss the improvement of microbial control and pest management through genetic recombinant technology. This review is not necessarily exhaustive, but it covers the essential groups of agents and the approaches that demonstrate the greatest potential for enhancing microbial control.

ENTOMOPATHOGENS

The entomopathogens that have been used in microbial control of insect pests include representatives of the five major groups of microorganisms: viruses, bacteria, fungi, protozoa, and nematodes. Those developed and used as microbial control agents are considered only as alternatives to conventional chemical insecticides (Waage, 1997), and their development and use in pest management systems have always been limited to techniques and methods employed with conventional chemical pesticides.

In terms of development and commercialization, the bacterium *Bacillus thuringiensis* Berliner is the most successful biopesticide. Copping (1999) stated that *B. thuringiensis* products represented 39 percent of the biopesticide market, with activity against lepidopteran larvae (primarily *B. thuringiensis* var. *kurstaki* and *B. thuringiensis* var. *azaiwai*), dipteran larvae *(B. thuringiensis* var. *israeliensis),* and coleopteran larvae *(B. thuringiensis* var. *tenebrionis).* Other entomopathogenic bacteria of note in terms of commercialization include *B. popilliae* Dutky, which infects soil-inhabiting coleopteran grubs, and *Bacillus sphaericus* Meyer and Neide, which infects mosquito larvae.

The entomopathogenic viruses are primarily the nuclear polyhedrosis viruses (NPVs) or baculoviruses, cytoplasmic polyhedrosis viruses (CPVs), and granulosis viruses (GVs). Of these, the CPVs are more numerous, with more than 200 isolated from various hosts. More than 125 NPVs have been isolated from lepidopteran, hymenopteran, dipteran, and orthopteran hosts, and only approximately 50 GVs have been isolated, primarily from lepidopteran larvae. The commercially successful products are restricted to the NPVs, with only 24 on the market (Copping, 1999). Yet, a number are economically viable products, including the *Heliothis zea* NPV and the *Spodoptera exigua* NPV, for vegetables, grapes, cotton, and other crops.

McCoy, Samson and Boucias (1986) noted that more than 750 species of fungi are entomopathogenic. The deuteromycetes (e.g., *Beauveria, Metarhizium, Verticillium, Nomuraea*) have the greatest potential for development and commercialization as microbial control agents, primarily because of their relatively broad host ranges. Although the fungi are highly dependent upon microclimate conditions (e.g., humidity, free moisture), forty-seven fungus-based products are available on the biopesticide market (Copping, 1999).

The activity of entomopathogenic nematodes against insect hosts is dependent upon symbiotic bacteria (family Enterobacteriaceae) that reside within the nematode gut. The bacteria are released into the host insect's hemocoel after successful parasitization by the nematode, causing rapid (e.g., twenty-four to forty-eight hours) septicemic death of the host (Kaya, 1985). The most promising successful candidates for development and commercialization are in either the family Heterorhabditidae or the family Steinernematidae and are used to control pests in soil or cryptic habitats.

The entomopathogenic protozoa are characterized by chronic rather than acute infections with low levels of pathogenicity. They remain relatively unattractive for commercialization, with only two products available as biopesticides worldwide (Copping, 1999). Their utility in pest management programs lies in their natural regulation of pest populations.

Opportunities exist in genetically improving entomopathogens for microbial control. To date, successes achieved in genetic improvement are directly related to available knowledge of the biology and pathogenicity of microorganisms as well as the ease of their genetic manipulation. Perhaps the greatest successes have been reported with the entomopathogenic baculoviruses, for several of which the genomic sequence is known (Ahrens, Russell and Rohrmann, 1977; Ayres et al., 1994; Gomi, Majima and Maeda, 1999). The insecticidal toxins produced by the entomopathogenic bacterium *B. thuringiensis* also have been extensively studied and manipulated for host range expansion, improved efficacy, and plant transgenics. Other groups of entomopathogens, including the fungi, nematodes, and protozoa, are far less studied and are not as amenable to enhancement via biotechnological methods.

Entomopathogenic Viruses

Among the entomopathogenic viruses, only the baculoviruses have been genetically engineered to enhance insecticidal performance. Approximately twenty have been successfully engineered to yield recombinant forms that significantly reduce the median lethal time (LT_{50}) of infected hosts or the damage caused by their feeding in comparison to their wild-type forms.

Biotechnological methods for producing recombinant baculoviruses are reviewed by Jarvis (1997). The most common recombinant methods involve plasmid transfer vectors, although methods for direct cloning (O'Reilly and Miller, 1991; Lu and Miller, 1996) and insertion of foreign genes (Vlak et al., 1990; Gritsun et al., 1997) into the baculovirus genome have been reported.

Recombinant baculoviruses engineered to express neurotoxic peptides possess perhaps the greatest potential for adoption and commercialization in microbial control. The sources of neurotoxins are mites (phylum Arthropoda: order Acari), sea anemones (phylum Cnidaria: class Anthozoa), scorpions (phylum Arthropoda: order Scorpionida), and spiders (phylum Arthropoda: order Araneida) (Hoover et al., 1995; Prikhod'ko, Dickson and Miller, 1996; Prikhod'ko et al., 1998; Hughes et al., 1997; Popham, Li and Miller, 1997; Popham, Prikhod'ko and Miller, 1998). These peptides are insect specific, have a rapid effect on the host, have no effect on the ability of the virus to infect or replicate, and are active at low concentrations (Possee et al., 1997).

Neurotoxins expressed by recombinant baculoviruses generally disrupt presynaptic, voltage-sensitive sodium channels in the cerebral and peripheral nervous systems of the insect host. They cause cessation of feeding and paralysis of the hosts and appear to act at multiple sites, with some being inhibitory and others excitatory in action. Thus, coexpression could improve and even synergize efficacy and performance (Prikhod'ko et al., 1998).

Baculoviruses also have been engineered to express factors that alter the physiological state of the target insect host. Physiological effectors inserted into the baculovirus genome exhibit deleterious effects upon water balance (Maeda, 1989), pheromone synthesis (Ma et al., 1998), ecdysis and development (Bonning et al., 1992, 1997), chitinase activity (Gopalakrishnan, Muthukrishnan and Kramer, 1995), and cell regulation (Lee et al., 1997). In general, baculoviruses with these insertions killed more hosts at a faster rate than their respective wild-type, nonrecombinant forms. O'Reilly and Miller (1991) improved the performance of a baculovirus against the fall armyworm,

Spodoptera frugiperda (J. E. Smith), by deleting the ecdysteroid glucosyl transferase (*egt*) gene from the virus genome. However, other effectors inserted into baculovirus genomes either had no effect or exhibited deleterious effects on the baculovirus performance (Eldridge et al., 1991, 1992; Bonning et al., 1992; O'Reilly, 1995; Sato et al., 1997).

Black et al. (1997) reviewed the potential for commercialization of recombinant and wild-type baculoviruses. It is generally acknowledged that recombinant baculoviruses expressing insect-specific neurotoxins have the greatest potential in insect pest management programs; however, regulatory considerations may favor those recombinant forms that express physiological factors. Given their specificity and relative safety, recombinant baculoviruses may even appear on the market as microbial insecticides for controlling lepidopteran pests of cotton and vegetables. Genetic engineering could further improve the host ranges of baculoviruses, identify or modify additional neurotoxins and physiological effectors, and alter the host immune system with immunosuppressant genes and peritrophic matrix-degrading enzymes.

Entomopathogenic Bacteria

Bacteria are excellent candidates for genetic enhancement as microbial control agents. They are relatively easy to culture with artificial media, and critical genes are often located on large, nonessential extrachromosomal plasmids that can be easily transferred among different species. Those bacterial species receiving the greatest attention are *B. thuringiensis, B. sphaericus,* and the symbionts from the family Enterobactericeae within the guts of entomopathogenic nematodes.

Bacillus thuringiensis

Bacillus thuringiensis produces several crystalline peptide toxins that accumulate in the cytoplasm. The most studied of these toxins are the *delta*-endotoxins. The crystalline toxins dissolve in the alkaline contents of the gut of the insect host and bind with receptors on the microvillar borders of the gut epithelium of susceptible hosts. The gut is paralyzed and the gut wall disintegrates, resulting in sepsis and death.

The genes encoding the toxins are termed *cry* genes and are mostly located on the extrachromosomal plasmids within the bacterial cell. The *cry* genes are grouped (*cryI-VI*) according to specificity of action and known sequence similarity (Rajamohan and Dean, 1996); functions have been assigned to some regions (Thompson, Schnepf and Feitelson, 1995). Discovery of new toxins and manipulation of the toxins via biotechnological methods show the greatest promise in terms of enhancing the activity and utility of *B. thuringiensis* in insect pest management programs.

Manipulation of the *cry* genes can improve the performance and host range of *B. thuringiensis*. Cloning vector systems are a means of recombination of the genes that encode the toxins produced by the entomopathogen (Carlton, 1996). This allows for the selection and cloning of novel combinations of insecticidal toxins within one strain.

Biotechnological methods have also yielded novel delivery systems for the toxins (Koziel et al., 1993). Alternative delivery systems address some factors limiting the activity of *B. thuringiensis* in the field (e.g., persistence). Toxins have been "biologically encapsulated" by genetic means, thereby creating a form of delivery that extends the field persistence while reducing the inoculum of bacterial endospores applied to the crop, a concern of regulatory agencies in several countries. For example, genes encoding *B. thuringiensis* toxins were cloned into *Pseudomonas fluorescens* Migula, a nonpathogenic bacterium. The *P. fluorescens* cells were subsequently killed, yielding *B. thuringiensis* toxins encapsulated within the cellular membranes (Carlton, 1996). Lereclus et al. (1995) also produced an encapsulated form of *B. thuringiensis* toxins by genetically modifying *B. subtilis* (Ehrenberg) to express the toxins.

In addition, *cry* genes have been engineered into epiphytic and endophytic bacteria. Genes encoding for the *delta*-endotoxins have been engineered into *P. fluorescens* (Watrud et al., 1985) and the xylem-limited bacterium *Clavibacter xyli* Davis (Tomasino et al., 1995). *P. fluorescens* is an epiphyte on corn roots, and *C. xyli* is an endophyte limited to the xylem of corn. Genetically modified forms of these bacteria enhanced the performance of *B. thuringiensis* toxins against soil-inhabiting coleopteran rootworms (*Diabrotica* spp.) and the stem-burrowing larvae of the European corn borer, *Ostrinia nubilalis* Hübner, respectively.

Scientists have also successfully engineered the genomes of a number of plants to express *delta*-endotoxin activity against a variety of susceptible insect pests. This is more fully discussed in the section titled "Transgenic Plants Expressing *Bacillus thuringiensis* Toxins."

Bacillus sphaericus

Bacillus sphaericus is being developed as a biopesticide for its larvicidal activity against mosquitoes and blackflies. The activity is largely via peptide toxins. Their action is not fully understood, but strains have been genetically modified to express toxins with enhanced activity against larvae (Thanabalu and Porter, 1995).

Symbiotic Enterobacteriaceae

Enterobacteriaceae bacteria that have symbiotic relationships with entomopathogenic nematodes exhibit antimicrobial as well as insecticidal activity. Mutagenic techniques are being used to study the genes that encode the antibiotics and to delineate further the nematode-bacteria relationship. Such studies may yield genes encoding insecticidal activity as well as possible pharmaceutical agents.

Entomopathogenic Nematodes

Genetic engineering may improve the pathogenicity, extend and enhance field stability, and expand the host range of entomopathogenic nematodes (Poinar, 1991). More than 250 species of insects are reportedly parasitized by nematodes in the Steinernematidae and Heterorhabditidae families. Several species and strains are commercially available in a number of niche markets, including those targeted to soil-inhabiting and other cryptic pests (Gaugler and Hashmi, 1996). Their utility, however, is limited by their susceptibility to temperature extremes, ultraviolet radiation, and desiccation.

Nematode pathogenicity is dependent upon a symbiotic relationship with bacteria in the family Enterobacteriaceae that reside within specialized vesicles in the nematode intestine. The relationship is specific in that each nematode species has its own unique bacterial species. When the infective nematode enters a susceptible insect host (usually through the mouth, anus, spiracles, or thin areas of the integument), the bacterium is released into the host hemocoel. The bacteria release extracellular enzymes that provide nutrients for the developing nematode. They also release antibiotic substances that limit microbial growth in the host cadaver, thereby providing a suitable growth environment without microbial competition. The nematode, in turn, provides protection and transportation for the bacteria between insect hosts.

Gaugler and Hashmi (1996) noted that nematode virulence, host-finding abilities, and other biological fitness factors have been improved by classical selection methods. However, little is actually known of the genetics of these entomopathogens, and scientists have used our extensive knowledge of the genetics of *Caenorhabditis elegans* (Maupas) Dougherty, a nonpathogenic nematode that is a model for studies in molecular genetics, for the foundation of molecular studies of the entomopathogenic nematodes.

Hashmi et al. (1998) engineered *Heterorhabditis bacteriophora* Poinar to express the heat-shock protein hsp70 of *C. elegans*. The transformed *H. bacteriophora* demonstrated increased thermotolerance, grew and developed normally, and showed no diminished effects in terms of biological fitness factors. Subsequent field evaluations demonstrated that the thermotolerance trait had no effect on infection rates (Gaugler, Wilson and Shearer, 1997).

To date, little is known about the genetics of the entomopathogenic nematodes. However, considerable potential exists for their genetic manipulation to improve their susceptibility to environmental stressors and to enhance their performance in biological control programs.

Entomopathogenic Fungi

Enhancement of the entomopathogenic fungi by genetic manipulation is only in its infancy because of the limited knowledge available on the molecular and biochemical factors involved in host cuticle penetration, pathogenesis, and host death. Genes regulating these processes must be identified for genetic engineering to be used in enhancing the performance of the fungi as microbial control agents (St. Leger, 1993).

The lack of an effective gene transfer system for the filamentous fungi has limited the genetic improvement of entomopathogenic fungi by molecular means. The most frequently used gene transfer methods have involved the isolation of protoplasts. However, transformation frequencies obtained with protoplasts are usually low for the entomopathogenic fungi (Sandhu et al., 2001). Some success is occurring with alternative transformation methods (Canton and Vandenburg, 1999; dos Reis et al., 2004).

The deuteromycete fungi appear to be the most suitable for genetic improvement as microbial control agents. These fungi are generally easy to cultivate on artificial media, and no sexual cycle has been identified for them. Fungal strains resistant to the commonly used fungicide benomyl have been developed by mutagenesis (Valadares-Inglis and Inglis, 1997; Sandhu et al., 2001).

The greatest success in genetic characterization and modification, however, has been for the process of host cuticle penetration by *Metarhizium anisopliae*. In a series of studies, St. Leger and various co-workers have elucidated the physiological and enzymatic processes involved in the penetration process (Joshi, St. Leger and Roberts, 1997; St. Leger, Joshi and Roberts, 1998) and have cloned various genes regulating the formation of the appressorium, a specialized fungal structure involved in cuticle penetration (St. Leger, 1993, 1995). In addition, the *Pr1* gene, which encodes a protease involved in cuticle penetration, was inserted into the genome of *M. anisopliae*. The *Pr1* protease produced by the transformed fungus was released into the hemocoel of the host, rather than during the cuticle penetration phase, resulting in a significantly reduced lethal time and lower feeding damage by infected lepidopteran larvae (St. Leger et al., 1996). These and other studies demonstrate the potential of genetic modification in improving the performance of entomopathogenic fungi against susceptible insect hosts.

BIOCONTROL AGENTS FOR PLANT DISEASES

Microbial control of plant diseases is not as advanced as the microbial control of insect pests; however, interest in commercializing microbial control products was rejuvenated in the 1990s as markets developed for products to control plant diseases biologically (Copping, 1999). Initially, microbial control of plant disease was restricted to wound protection and healing (Corke and Rishbeth, 1981), but more recently successes have been realized with the biocontrol of plant disease microbes in soils and on aerial structures and portions of plants.

Commercially produced yeasts and fungi are now available worldwide for vegetable producers to manage fungal diseases (Punja and Utkhede, 2003). Several *Trichoderma* spp. are being used or developed as anti fungal agents in controlling such diseases as soil-borne blights, damping-off, and take-all (Yedidia et al., 2000; Viterbo et al., 2001). Yeast antagonists show promise in limiting the activity of, and the subsequent losses caused by, postharvest diseases of fresh fruits and vegetables (Karabulut and Baykal, 2004).

The primary mechanisms of action include production of hydrolytic enzymes and antibiotics, competition for nutrients and colonization sites, induction of plant defense mechanisms, and interference with the pathogenesis of the disease. Molecular and genetic techniques have elucidated these mechanisms for several agents (Arisan-Atac, Heidenreich and Kubicek, 1995); others are currently being identified or clarified. Biotechnology is also responsible for identifying, isolating, and sequencing genes that encode metabolites responsible for the modes of action (Hayes et al., 1994; Donzelli et al., 2001). Such studies will provide new opportunities to improve the performance of biocontrol microorganisms (Thomashow, 1996).

Genetic marker and biotechnological techniques also assist in examining the spread and epidemiology of plant diseases (Punja and Utkhede, 2003). As with several entomopathogens, genetic transformations have been used to improve the delivery of plant disease biocontrol agents. In one notable example, the plant root-colonizing bacterium *P. fluorescens* was genetically transformed, applied to wheat seed, and successfully suppressed Rhizoctonia root rot (Huang et al., 2004). Additional efforts will yield improved agents as we acquire more knowledge of the mode of action and the genetics of these biocontrol agents.

TRANSGENIC PLANTS EXPRESSING BACILLUS THURINGIENSIS *TOXINS*

Genetic engineering of plants to express the *delta*-endotoxins of *B. thuringiensis* was recently reviewed by O'Callaghen et al. (2005) and Babu et al. (2003), with previous reviews by Koziel et al. (1993)

and Peferoen (1997). The increasing use of genetically modified crops worldwide underscores the effectiveness of the strategy in managing insect pests and its acceptance by growers and selected markets. Yet public concern remains with respect to human health, development of insect resistance, and environmental effects.

Four *cry* genes have been inserted into the genomes of commercially available cotton and maize (Shelton, Zhao and Roush, 2002). Other crops, including potatoes, tomatoes, cereals, rice, and trees are also being engineered to express the *B. thuringiensis* endotoxins. An additional peptide toxin (not a *delta*-endotoxin) from *B. thuringiensis* is also being field-tested in transgenic cotton (OGTR, 2003).

Insect targets of this pest management strategy are primarily lepidopterans and coleopterans. In cotton, the major targets are *Helicoverpa zea* (Boddie), *Helicoverpa armigera* (Hübner) and *Heliothis virescens* (F.). Transgenic maize and sweet corn are used against lepidopteran foliage, silk feeders, and stem borers, including *Helicoverpa zea, O. nubilalis, Diatraea grandiosella* Dyar, and *Spodoptera* spp. Potatoes have been engineered for resistance to the Colorado potato beetle, *Leptinotarsa decemlineata* L., as well as lepidopteran borers that attack the stored tubers. Other insect targets include stem borers (*Chilo suppressalis* Walker and *Scirpophaga incertulas* [Walker]), leafrollers (*Cnaphalocrocis medinalis* [Guenee] and *Marasmia patnalis* Bradley), and fruit and foliage feeders (e.g., *Manduca sexta* L.) (Koziel et al., 1993; Bennett et al., 1997; Schuler et al., 1999).

Whereas the isolation of the *cry* genes was a relatively simple task, the engineering of plants to express an adequate level of *delta*-endotoxins proved more difficult. Alterations in the protein sequence and in the *cry* gene sequence proved successful in elevating the expression of the toxins to desirable and effective levels in the modified plants.

As adoption of transgenic plants for pest management increases, concerns about human health and potential environmental effects are being addressed. Concerns of possible negative effects on pollinators and nontarget lepidopterans have proven unfounded (Wraight et al., 2000; Oberhauser et al., 2001; Zangerl et al., 2001; Shelton, Zhao and Roush, 2002). Studies thus far demonstrate that the endotoxins have little to no direct or indirect impact on beneficial natural enemies, yet this remains a concern as additional data are collected (Schuler et al.,

1999; Shelton, Zhao and Roush, 2002). The U.S. Environmental Protection Agency and the U.S. Food and Drug Administration have concluded that *B. thuringiensis*—expressing maize and potatoes pose no threats to human and animal health and nutrition (Carozzi and Koziel, 1997; Feldman and Stone, 1997). Transgenic plants are revolutionizing agriculture and insect pest management.

CONCLUSION

Biotechnology is revolutionizing many aspects of our lives, including the study and cure of human health issues, food production and security, production of pharmaceuticals, and even the management of pests that are nuisances, disease vectors, or damaging to crops and stored products. Genetic techniques offer much in terms of understanding the mechanism of action and epidemiology of microorganisms that infect and naturally regulate insect pests and plant diseases. Molecular methods have allowed us to identify, isolate, sequence, and clone genes that encode or regulate the activity of these microbial agents against pests. Genetic engineering and recombinant techniques have improved or enhanced the activity of selected microbial control agents. Although all of these are valuable tools in microbial control, genetic engineering is not the panacea of pest management. Genetic techniques can be used to improve or enhance the activity or utility of microbial control agents; however, microbial control is only one component of pest management. We must continue to understand the natural epidemiology of microbial control agents, the benefits and risks associated with their use, and the potential markets for their development and use.

REFERENCES

Ahrens, C.H., R.L.Q. Russell and G.F. Rohrmann. (1997). The sequence of the *Orgyia pseudotsugata* multinucleocapsid nuclear polyhedrosis virus genome. *Virology* 229: 381-399.

Arisan-Atac, I., E. Heidenreich and C.P. Kubicek. (1995). Randomly amplified polymorphic DNA fingerprinting identifies subgroups of *Trichoderma viride* and other *Trichoderma* sp. capable of chestnut blight biocontrol. *FEMS Microbiology Letters* 126: 249-255.

Ayres, M.D., S.C. Howard, J. Kuzio, M. Lopez-Ferber and R.D. Possee. (1994). The complete sequence of *Autographa californica* nuclear polyhedrosis virus. *Virology* 202: 586-605.

Babu, R.M., A. Sajeena, K. Seetharaman and M.S. Reddy. (2003). Advances in genetically engineered (transgenic) plants in pest management—An overview. *Crop Protection* 22: 1071-1086.

Baker, K.F. and R.J. Cook. (1974). *Biological Control of Plant Pathogens,* W.H. Freeman and Co., San Francisco, p. 433.

Bennett, J., M.B. Cohen, S.K. Katiyar, B. Ghareyazie and G.S. Khush. (1997). Enhancing insect resistance in rice through biotechnology. In *Advances in Insect Control: The Role of Transgenic Plants,* N. Carozzi and M.G. Koziel, eds. Taylor and Francis, London, pp. 75-93.

Black, B.C., L.A. Brennan, P.M. Dierks and I.E. Gard. (1997). Commercialization of baculoviral insecticides. In *The Baculoviruses,* L.K. Miller, ed. Plenum Press, New York, pp. 447-462.

Bonning, B.C., M. Hirst, R.D. Possee and B.D. Hammock. (1992). Further development of a recombinant baculovirus insecticide expressing the enzyme juvenile hormone esterase from *Heliothis virescens. Insect Biochemistry and Molecular Biology* 22: 453-458.

Bonning, B.C., V.K. Ward, M.V. vanMeer, T.P. Booth and B.D. Hammock. (1997). Disruption of lysosomal targeting is associated with insecticidal potency of juvenile hormone esterase. *Proceedings of the National Academy of Sciences USA* 94: 6007-6012.

Burges, H.D. (ed.). (1981). *Microbial Control of Pests and Plant Diseases 1970-1980,* Academic Press, London, p. 949.

Burges, H.D. and N.W. Hussey (eds.). (1971). *Microbial Control of Insects and Mites,* Academic Press, London, p. 861.

Canton, F.A. and J.D. Vandenburg. (1999). Genetic transformation and mutagenesis of the entomopathogenic fungus *Paecilomyces fumosoroseus. Journal of Invertebrate Pathology* 74: 281-288.

Carlton, B.C. (1996). Development and commercialization of new and improved biopesticides. In *Engineering Plants for Commercial Products and Applications,* G.B. Collins and R.J. Shepherd, eds. New York Academy of Sciences, New York, pp. 154-163.

Carozzi, N. and M.G. Koziel. (1997). Transgenic maize expressing a *Bacillus thuringiensis* insecticidal protein for control of European corn borer. In *Advances in Insect Control: The Role of Transgenic Plants,* N. Carozzi and M.G. Koziel, eds. Taylor and Francis, London, pp. 63-74.

Copping, L. (1999). *The Biopesticide Manual,* British Crop Protection Council, Farnham, p. 333.

Corke, A.T.K. and J. Rishbeth. (1981). Use of microorganisms to control plant diseases. In *Microbial Control of Pests and Plant Diseases,* H.D. Burges, ed. Academic Press, London, pp. 717-736.

Donzelli, B.G., M. Lorito, F. Scala and G.E. Harman. (2001). Cloning, sequence and structure of a gene encoding an antifungal glucan 1,3-β-glucosidase from *Trichoderma atroviride (T. harzianum). Gene* 277: 199-208.

dos Reis, M.C., M.H.P. Fungaro, R.T.D. Duarte, L. Farlanto and M.C. Furlaneta. (2004). *Agrobacterium tumefaciens*-mediated genetic transformation of the entomopathogenic fungus *Beauveria bassiana*. *Journal of Microbiological Methods* 58: 197-202.

Eldridge, R., F.M. Horodyski, D.B. Morton, D.R. O'Reilly, J.W. Truman, L.M. Riddiford and L.K. Miller. (1991). Expression of an eclosion hormone gene in insect cells using baculovirus vectors. *Insect Biochemistry* 21: 341-351.

Eldridge, R., D.R. O'Reilly, B.D. Hammock and L.K. Miller. (1992). Insecticidal properties of genetically engineered baculoviruses expressing an insect juvenile hormone esterase gene. *Applied Environmental Microbiology* 58: 1583-1591.

Feldman, J. and T. Stone. (1997). The development of a comprehensive resistance management plan for potatoes expressing the Cry3A endotoxin. In *Advances in Insect Control: The Role of Transgenic Plants,* N. Carozzi and M.G. Koziel, eds. Taylor and Francis, London, pp. 49-61.

Frey, P.M. (2001). Biocontrol agents in the age of molecular biology. *Trends in Biotechnology* 19: 432-433.

Gaugler, R. and S. Hasnmi. (1996). Genetic engineering of an insect parasite. *Genetic Engineering: Principles and Methods* 18: 135-141.

Gaugler, R., M. Wilson and P. Shearer. (1997). Field release and environmental fate of a transgenic entomopathogenic nematode. *Biological Control* 9: 75-80.

Georgis, R. (1997). Commercial prospects of microbial insecticides in agriculture. In *British Crop Protection Council Symposium Proceedings No. 68. Microbial Insecticides: Novelty or Necessity,* H.F. Evans, ed. British Crop Protection Council, Farnham, pp. 241-252.

Gomi, S., K. Majima and S. Maeda. (1999). Sequence analysis of the genome of *Bombyx mori* nucleopolyhedrovirus. *Journal of Genetic Virology* 80: 1323-1337.

Gopalakrishnan, B., S. Muthukrishnan and K.J. Kramer. (1995). Baculovirus-mediated expression of a *Manduca sexta* chitinase gene: Properties of the recombinant protein. *Insect Biochemistry and Molecular Biology* 25: 255-262.

Gritsun, T.S., M.V. Mikhailov, P. Roy and E.A. Gould. (1997). A new, rapid and simple procedure for direct cloning of PCR products into baculoviruses. *Nucleic Acid Research* 24: 1864-1872.

Hashmi, S., G. Hashmi, I. Glazer and R. Gaugler. (1998). Thermal response of *Heterorhabditis bacteriophora* transformed with the *Caenorhabditis elegans hsp70* encoding gene. *Journal of Experimental Zoology* 281: 164-170.

Hayes, C.K., S. Klemsdal, M. Lorito, A.D Pietro, C. Peterbauer, J.P Nakas, A. Tronsmo and G.E. Harman. (1994). Isolation and sequence of an endochitinase-encoding gene from a cDNA library of *Trichoderma harzianum*. *Gene* 138: 143-148.

Hoover, K., C.M. Schultz, S.S. Lane, B.C. Bonning, S.S. Duffey and B.D. Hammock. (1995). Reduction in damage to cotton plants by a recombinant baculovirus that causes moribund larvae of *Heliothis virescens* to fall off the plant. *Biological Control* 5: 419-426.

Huang, Z., R.F. Bonsall, D.V. Mavrodi, D.M. Weller and L.S. Thomashow. (2004). Transformation of *Pseudomonas fluorescens* with genes for biosynthesis of phenazine-1-carboxylic acid improves biocontrol of rhizoctonia root rot and *in situ* antibiotic production. *FEMS Microbiology Ecology* 49: 243-251.

Huffaker, C.B., P.S. Messenger and P. DeBach. (1971). The natural enemies component in natural control and the theory of biological control. In *Biological Control,* C.B. Huffaker, ed. Plenum Press, New York, pp. 16-67.

Hughes, P.R., H.A. Wood, J.P. Breen, S.F. Simpson, A.J. Duggan and J.A. Dybas. (1997). Enhanced bioactivity of recombinant baculoviruses expressing insect-specific spider toxins in lepidopteran crop pests. *Journal of Invertebrate Pathology* 69: 112-118.

Ignoffo, C.M. (ed.) (1988). *Handbook of Natural Pesticides, Vol V: Microbial Insecticides,* CRC Press, Boca Raton, Florida, p. 436.

Jarvis, D.L. (1997). Baculovirus expression vectors. In *The Baculoviruses,* L.K. Miller, ed. Plenum Press, New York, pp. 389-431.

Joshi, L., R.J. St. Leger and D.W. Roberts. (1997). Isolation of a cDNA encoding a novel subtilisin-like protease (PR1B) from the entomopathogenic fungus, *Metarhizium anisopliae* using differential display-RT-PCR. *Gene* 197: 1-8.

Karabulut, O.A. and N. Baykal. (2004). Integrated control of postharvest diseases of peaches with a yeast antagonist, hot water and modified atmospheric packaging. *Crop Protection* 23: 431-435.

Kaya, H.K. (1985). Entomogenous nematodes for insect control in IPM systems. In *Biological Control in Agricultural IPM Systems,* M.A. Hoy and D.C. Herzog, eds. Academic Press, Orlando and London, pp. 283-302.

Koziel, M.G., N.B. Carozzi, T.C. Curriei, G.W. Warren and S.V. Evola. (1993). The insecticidal crystal proteins of *Bacillus thuringiensis*: Past, present and future uses. *Biotechnological and Genetic Engineering Review* 11: 171-228.

Lacey, L.A. and M.S. Goettel. (1995). Current developments in microbial control of insect pests and prospects for the early 21st century. *Entomophaga* 40: 3-27.

Lee, S.Y., X. Qu, W. Chen, A. Poloumienko, N. MacAfee, B. Morin and C. Lucarotti. (1997). Insecticidal activity of a recombinant baculovirus containing an antisera-*c-myc* fragment. *Journal of Genetic Virology* 78: 273-281.

Lereclus, D., H. Agaissi, M. Gominet and J. Chaufaux. (1995). Overproduction of encapsulated insecticidal crystal proteins in a *Bacillus thuringiensis spo0A* mutant. *Biotechnology* 13: 67-71.

Lu, A. and L.K. Miller. (1996). Generation of recombinant baculoviruses by direct cloning. *Biotechniques* 26: 63-70.

Ma, P.W.K., T.R. Davis, H.A. Wood, D.C. Knipple and W.L. Roelofs. (1998). Baculovirus expression of an insect gene that encodes multiple neuropeptides. *Insect Biochemistry and Molecular Biology* 28: 239-249.

Maeda, S. (1989). Increased insecticidal effect by a recombinant baculovirus carrying a synthetic diuretic hormone gene. *Biochemical and Biophysiological Research Communications* 165: 1177-1179.

McCoy, C.W., R.A. Samson and D.G. Boucias. (1986). Entomogenous fungi. In *Handbook of Natural Pesticides. Volume 5: Microbial Pesticides,* C.M. Ignoffo and N.B. Mandaira, eds. CRC Press, Baco Raton, Florida, pp. 151-236.

Oberhauser, K.S., M.D. Prysby, H.R. Mattila, D.E. Stanley-Horn, M.K. Sears, G.P. Dively, E. Olson, J.M. Pleasants, W.K.F. Lam and R.L. Hellmich. (2001).

Temporal and spatial overlap between the monarch larvae and corn pollen. *Proceedings of the National Academy of Sciences USA* 98: 1444-1454.

O'Callaghan, M.O., T.R. Glare, E.P.J. Burgess and L.A. Malone. (2005). Effects of plants genetically modified for insect resistance on nontarget organisms. *Annual Review of Entomology* 50: 271-292.

Office of the Gene Technology Regulator. (2003). Intentional release. DIR034/2003. The evaluation of transgenic cotton plants expressing the VIP gene. http://www.ogtr.gov.au/ir/dir034.htm.

O'Reilly, D.R. (1995). Baculovirus-encoded ecdysteroid UDP-glucosyltransferases. *Insect Biochemistry and Molecular Biology* 25: 541-550.

O'Reilly, D.R. and L.K. Miller. (1991). Improvement of a baculovirus pesticide by deletion of the EGT gene. *Biotechnology* 9: 1086-1089.

Peferoen, M. (1997). Insect control with transgenic plants expressing *Bacillus thuringiensis* crystal proteins. In *Advances in Insect Control: The Role of Transgenic Plants,* N. Carozzi and M.G. Koziel, eds. Taylor and Francis, London, pp. 21-48.

Poinar, G.O. (1991). Genetic engineering of nematodes for pest control. In *Biotechnology for Biological Control of Pests and Vectors,* K. Maramorosch, ed. CRC Press, Boca Raton, Florida, pp. 77-93.

Popham, H.J.R., Y. Li and L.K. Miller. (1997). Genetic improvement of *Helicoverpa zea* nuclear polyhedrosis virus as a biopesticide. *Biological Control* 10: 83-91.

Popham, H.J.R., G.G. Prikhod'ko and L.K. Miller. (1998). Effect of deltamethrin treatment on lepidopteran larvae infected with baculoviruses expressing insect-specific toxins mu-Aga-IV, AsII, or ShI. *Biological Control* 12: 79-87.

Possee, R.D., A.L. Barnett, R.E. Hawtin and L.A. King. (1997). Engineered baculoviruses for pest control. *Pesticide Science* 51: 462-470.

Prikhod'ko, G.G., M. Dickson and L.K. Miller. (1996). Properties of three baculovirus-expressing genes that encode insect-selective toxins: mu-Aga-IV, As II, and ShI. *Biological Control* 7: 236-244.

Prikhod'ko, G.G., H.J.R. Popham, J.J. Felcetto, D.A. Ostlind, V.A. Warren, M.M. Smith and V.M. Garsky. (1998). Effects of simultaneous expression of two sodium channel toxin genes on the properties of baculoviruses as pesticides. *Biological Control* 12: 66-78.

Punja, Z.K. and R.S. Utkhede. (2003). Using fungi and yeasts to manage vegetable crop diseases. *Trends in Biotechnology* 21: 400-407.

Rajamohan, F. and D.H. Dean. (1996). Molecular biology of bacteria for the biological control of insects. In *Molecular Biology of the Biological Control of Pests and Diseases of Plants,* M. Gunasekaran and D.J. Weber, eds. CRC Press, Boca Raton, Florida, pp. 50-66.

Sandhu, S.S., S.E. Unkles, R.C. Rajak and J.R. Kinghorn. (2001). Generation of benomyl resistant *Beauveria bassiana* strains and their infectivity against *Helicoverpa armigera. Biocontrol Science and Technology* 11: 250-254.

Sato, K., M. Komoto, T. Sato, H. Enei, M. Kobayashi and T. Yaginuma. (1997). Baculovirus-mediated expression of a gene for trehalase of the mealworm beetle, *Tenebrio molitor,* in insect cells, SF-9, and larvae of the cabbage armyworm, *Mamestra brassicae. Insect Biochemistry and Molecular Biology* 27: 1007-1016.

Schuler, T.H., G.M. Poppy, B.R. Kerry and I. Denholm. (1999). Potential side effects of insect resistant transgenic plants on arthropod natural enemies. *Trends in Biotechnology* 17: 210-216.

Shelton, A.M., J.Z. Zhao and R.T. Roush. (2002). The economic, ecological, food safety and social consequences of the deployment of Bt transgenic plants. *Annual Review of Entomology* 47:445-481.

St. Leger, R.J. (1993). Biology and mechanisms of insect-cuticle invasion by deuteromycete fungal pathogens. In *Parasites and Pathogens of Insects,* N.E. Beckage, S.N. Thompson and B.A. Federici, eds. Academic Press, New York, pp. 211-216.

St. Leger, R.J. (1995). The role of cuticle-degrading proteases in fungal pathogenesis of insects. *Canadian Journal of Botany* 73 (Suppl. 1): S1119-S1125.

St. Leger, R.J., L. Joshi, M.J. Bidochka and D.W. Roberts. (1996). Construction of an improved mycoinsecticide overexpressing a toxic protease. *Proceedings of the National Academy of Sciences USA* 93: 6349-6354.

St. Leger, R.J., L. Joshi and D.W. Roberts. (1998). Ambient pH as a major determinant in the expression of cuticle-degrading enzymes and hydrophobin by *Metarhizium anisopliae. Applied Environmental Microbiology* 64: 709-713.

Steinhaus, E.A. (1975). *Disease in a Minor Chord,* Ohio State University Press, Columbus, p. 116.

Thanabalu, T. and A.G. Porter. (1995). Efficient expression of a 100-kilodalton mosquitocidal toxin in protease-deficient recombinant *Bacillus sphaericus. Applied Environmental Microbiology* 61: 4031-4036.

Thomashow, L.S. (1996). Biological control of plant root pathogens. *Current Opinion in Biotechnology* 7: 343-347.

Thompson, M.A., H.E. Schnepf and J.S. Feitelson. (1995). Structure, function and engineering of *Bacillus thuringiensis* toxins. *Genetic Engineering, Principles and Methods* 17: 99-117.

Tomasino, S.F., R.T. Leister, M.B. Dimock, R.M. Beach and J.L. Kelly. (1995). Field performance of *Clavibacter xyli* subsp. *Cynodontis* expressing the insecticidal protein gene cry1A(c) of *Bacillus thuringiensis* against European corn borer in field corn. *Biological Control* 5:442-456.

Valadares-Inglis, M.C. and P.W. Inglis. (1997). Transformation of the entomopathogenic fungus, *Metarhizium flavoviride* strain CG423 to benomyl resistance. *FEMS Microbiological Letters* 155: 199-202.

Viterbo, A., S. Haran, D. Friesem, O. Ramot and I. Chet. (2001). Antifunfal activity of a novel endochitinase gene (chit36) from *Trichoderma harzianum Rifai* TM. *FEMS Microbiology Letters* 200: 169-174.

Vlak, J.M., A. Schouten, M. Usmany, G.J. Belsham, E.C. Kinge-Roode, A.J. Maule and J.W.M. vanLent. (1990). Expression of cauliflower mosaic virus gene I using a baculovirus vector based on the p10 and a novel selection method. *Virology* 179: 312-320.

Waage, J.K. (1997). Biopesticides at the cross-roads: IPM products or chemical clones. In *British Crop Protection Council Symposium Proceedings No. 68. Microbial Insecticides: Novelty or Necessity,* H.F. Evans, ed. British Crop Protection Council, Farnham, pp. 11-19.

Watrud, L.S., F.J. Peralk, M. Tran, K. Kusano, E.J. Mayer, M.A. Miller-Wideman, M.G. Obukowicz, D.R. Nelson, J.P. Kreitinger and R.J. Kaufman. (1985). Cloning of *Bacillus thuringiensis* subsp. *Kurstaki* delta-endotoxin gene into *Pseudomonas fluorescens*: molecular biology and ecology of an engineered microbial pesticide. In *Engineered Organisms in the Environment,* H.O. Harlvorson, D. Poinar and M. Rogul, eds. American Society for Microbiology, Washington, DC, pp. 56-77.

Wraight, L.C., A.R. Zangerl, M.J. Carroll and M.R. Berenbaum. (2000). Absence of toxicity of *Bacillus thuringiensis* pollen to black swallowtails under field conditions. *Proceedings of the National Academy of Sciences USA* 97: 7700-7703.

Yedidia, I., N. Benhamou, Y. Kapulnik and I. Chet. (2000). Induction and accumulation of PR proteins activity during early stages of root colonization by the mycoparasite *Trichoderma harzianum* strain T-203. *Plant Physiology and Biochemistry* 38: 863-873.

Zangerl, A.R., D. McKenna, C.K. Wraight, M. Carroll, P. Picarello, R. Warner and M.R. Berenbaum. (2001). Effects of exposure to Event 176 *Bacillus thuringiensis* corn pollen on monarch butterflies and black swallowtails. *Proceedings of the National Academy of Sciences USA* 98: 11908-11912.

Chapter 5

Control and Management of Plant-Parasitic Nematodes

Franco Lamberti
Nicola Greco
Alberto Troccoli

INTRODUCTION

Since the report of the first plant-parasitic nematode, *Anguina tritici,* in 1743 by Needham (Poinar, 1983), approximately 3,000 species of plant-parasitic or candidate plant parasitic nematodes have been described within some 150 genera. However, the economically important plant parasitic nematodes can be included in a list of 150 or so species belonging to some forty-five to fifty genera.

Nematodes may damage plants directly, causing disruption or necrosis of infected plant tissues, and indirectly by interacting with other organisms, thus increasing the severity of the damage they cause (Khan, 1993; Abawi and Chen, 1998) and reducing *Rhizobium* nodulation on the roots of leguminous plants (Riggs and Niblack, 1993; Vovlas, Castillo and Troccoli, 1998; Sikora, Greco and Silva, 2005). Other species may affect plants by transmitting plant viruses to them (Santos et al., 1997; Taylor and Brown, 1997; Brown, Zheng and Zhou, 2004). Moreover, several nematode species survive on or within planting material, with which they can be spread over long distances; therefore, they are included in lists of quarantine pests by

The authors wish to thank Dr. Ken Evans for reviewing the manuscript, and all colleagues who have supplied photographs.

Management of Nematode and Insect-Borne Plant Diseases

doi:10.1300/5754_05

many countries. As most plant-parasitic nematodes attack roots, the symptoms they cause in the aerial plant parts usually are not typical but simply those of plants with damaged roots. Only for a limited number of nematodes are the symptoms so typical that they indicate infection by a given species.

The extent of yield loss caused by nematodes can be so severe as to make crops uneconomic and, therefore, to warrant the use of control means. However, for a sound approach to nematode management, information is necessary on nematode species, biology, population density at planting, population dynamics, economic thresholds of the host plants, pedo-climatic conditions (which may or may not favor nematode development), and the effectiveness, availability, and possible side effects of different control options. Such information, where available, will be given below for each of the most important nematode groups and species.

KNOWN PLANT-PARASITIC NEMATODES

For the most part, clear evidence of damage to plants and crops is available in the literature for species included in thirty-two genera (Table 5.1).

TABLE 5.1. Genera of plant-parasitic nematodes containing species for which evidence exists that they affect plants and crops.

Class	Order	Family	Genus	Authority
Adenophorea	Dorylaimida[a]	Longidoridae	*Longidorus*	Mycoletzky
			Paralongidorus	Siddiqi, Hooper, and Khan
			Xiphinema	Cobb
	Triplonchida[a]	Trichodoridae	*Paratrichodorus*	Siddiqi
			Trichodorus	Cobb
Secernentea	Aphelenchidae[b]	Aphelenchoididae	*Aphelenchoides*	Fischer
			Bursaphelenchus	Fuchs
			Rhadinaphelenchus	J.B. Goodey
	Tylenchida[c]	Anguinidae	*Anguina*	Scopoli
			Ditylenchus	Filipjev
		Belonolaimidae	*Belonolaimus*	Steiner

TABLE 5.1 *(continued)*

Class	Order	Family	Genus	Authority
		Criconematidae	*Hemicriconemoides*	Chitwood and Birchfield
			Mesocriconema	Andrassy
		Dolichodoridae	*Dolichodorus*	Cobb
		Hemicycliophoridae	*Hemicycliophora*	de Man
		Heteroderidae	*Globodera*	Skarbilovich
			Heterodera	Schmidt
		Hoplolaimidae	*Helicotylenchus*	Steiner
			Hoplolaimus	von Daday
			Rotylenchus	Filipjev
			Scutellonema	Andrassy
		Meloidogynidae	*Meloidogyne*	Goeldi
		Paratylenchidae	*Cacopaurus*	Thorne
			Gracilacus	Raski
			Paratylenchus	Micoletzky
		Pratylenchidae	*Hirschmanniella*	Luc and Goodey
			Nacobbus	Thorne and Allen
			Pratylenchus	Filipjev
			Radopholus	Thorne
		Rotylenchulidae	*Rotylenchulus*	Linford and Oliveira
		Telotylenchidae	*Tylenchorhynchus*	Cobb
		Tylenchulidae	*Tylenchulus*	Cobb

[a]According to Hunt, 1993.

[b]According to Maggenti, 1991.

[c]According to Siddiqi, 2000.

Longidorids

Species of longidorids are of interest as pests of both perennial and annual crops and as vectors of plant nepoviruses (Brown and Trudgill, 1997; Brown, Zheng and Zhou, 2004). They are among the largest plant parasitic nematodes (up to 10-12 mm long and 30-35 μm diameter at midbody). The most important genera are *Longidorus* and *Xiphinema* (Figure 5.1A, C, and D). Within the genus *Longidorus, Longidorus elongatus* (de Man) Thorne and Swanger and *Longidorus macrosoma* Hooper appear to be the most widespread. Among the species of *Xiphinema,* the most common are *Xiphinema diversicaudatum* (Micoletzky) Thorne, *Xiphinema index* Thorne and Allen, *Xiphinema italiae* Meyl, and the *Xiphinema americanum* group. *X. index* is of worldwide importance as the natural vector of the grapevine fan leaf virus (GFLV) complex (Figure 5.1F), and species of the *X. americanum* group are vectors of tomato ringspot, tobacco ringspot, peach rosette mosaic virus, and cherry rasp leaf virus. The *X. americanum* group is of quarantine concern for several countries. Although longidorids are adapted to a wide variety of environmental conditions, they prefer sandy or coarse soils. Fluctuation in soil moisture content affects their population dynamics, with dry soils suppressing population densities severely. Generally, species of *Longidorus* are mostly associated with herbaceous plants, whereas *Xiphinema* spp. are associated with trees (Hunt, 1993). They are ectoparasites that feed mostly on root tips, causing severe gall-like swellings (Figure 5.1I). They develop better at temperatures in the range 20-25°C, and those attacking trees can be found at depths of as much as 1 m or wherever there are roots. The life cycle may take from a few months to one year to be completed, and adult specimens may survive in the soil for several years, retaining viruses that they can transmit. Because of their habit, they are very sensitive to synthetic nematicides, a characteristic that makes chemical control a feasible and effective practice.

Damage by longidorids is more severe on perennial than on annual crops because their long life cycle (more than one year in many instances) is affected by the frequent drastic cultural practices used for annual crops, such as deep ploughing and weeding, and by the senescence and maturity of the crops. Rotation of annual crops can be a convenient control option and sometimes is useful to solve a replant or soil sickness problem.

Fumigant nematicides are generally more efficient (Figure 5.1H) than nonvolatile compounds because they disperse more readily and thoroughly through the soil profile, whereas nonvolatile compounds depend very much on the water present in the soil for their movement. Depending on the rate of application, fumigants can reduce nematode

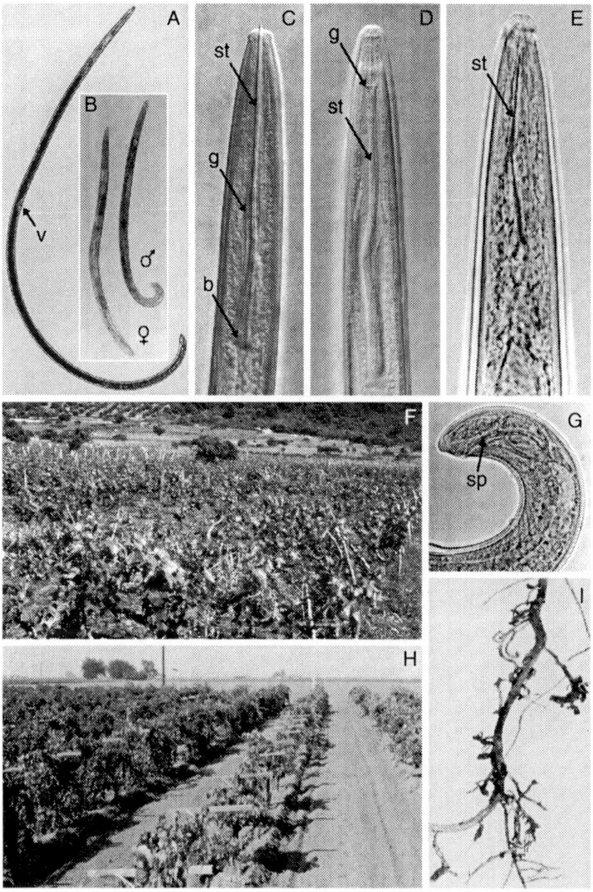

FIGURE 5.1. Longidorid and trichodorid nematodes. (A) Entire female of *Xiphinema index* showing vulva (v) at mid-body length; (B) male and female specimens of *Trichodorus variabilis* Roca; (C) anterior region of *Xiphinema* sp. (note the long stylet [st] with a basal bulb [b] and the position of the guide ring [g] near the middle of the stylet); (D) anterior region of *Longidorus* sp. showing absence of basal knobs on the stylet (st) and the position of the guide ring (g) in the anterior part of the stylet; (E) specimen of *T. variabilis* (note the slightly arcuate shape of the stylet); (F) vineyard showing a patch with poorly growing and yellowing plants caused by *X. index* and the GFLV; (G) tail region of a male *T. variabilis* showing spicules (sp); (H) vineyard infested with *Xiphinema americanum* in California (United States): treated with D-D (left) and untreated (right); (I) roots of grapes showing tip galls caused by the feeding of *X. index*. (photographs B, E, and G courtesy of Dr. Francesco Roca).

populations down to 90 cm depth, but the effects of nonvolatile nematicides are largely restricted to the upper 30-40 cm of the soil.

The control of longidorid species that act as vectors of plant nepoviruses should aim for as close as possible to complete eradication, as just a few surviving nematodes can still efficiently transmit the virus (Thomason and McKenry, 1975) and their populations will increase again within a few months or years (Coiro, Taylor and Lamberti, 1987). Eradication of soil populations of virus vector nematodes is almost impossible, and any attempt is usually very expensive because very high doses of fumigant nematicides have to be injected into the soil by appropriate machinery to achieve satisfactory results.

High doses of D-D (1,2-dichloropropane-1,3-dichloropropene; Figure 5.1H) or methyl bromide have been used extensively in grapevine and fruit tree replanting in the past with some success. However, these and similar chemicals were recently more or less banned owing to their pollutant effects. One alternative that is still available is 1,3-dichloropropene (1,3-D).

Crop rotation often is an effective method to control both soil-borne viruses and the vector nematodes. However, it requires a minimum break of five years before replanting with healthy plants (Raski et al., 1965). To make a rotation effective, it is necessary to control weeds and any surviving plants of the previous crop as they can act as hosts for the nematode or a reservoir for the virus.

Perhaps the most convenient way to control longidorid vectors of plant viruses is a combination of a two- to three-year rotation and soil fumigation. Fumigants available on the market in developed countries are 1,3-D, chloropicrin, and metham-sodium. Methyl bromide will be available in developing countries only for one more decade. Spraying systemic herbicides before pulling out the viruliferous host plants, to kill their roots and thus eliminate the virus reservoir for the nematode, combined with crop rotation and possibly with a chemical soil treatment has also been recommended (Greco and Esmenjaud, 2004).

Trichodorids

Several species of *Paratrichodorus* and *Trichodorus* (Figure 5.1B, E, and G) transmit tobraviruses (Brown, Zheng and Zhou, 2004). They are rather small in size (0.5-0.8 mm long) and also cause direct damage to crops (Ploeg and Brown, 1997). Among the most widespread species are *Trichodorus primitivus* (De Man) Micoletzky, *Trichodorus viruliferus* Hooper, *Paratrichodorus minor* (Colbran) Siddiqi, and *Paratrichodorus porosus* (Allen) Siddiqi. Trichodorids are better adapted to rather sandy soils and are susceptible to soil moisture stress and mechanical disturbance (Hunt, 1993). As they are ectoparasites, trichodorids are also very sensitive to nematicides. Thus, fumigants such as 1,3-D or chloropicrin or nonvolatile chemicals may satisfactorily control both the direct damage to plants and virus dissemination. Trichodorids have short life cycles of a few months and cause damage mainly to annual herbaceous plants. Therefore, crop rotation and soil tillage to enhance nematode desiccation normally give acceptable results in the control of these parasites. However, virus transmission rates can remain high even with few nematodes owing to the efficiency of the vectors (Decramer, 1995).

Aphelenchidae

Within the family Aphelenchidae, nematodes may cause different problems according to the genus and sometimes to the species within a genus. Plant-parasitic species can be as thin as 12 μm and rather short (0.5-1.1 mm). They have short life cycles (ten to twenty days) and can withstand desiccation in planting materials, with which they are usually dispersed (Hunt, 1993). Therefore, most of the species in this group are considered quarantine microorganisms.

One of the most important genera is *Aphelenchoides* (Figure 5.2E). Among the several species are the leaf and bud nematodes, *Aphelenchoides fragariae* (Ritzema Bos) Christie and *Aphelenchoides ritzemabosi* (Schwartz) Steiner and Buhrer. They reproduce and cause damage most severely when the temperature is around 20-25°C and the surfaces of host plants remain wet for several hours a day. These two species cause severe damage mostly to strawberry and chrysanthemum (Figure 5.2B), respectively, but, as they have wide host ranges (Hunt, 1993), they also damage a number of broad-leaved crops, including various vegetables (Figure 5.2A) and flowers. Planting materials should be free of them or previously treated by immersion in hot water at 46°C for five minutes. Care is necessary as higher temperatures can damage canes and runners. Systemic nematicides, applied as foliar sprays (if permitted) or granules to the soil, also give good results.

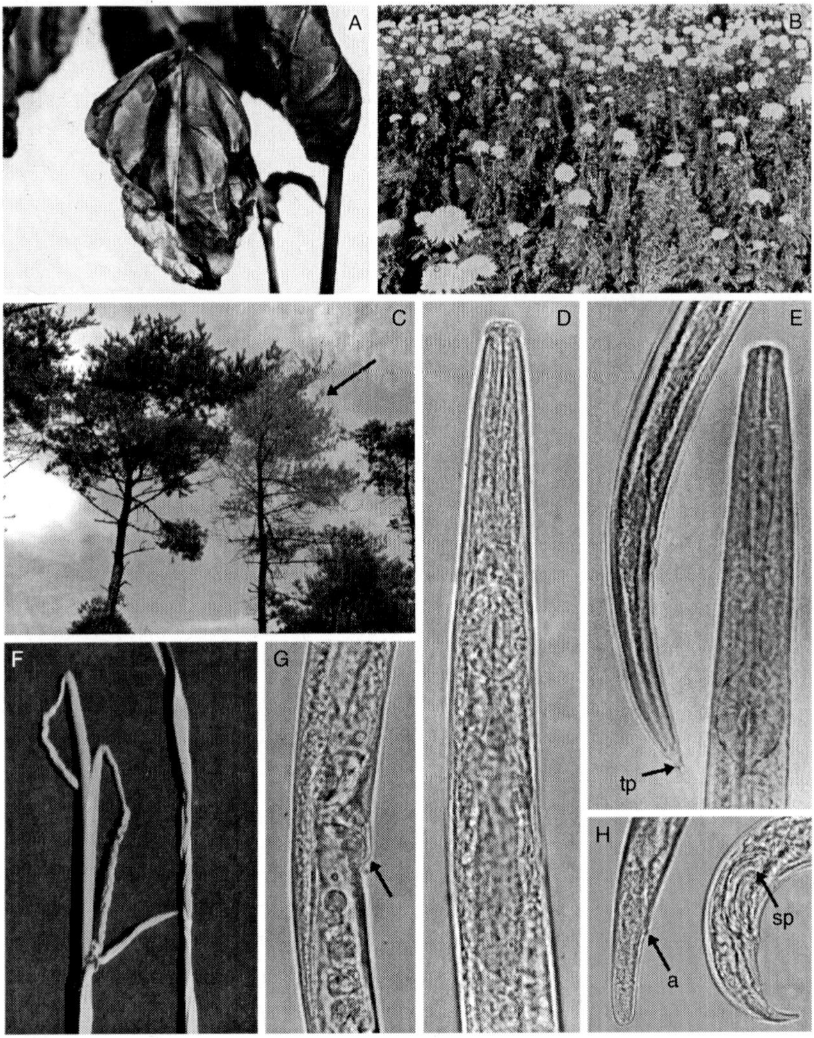

FIGURE 5.2. Aphelenchoidinae. (A) Necrotic leaves of basil; (B) A chrysanthemum crop showing plants growing poorly and with necroctic leaves caused by infection by *Aphelenchoides ritzemabosi*; (C) A pine (arrow) killed by *Bursaphelenchus xylophilus* in Japan; (D) Anterior region of *B. xylophilus*; (E) anterior region (right) and tail region (left) of a female *A. ritzemabosi* showing the mucronate tail tip (tp); (F) wrinkled leaves of rice caused by *A. besseyi* in India; (G) vulval region (arrow) of *B. xylophilus*; (H) tail regions of a female (left), showing anus (a) and male (right), showing spicules (sp) of *B. xylophilus*.

Aphelenchoides arachidis Bos and *Aphelenchoides besseyi* Christie are seed-borne parasites of peanuts and rice (Figure 5.2F), respectively, and the latter species can also be dispersed with flood water. They require temperatures of around 25°C. Infested dormant seeds of peanuts should be immersed for five minutes in four times their volume of water heated to 60°C (Minton and Baujard, 1990; Hunt, 1993). For *A. besseyi,* presoaking rice seeds for eighteen to twenty-four hours in cold water followed by immersion for ten to fifteen minutes in water at 51-53°C is required to control the nematode. Also, some cultivars of rice resistant to *A. besseyi* are known (Bridge, Luc and Plowright, 1990).

Aphelenchoides composticola often damages mushrooms under cultivation. Strict hygiene (cleaning the trays), heating the compost to 60°C for eight hours, or treating with low-persistence nematicides at the start of cultivation should successfully control the problem (Richardson and Grewal, 1993).

The genus *Bursaphelenchus* comprises several species that damage *Pinus* spp. The most important is the pine wilt nematode, *Bursaphelenchus xylophilus* (Steiner and Buhrer) Nickle (Figure 5.2D, G, and H), a destructive pest of pine (Figure 5.2C) transmitted by Cerambycidae beetles, mainly *Monochamus alternatus* in Japan and *M. carolinense* in the United States (Hunt, 1993; Sutherland and Webster, 1993). In Europe the nematode has been found in Portugal (Mota et al., 1999). *B. xylophilus* is favored by hot temperatures; its life cycle can be completed in twelve days at 15°C and in only three days at 30°C. A susceptible pine can be killed within thirty to forty days from first infection.

Good control of this nematode can be achieved only by preventing emergence of its vector from dead pine trees and branches (Mamiya, 1984). These should be burned or have their bark removed and be sprayed with high concentrations of insecticide emulsions. Timber sawdust should always be heat-treated at 60-80°C for two to three hours or fumigated in closed containers. Some *Pinus* species are resistant to *B. xylophilus* (Mamiya, 1984). A priority objective in control work is the search for natural antagonists of both the nematode and its vectors.

Rhadinaphelenchus cocophilus (Cobb) Goodey, the causal agent of red ring disease of coconut, which causes plant death, is also naturally spread by a coleopteran vector, of the Curculionidae, *Rhynchophorus palmarum* L. (Hunt, 1993). The nematode is widespread in the Caribbean and in Latin American countries, where it causes a 10-15 percent yield reduction. Coconut palms are more susceptible to the nematode when they are three to ten years old. There are no effective means for controlling the nematode in living palms. Burning of the dead or yellow plants and stringent insecticide spraying programs to control the vector are the only way to obtain satisfactory results (Griffith and Koshy, 1990).

Anguinidae

In the past, *A. tritici* (Steinbuch) Filipjev caused severe damage to various wheat species and barley (Figure 5.3B), inducing seed and leaf galls. This was the first plant-parasitic nematode to be reported in the literature. It survives within seed kernels as the second stage juvenile. After infested seeds are sown, second-stage juveniles (Figure 5.3A) become active, infect the seedlings, and complete their life

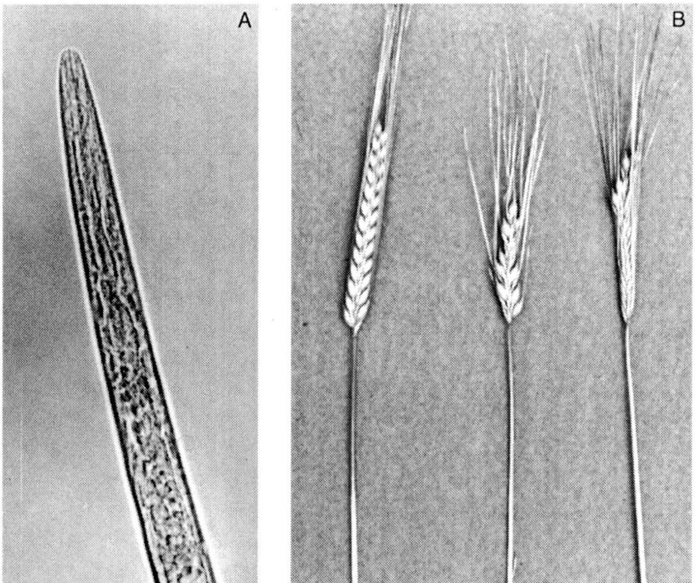

FIGURE 5.3. (A) Infective second-stage juveniles of *A. tritici;* (B) damage to barley in Syria: healthy (left) and infected (right) ears showing lack of kernel formation because of infection by the nematode.

cycle in the new kernels. Only one generation per year is completed. Infested seeds are lighter than healthy ones, and modern seed-cleaning techniques have greatly reduced the economic importance of this nematode, the causal agent of earcockle disease. However, *A. tritici* is still rather common in countries of the Middle East and on the Indian subcontinent. One year's crop rotation and cleaning seeds by winnowing or floating in 20 percent brine can effectively control this disease. Also, presoaking cereal seeds in water to reactivate nematode juveniles and then soaking them at 50-56°C for thirty to thirty-five minutes before sowing is quite satisfactory (Swarup and Sosa-Moss, 1990). Some Indian varieties were reported to be resistant to the nematode (Parveen et al., 2003).

Ditylenchus is a nonhomogeneous genus in terms of its bionomics and behavior, such that different populations of *Ditylenchus dipsaci* (Kühn) Filipjev are considered to be separate races.

Ditylenchus dipsaci (Figure 5.4A and B) is a migratory endoparasite that has a rather short life cycle (nineteen to twenty-three days at 15°C) and a wide host range, being able to infect more than 450 plant species, including weeds and grasses, of which it attacks aerial plant parts and subterranean reproduction organs but not roots. Among the most damaged crops are bulbous vegetables (Figure 5.4C-E) and flowers, alfalfa, broad bean (Figure 5.4G), clovers, strawberry (Figure 5.4H), carrot (Figure 5.4F), and, in a few countries, oats, corn, and sugar beet. The symptoms the nematodes cause on most of these crops are rather typical and consist mainly of leaf, stem, and bulb deformations (Figure 5.4D, E, and G) and rotting (Figure 5.4F). However, each of the approximately thirty known host races of the nematode (Sturhan and Brzeski, 1991) reproduces on and damages a more restricted range of plant species. Only the giant race, affecting broad bean mostly in North African countries, can be identified by molecular techniques (Esquibet et al., 1998). The nematode survives during unsuitable environmental conditions, such as dry and hot or cool periods and plant senescence, in soil and plant parts as a preadult quiescent stage. In the field, survival is better in clay than in sandy soils. Nematode infection and damage is favored by environmental conditions that leave the host plant wet, such as rain, fog, dew, and sprinkler irrigation, with reproduction rates that can exceed 1,000-fold. Severe damage to onion may occur in fields infested with five

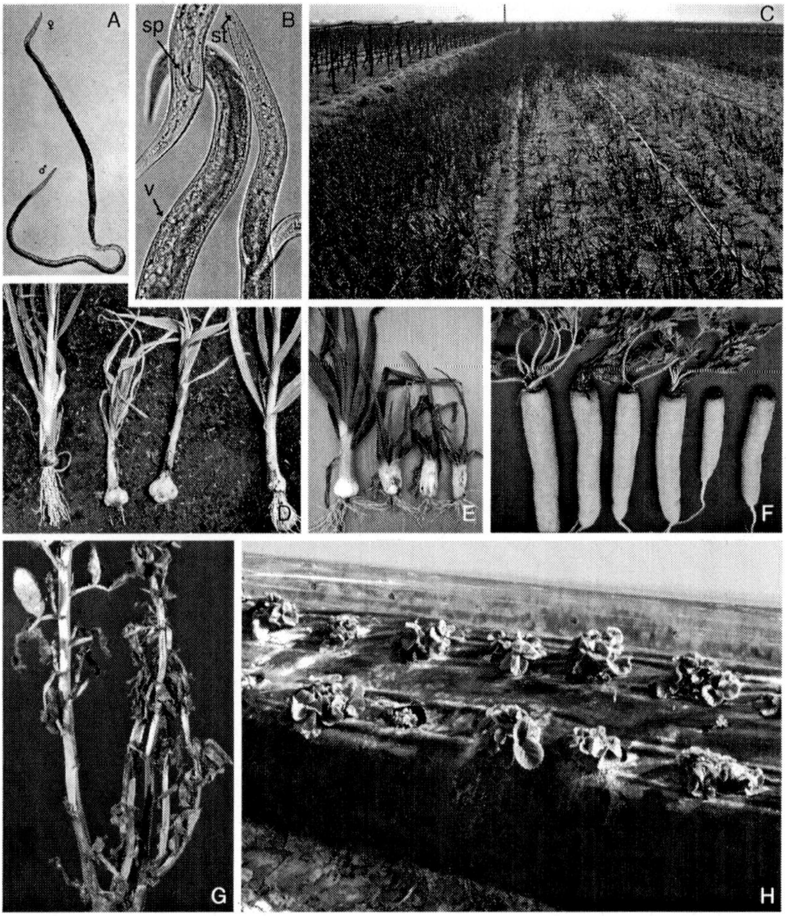

FIGURE 5.4. *D. dipsaci.* (A) Adult female and male; (B) posterior regions of a male showing spicules (sp) and of a female showing vulva (v) and anterior region with the stylet (st); damage to: (C, E) onion, (D) garlic, (F) carrot, (G) broad bean, and (H) strawberry.

nematodes per 500 cm^3 soil. Temperatures in the range 15-20°C are the most suitable for development of this nematode.

Ditylenchus dipsaci is disseminated with seeds (alfalfa, broad bean, onion), bulbs, rhizomes, cloves (garlic), taproots (carrots, sugar beet), runners, and other infested planting material and, therefore, is considered a quarantine pest. The main rule to avoid spreading this pest is selection of healthy planting material. Sanitation schemes (Tacconi, 1989) may help to eradicate the problem from some regions. Also very satisfactory are hot water treatments of vegetable and flower bulbs for three to four hours at 44-45°C, especially if combined with certain additives (Roberts and Matthews, 1995). Lightly infested seeds (broad bean and alfalfa) can be successfully treated with methyl bromide in closed containers at the rate of 1,000 mg/h/l. Higher rates (3,000 mg/h/l) can disinfest even highly infested broad bean seeds but also reduce seed germination (Powell, 1974). When soil is infested, preplant treatments with fumigant and nonvolatile systemic nematicides can give very good results (Whitehead, 1998). Soil solarization (Siti et al., 1982; Greco, Brandonisio and Elia, 1985) can also be applied with success. Crop rotation requires at least three to four years to reduce soil populations and must be combined with strict weed control, as many weed species are good hosts for the nematode. For some crops, including alfalfa, clovers, and oats, resistant cultivars are available (Whitehead, 1998).

Ditylenchus destructor Thorne is similar to *D. dipsaci* but attacks only above-ground plant parts. Potato is among the crops most damaged; tubers may rot because of subsequent bacterial invasion. It is considered a quarantine pest. The use of healthy potato tubers is a prerequisite to avoid yield loss and, combined with crop rotation, may solve the problem completely (Jatala and Bridge, 1990; Brodie, Evans and Franco, 1993).

Ditylenchus angustus (Butler) Filipjev is a severe pest of deep-water and lowland rice in several Asian countries, especially when temperatures are in the range 27-30°C. The nematode is ectoparasitic on young leaves and survives within the dried glumes of spikelets, to resume activity in the water of the next crop. It is considered a quarantine pest. Good results in the control of *D. angustus* can be achieved by destruction of crop residues, use of crop rotation, control of weeds, volunteer rice, water flow, and use of resistant cultivars (Bridge, Luc and Plowright, 1990).

Ditylenchus africanus Wendt, Swart, Vrain, *et* Webster, reported until 1994 as *D. destructor*, is a parasite of peanut in South Africa (Venter et al., 1995) that causes pod and seed deformation and seed pregermination before harvest. The nematode should be considered a quarantine pest. The use of healthy seeds and soil treatments should provide good control.

Hygiene measures should be applied in the preparation of mushroom beds and composts to prevent establishment and attacks of *Ditylenchus myceliophagus* Goodey. Compost and casing material should be heated up to 60°C (Hesling, 1974; Richardson and Grewal, 1993) for about eight hours.

Belonolaimidae

Belonolaimus longicaudatus Rau and *Belonolaimus gracilis* Steiner are root ectoparasites on various plants, including pines (Smart and Nguyen, 1991). They can be well controlled with nematicides or crop rotation.

Criconematidae

Also known as ring nematodes, Criconematidae are root ectoparasites of both herbaceous (Figure 5.5E) and woody plants and have a rather typical cuticle (Figure 5.5A-D). *Hemicriconemoides mangiferae* Siddiqi has been reported as a pest of various tropical fruit trees, including mango (Cohn and Duncan, 1990). However, no attempts have ever been made to determine the amount of damage caused or to test control measures.

Mesocriconema (Criconemella) xenoplax (Raski) Loof is considered one of the major causal agents of peach tree short life in the southeastern United States (Nyczepir and Halbrendt, 1993). It can be controlled by soil fumigation; however, investigations aimed at developing nonchemical means of nematode control indicated that interplanting *Paspolum* spp. in the tractor rows of peach plantations would suppress populations of *M. xenoplax* and increase yields. Also, wheat seems to suppress populations of *M. xenoplax*. Lovell peach is partially tolerant of this nematode.

Dolichodoridae

Dolichodorus heterocephalus Cobb can cause severe damage on celery, corn, and other plants. Nematicides are very effective in controlling this ectoparasitic nematode (Smart and Nguyen, 1991).

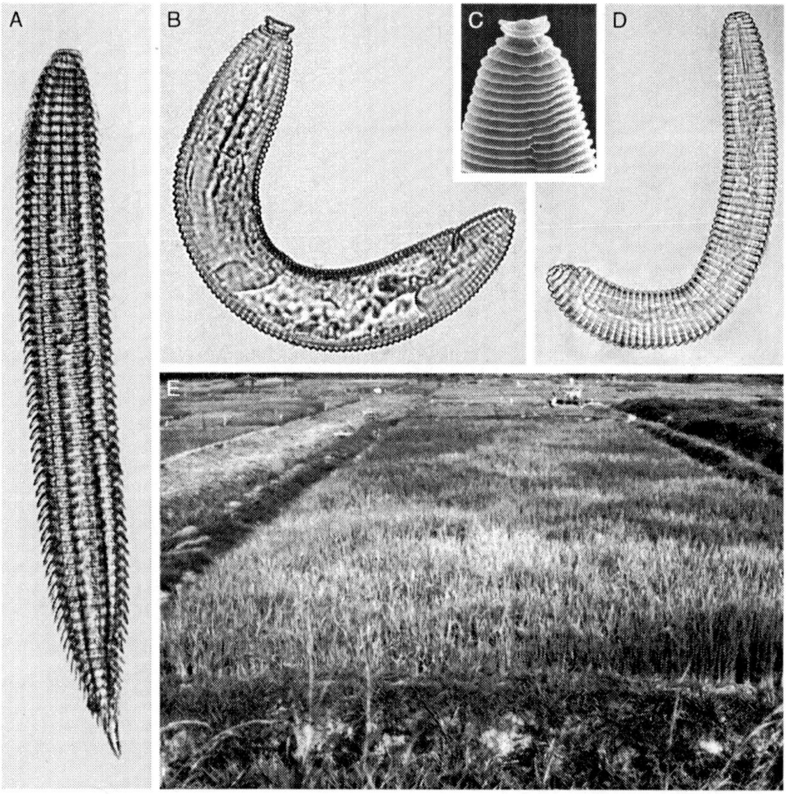

FIGURE 5.5. Specimens of (A) *Ogma araguaensis* Crozzoli *et* Lamberti; (B) *Discocriconemella* sp. with SEM view of the (C) expanded labial disc and (D) *Mesocriconema sphaerocephala* (Taylor) Loof. Note the rather thick and strongly annulated cuticle in A, B, and D. (E) A rice crop in Liberia damaged by *Mesocriconema curvatum* (Rasky) Loof *et* De Grisse (photographs A-D courtesy of Dr. Renato Crozzoli).

Hemicycliophoridae

Hemicycliophora arenaria Raski constitutes a threat to the citrus industry in southern California. Infected roots show galls. Sanitation of planting material can be achieved by hot-water dips (10 minutes at 46°C). Soil fumigation solves preplant problems; resistant rootstocks are also available (Duncan and Cohn, 1990).

Heteroderidae

The Heteroderidae include several of the most damaging and widespread nematode species (Baldwin and Mundo-Ocampo, 1991; Evans and Rowe, 1998). The species of this family are typical as they show marked sexual dimorphism, having wormlike adult males and globose (Figure 5.6B) or lemon-shaped (Figure 5.7A and B) adult females. The cyst, which is the final life stage, contains many eggs (up to 500) and can survive in the soil and on infected plant parts for several years (up to twenty years, depending on the species). Generally, eggs within cysts are stimulated to hatch only by root exudates of host plants, provided that suitable temperature and soil moisture conditions also occur. Moreover, eggs within cysts may undergo a diapause or dormancy period. The second-stage juvenile emerging from the egg is the only infective stage and penetrates young roots near to

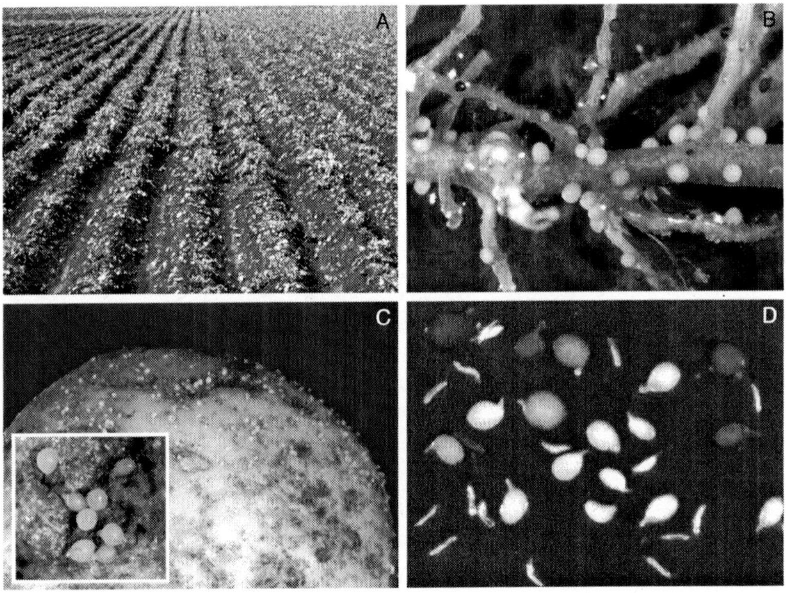

FIGURE 5.6. Potato cyst nematode, *G. pallida*. (A) Potato crop severely damaged; (B) potato roots showing many white, globose females; (C) potato tuber severely infested at harvest; (D) different developmental stages extracted from infected potato roots.

FIGURE 5.7. Cyst-forming nematodes, *Heterodera* spp. Roots of (A) chickpea bearing lemon-shaped females (f) of *H. ciceri* and of (B) carrot severely infested by *H. carotae*. Note the absence (A) and the presence (B) of egg sacs (es) containing many eggs. (C) Tranverse section of a root of chickpea infected by *H. ciceri* and showing the large syncitial cells (sc) disrupting the xylem. (D) Extensive yellowing and poor growth of pea infested by different population densities of *H. goettingiana* in microplots, (E) hard wheat by *H. avenae,* (F) sugar beet by *H. schachtii* (close-up shows white females on feeder roots) and (G) soybean by *H. glycines* in Tennessee (United States).

their tips. As these nematodes are sedentary endoparasites, the juveniles quickly select a feeding site within the central cylinder of the roots and induce the formation of a multicellular syncytium on which they feed. Syncytial cells cause interruption of the vascular tissue (Figure 5.7C), thus impairing root function. The second-stage juveniles undergo three moults to the third, fourth, and adult stages (Figure 5.6D). As they molt, female stages swell and, when adult, they rupture the root cortex and become visible from outside, although the head region remains inserted in the root. Mature females and cysts are approximately 0.4-0.8 mm long and in diameter and can be seen even with the naked eye by an experienced person. However, observation under a dissecting microscope at low magnification (6-10×) is suggested for the observation of younger developmental stages (Figure 5.6D). Second-stage juveniles are transparent, 400-600 μm long, and 18-24 μm wide, and adult males are 1,000-1,300 μm long and 30-40 μm wide. Generally, the eggs, which are approximately 50 μm wide and 100 μm long, are retained within the female body; at the end of the life cycle, the thickness of the cuticle of the female body increases to become a brown cyst. Females of several cyst nematode species, before becoming cysts, also lay eggs within a gelatinous matrix or egg sac (the egg mass). Eggs deposited in the egg sac normally do not undergo diapause or dormancy and are much less dependent on root exudates to hatch. Third- and fourth-stage males are also swollen, but much less so than females, and, when they reach adulthood, are wormlike, migratory, and not pathogenic.

Cyst nematodes have rather narrow host ranges, with sometimes only one or just a few crop plants attacked by the same species. The most important genera of this family are *Globodera* and *Heterodera*.

The genus *Globodera* includes species with almost spherical females and cysts (Figure 5.6B-D). Of the fourteen species known, the following are the most important.

The potato cyst nematodes, *Globodera pallida* (Stone) Behrens and *Globodera rostochiensis* (Wollenweber) Behrens, are among the most destructive nematode pests of agriculture (Figure 5.6A; Marks and Brodie, 1998). They develop better at soil temperatures in the range 15-25°C, with the optimum around 20-21°C. *G. pallida* appears to be adapted better than *G. rostochiensis* to lower temperatures. Both species infect roots (Figure 5.6B), stolons, and tubers (Figure 5.6C), through which they can be dispersed. Therefore, they are quarantine pests. In temperate areas, eggs in cysts undergo diapause, whereas in tropical areas they do not. Usually, there is only one generation of *G. rostochiensis* per crop cycle of potato, during which the nematode population may increase between 9-fold (during cool growing periods; Greco and Moreno, 1992) and 65-fold (on spring-sown crops; Greco et al., 1982). The annual decline of the nematode population is estimated at 50 percent in temperate regions and at up to 70-90 percent in warm and dry regions. The potato cyst nematodes have a worldwide distribution and can cause total crop loss (Marks and Brodie, 1998). The tolerance limit of potato to the nematodes is approximately 1.9 eggs/g soil, with 32 eggs/g soil causing 50 percent yield loss and 128 eggs/g soil nearly complete crop failure (Greco et al., 1982; Seinhorst, 1982; Table 5.2). The genetics of these nematodes is rather complex, as five pathotypes have been reported for *G. rostochiensis* and three for *G. pallida,* with differences occurring between Andean and European pathotypes (Fleming and Powers, 1998).

The control of potato cyst nematodes must be based primarily on the use of seed tubers free of the nematodes. Good control has often been attained with the application of fumigant nematicides (Figure 5.14A) such as 1,3-D or metham-sodium or of nonfumigant compounds such as the organophosphates fensulphothion, fenamiphos, and ethoprophos or the carbamates aldicarb and oxamyl (Whitehead and Turner, 1998). However, chemical treatments are not always fully successful and in many cases are too expensive for protecting a relatively low-value cash crop such as field-scale production of potatoes. Soil solarization for four to eight weeks is also effective in warm areas, but it may not be an economic proposition. In terms of costs, crop rotation would be more appropriate, but break periods with nonhosts should be around five to six years in temperate countries and three to four years in dry and warm areas, provided any volunteer potatoes are well controlled.

Fortunately, several resistant and/or tolerant potato cultivars are available, but they must be chosen with care, taking into consideration the nematode species and pathotypes. In fact, no potato cultivar is fully resistant to both species of potato cyst nematodes. Cultivars usually show resistance to just one pathotype of any given species and only exceptionally to more. Moreover, cultivars with partial resistance

TABLE 5.2. Tolerance limits of some crop plants to phytoparasitic nematodes and estimated soil population densities of these parasites that may cause given levels of yield losses (mostly abstracted from reports by Lamberti and Greco, 1989; Sasanelli, 1994).

		Nematodes/g or cm^3 of soil			
			Causing yield loss		
Nematode	**Crop plant**	**Tolerance limit**	**20 percent**	**50 percent**	**80 percent**
Globodera rostochiensis	Potato	1.9	10	34	100
G. pallida	Potato	1.7	10	28	80
H. avenae	Wheat	0.3-1	2.3-5	13-64	60?
H. carotae	Carrot	0.3-0.8	7	20	55
H. ciceri	Chickpea	1	6	10	32
	Lentil	2.5	20	128	
H. goettingiana	Broad bean	0.8	5	15	32
	Pea	0.5	3	8	16
	Vetch	2.0	20	78	–
H. schacthii	Sugar beet	1.5-1.8	9	28	66
Meloidogyne artiellia	Winter chickpea	0.14	1.0	2.5	5.5
	Spring chickpea	0.01	0.11	0.26	0.7
	Wheat	0.43	2.5	8	20
M. incognita	Cantaloupe	0.19	1.0	2.5	5
	French bean	0.03-0.25	0.2-5	0.8-16	?-36
	Pepper	0.16	0.8	2.5	16
	Sugar beet	1.1	6	14	55
	Tobacco	2	10	32	70
	Tomato	0.55	2.5	7	18
Pratylenchus thornei	Winter chickpea	0.03	0.25	1.0	–

alone are available to control *G. pallida*. However, the continuous use of resistant cultivars should be avoided as it may select for different, more virulent pathotypes.

Integrated control, the use of two or more effective measures of control, is perhaps the correct answer to solve problems such as that of potato cyst nematodes in extensive agricultural systems.

Species of the *Globodera tabacum* (Lownsbery *et* Lownsbery) Behrens complex are adapted to warm seasons and damage tobacco, mainly in the United States and China, although they have been reported from other countries too.

Within the genus *Heterodera,* more than sixty-five species have been described. Among them, attention has been given to the most damaging species, some of which have been known for a long time as economically important agricultural pests, whereas others have only recently been described and have more restricted distributions.

Heterodera avenae Wollenweber is a parasite of cereals and grasses of worldwide distribution (Swarup and Sosa-Moss, 1990; Rivoal and Cook, 1993). It attacks mainly oats, barley, and wheat (both *Triticum aestivum* [Figure 5.7E] and *Triticum durum*), but affects rye less. Damage is more severe on spring cereals than on winter cereals (Whitehead, 1998) and in sandy soil, whereas nematode populations may remain at nondamaging levels in rather heavy soils. The nematode develops only one generation per crop cycle and reproduces better when soil temperature is in the range 15-20°C. In temperate regions, eggs in cysts will not hatch before a two-month period at 10°C, whereas in Mediterranean areas and in Australia this cool period is not necessary and eggs will hatch as soon as the temperature drops to 15°C and the first autumn rains occur, even in the absence of a host crop. The tolerance limit of wheat to the nematode was estimated at about 0.3-1 eggs/g soil (Meagher and Brown, 1974; Greco and Brandonisio, 1987) (Table 5.2).

The former Gotland strain of *H. avenae,* now considered as *H. filipjevi* (Madzhidov) Stelter, occurs in several European and Mediterranean countries and probably is more widely distributed than previously thought. *Heterodera latipons* Franklin is rather common in southern Mediterranean countries. Both species attack winter cereals and behave similarly to *H. avenae.*

Extensive nematicidal treatments would be too expensive for cereals. However, in Australia low rates of nematicides placed in the sowing row are recommended (Brown, 1987), with the decision to apply based on a bioassay (Simon, 1980). Nevertheless, three- to four-year rotations with noncereal crops remain the most suitable and economical means of controlling this nematode (Figure 5.15A). Resistant cultivars are available, but the occurrence of races (Rivoal and

Cook, 1993) complicates their use (Whitehead, 1998). *Heterodera cajani* Koshy attacks various Leguminosae, mainly in India. The main host plants are pigeon pea, chickpea, soybean, and French bean. The optimum temperature for nematode development is in the range 25-29°C, and each of the several generations per crop cycle may take approximately 16 days at 29°C. Eggs are also laid in an egg sac. The tolerance limit of pigeon pea to *H. cajani* was estimated at 2.6 eggs and juveniles/cm^3 soil (Sharma et al., 1993). Crop rotation is the only feasible measure to control this nematode economically, although soil solarization and nematicides are effective. No cultivar resistant to the nematode is available.

Heterodera carotae Jones parasitizes only carrot (Figure 5.7B), causing severe damage especially in sandy soils in Europe (Greco, 1986). It also occurs in Cyprus and in Michigan. *H. carotae* prefers temperatures in the range 15-20°C. The species typically produces nearly half of its eggs in the egg sac (up to 150 per egg sac; Figure 5.7B). Therefore, these eggs must be considered when estimating soil nematode population density. Usually, only one generation from cyst to cyst develops on a crop. However, as the nematode also lays eggs in the egg sacs, two or three generations per carrot cycle can develop. The tolerance limit of carrot to the nematode is 0.3-0.8 eggs/cm^3 soil (Table 5.2), and complete crop failure may occur in fields infested with 64 eggs/cm^3 soil.

Carrots are often economically important vegetable crops, and in such cases the use of nematicides is justified. Applications of fumigants, such as 1,3-D (100-200 kg/ha) (Figure 5.14B) or methyl isothiocyanate-releasing compounds at higher rates, give excellent control of *H. carotae*. Good results, but less dramatic, are also obtained with systemic nonvolatile nematicides such as aldicarb, fenamiphos, and oxamyl. As carrot is the only host crop for *H. carotae,* crop rotation can be an efficient method of control provided that the soil is maintained clean of volunteers. Unfortunately, four to five years of carrot rotation is necessary to obtain acceptable results (Greco, 1986). Periods of six to eight weeks of soil solarization during the summer generally provide satisfactory control of *H. carotae* in southern Italy (Greco, 1986), especially if combined with reduced rates of fumigants, such as 1,3-D injected at 30-35 cm depth or dazomet at 100-200 kg/ha (Figure 5.14D).

Heterodera ciceri Vovlas, Greco, *et* Di Vito is a serious pest of chickpea (Figure 5.7A) and lentil in Jordan, Lebanon, Syria, and Turkey. The nematode also affects peas and grass pea (*Lathyrus sativus* L.), with some populations also reproducing on alfalfa. *H. ciceri* develops from late fall to mid-spring and completes only one generation per crop cycle. The optimum temperature for development is approximately 20°C. Infected plants show poor *Rhizobium* nodule development and suffer water stress, as infected root systems are rather shallow. Infested crops show patchy growth and yellowing in April-May. The tolerance to the nematode is 1 egg/g soil for chickpea and 2.5 eggs/g soil for lentil (Table 5.2). Complete chickpea failure would occur in soil infested with sixty-four eggs/g soil. Soil solarization (Figure 5.14E) and crop rotation (Figure 5.15B) for three to four years are very good control measures for this nematode. Split application of 5 kg aldicarb/ha at sowing and one month after seedling emergence is also effective on spring chickpea but not on winter chickpea (Sikora, Greco and Silva, 2005). However, neither soil solarization nor nematicides are economic means of control. Resistance to *H. ciceri* has been detected in some wild *Cicer* sp. lines, but resistant cultivars are not yet available (Sikora, Greco and Silva, 2005).

Heterodera cruciferae Franklin is not a common species but is a major pest of various crucifer crops when it occurs (Whitehead, 1998). A four- to five-year crop rotation is feasible to control this nematode, but when a cash crop, such as Brussels sprouts must be planted in infested soil, fumigation with 1,3-D may be appropriate.

The soybean cyst nematode, *Heterodera glycines* Ichinohe, is one of the most destructive pests of soybean (Figure 5.7G) and may also damage several warm-season leguminous crops in the Americas and several Asian countries, including China, Japan, and South Korea (Schmitt, Wrather and Riggs, 2004). On the basis of four soybean differentials, sixteen different races can be distinguished. The optimal temperature for development is 25°C, at which the life cycle can be completed in twenty-one to twenty-five days and several generations per crop cycle can develop. *H. glycines* lays eggs in an egg sac. Egg sac eggs hatch promptly, whereas cyst eggs may (early generations) or may not (late generations), as they undergo diapause. Several resistant cultivars of soybean are available; however, their use is sometimes limited by the occurrence of races of the nematode virulent toward the available resistance or by the presence of other nematode pests pathogenic on soybeans, such as *Meloidogyne incognita* and *Rotylenchulus reniformis*. Relatively short crop rotations (three years) with nonhost plant species such as maize, sorghum, or cotton may reduce the soil population of *H. glycines* to nondamaging levels. Nematicides, mainly fumigants such as 1,3-D, are effective but not always economical. Integration of a four-year rotation with resistant cultivars is the usual approach to control of *H. glycines* (Schmitt, Wrather and Riggs, 2004).

Heterodera goettingiana Liebscher is a damaging parasite of peas (Figure 5.7D), broad beans, grass pea, and vetch in Europe, Mediterranean countries, and Washington state (Di Vito and Greco, 1986; Sikora, Greco and Silva, 2005). Other leguminous plants are poor hosts or nonhosts. The environmental requirements are similar to those for *H. carotae* and *H. ciceri*. When soil temperatures do not exceed 15°C and soil moisture content is good, the nematode also lays eggs in an egg sac. Eggs in newly formed cysts undergo to a two- to three-month dormancy period, whereas eggs in egg sacs do not. Therefore, the nematode can develop up to three generations per crop cycle if egg sacs are formed (crops sown in mid-fall and harvested the next spring) but only one if cysts, but no egg sacs, are formed (peas sown in mid-winter). The tolerance limits to the nematode are 0.5, 0.8, and 2.2 eggs/g soil for pea, broad bean, and vetch, respectively (Greco, Ferris and Brandonisio, 1991) (Table 5.2).

Heterodera goettingiana can be efficiently controlled by crop rotation with nonleguminous plants. However, it may take at least four years to reduce soil population densities to nondamaging levels (Di Vito and Greco, 1986). In southern Europe, the garden pea is considered a cash crop that justifies the use of nematicides in heavily infested soils. Fumigant nematicides such as 1.3-D (150-200 kg/ha) or dazomet (500 kg/ha) are effective, as are nonvolatile compounds such as aldicarb, carbofuran, oxamyl, and fenamiphos (at about 10 kg active ingredient/ha).

The beet cyst nematode, *Heterodera schachtii* Schmidt, has been considered for more than a century the most important nematode pest of sugar beet (Figure 5.7F; Baldwin and Mundo-Ocampo, 1991; Cooke, 1993). It is found mainly in European countries and, to some extent, also in North America. The nematode produces a small egg sac that may contain only a few eggs. The optimum temperature for egg hatching and development is 25°C, and an average of three generations for the sugar beet cycle have been reported on spring-sown crops and up to five on fall-sown crops in southern California. Infested sugar beet may show patchy growth and/or wilting during sunny times of day. In these fields, taproots show a bearded appearance because of an apparently supranormal production of rootlets, which bear white, lemon-shaped females (Figure 5.7F). Usually, developed sugar beet taproots are not infested, and seldom are severe infestations reported. The tolerance limit of sugar beet to *H. schachtii* under field conditions was estimated at 1.5-1.8 eggs/g soil (Table 5.2).

H. schachtii reproduces on a wide range of plants, mainly within the families Chenopodiaceae and Cruciferae, so much care must be taken when selecting alternative crops for rotation schemes. Effective rotations include sugar beet or host plants only once in five years in heavily infested soil (Cooke, 1993; Whitehead, 1998), and good weed control must be practiced. Soil population densities can be reduced to nondamaging levels by cultivating resistant oil radish and mustard cultivars in summer, soon after the harvest of a summer-harvested crop. These resistant crops stimulate eggs to hatch but do not act as hosts for the hatched juveniles.

Soil fumigation with 1,3-D or dazomet gives very good yield increases of sugar beet in infested soil (Cooke, 1993; Whitehead, 1998), as does incorporation into the soil of nonfumigant nematicides, such as aldicarb or oxamyl. However, such treatments are not always economical. Recently, cultivars of sugar beet resistant to *H. schachtii* have been released, although they do not appear to be as productive as susceptible ones.

Heterodera elachista Oshima in Japan, *Heterodera oryzae* Luc *et* Berdon-Brizuela in the Ivory Coast, Senegal, and Bangladesh, *Heterodera oryzicola* Rao *et* Jayaprakash in India, and *Heterodera sacchari* Luc *et* Merny in western Africa are all parasites of rice with local importance. However, the two former species also attack banana, and *H. sacchari* attacks sugarcane and a number of indigenous Cyperaceae and Gramineae (Luc, 1986). They are best controlled by crop rotation.

Heterodera trifolii Goffart, the clover cyst nematode, is mainly a parasite of clovers in many countries (Baldwin and Mundo-Ocampo, 1991). However, the nematode may infect and damage many leguminous and some nonleguminous plants, such as carnations. The yellow beet cyst nematode, which damages sugar beet in several central European countries and was earlier considered a biotype of *H. trifolii,* was recently described as *Heterodera betae* Wouts, Rumpenhorst, and Sturhan. *H. trifolii* can be controlled by crop rotations that exclude leguminous crops. However, for high-profit crops such as carnations, chemical and other expensive means of control can be used.

Heterodera zeae Koshy, Swarup, and Sethi attacks corn and most summer and winter cereals in India, Pakistan, Egypt, and some states in the United States. It is favored by high temperature (25-39°C). It is controlled mainly by crop rotation, although it has been shown that dressing maize seeds with aldicarb or applying low dosages of aldicarb, carbofuran, or fenamiphos to the soil eight days before sowing reduces crop losses (Whitehead, 1998).

Hoplolaimidae

Members of the genus *Helicotylenchus* are semiendoparasites. The most common species is *Helicotylenchus dihystera,* of worldwide distribution on a wide range of crops (Whitehead, 1998). Small doses of nonfumigant nematicides such as oxamyl, fensulphothion, ethoprophos, aldicarb, or carbofuran can satisfactorily control this pest.

Helicotylenchus multicinctus multiplies on many crop plants but is the most common nematode pest on bananas (Whitehead, 1998). Paring banana corms to remove infested tissue and dipping suckers in nematicide solutions for 30 minutes before planting are effective sanitation practices. Soil applications of systemic nematicides, such as oxamyl, carbofuran, aldicarb, or fenamiphos, reduce nematode populations and improve yields of banana and plantains both qualitatively and quantitatively. Such treatments are usually repeated every four to five months.

A few species of *Hoplolaimus* feed ecto- or semiendoparasitically on various crops, such as *Hoplolaimus columbus* on soybean and cotton, *Hoplolaimus pararobustus* on banana, *Hoplolaimus seinhorsti* on cowpea and various vegetable crops, and *Hoplolaimus indicus* on a wide range of plants (Whitehead, 1998). Depending on the situation, preplant applications of low doses of fumigant or nonvolatile nematicides can give good results.

Among the *Rotylenchus* species (Castillo and Vovlas, 2005), *Rotylenchus robustus* is known to cause damage on carrot and lettuce. Crop rotation or soil application of nematicides can effectively control this nematode.

Scutellonema bradys is a nematode pest infesting tubers of yams (*Dioscorea* spp.) (Whitehead, 1998). Control of the nematode can be achieved by rotation with nonhost plants such as groundnut, chillies, and tobacco or by application of systemic nematicides such as aldicarb, carbofuran, or oxamyl, even after planting.

Meloidogynidae

The most important genus of the Meloidogynidae family is *Meloidogyne,* which comprises approximately eighty species. These species cause the typical symptoms of galls (Figure 5.8A and B), whose size (2-20 mm diameter) and number (the roots can be transformed into an assembly of just galls) depend on the number of nematodes in the roots, the nematode and plant species, and the time elapsed from infection. On account of this symptom, they are known as root-knot nematodes and are the most important nematode pests of agricultural crops worldwide (Lamberti and Taylor, 1979; Sasser, 1979; Sasser and Carter, 1985; Eisenback and Hirschmann Triantaphyllou, 1991; Whitehead, 1998) (Figure 5.9A-E). They also reproduce on a wide range of wild plants.

Behavior, economic importance, crops attacked, and control measures used for *Meloidogyne* species vary according to the climatic regions where the problem occurs and the species involved.

Root-knot nematodes are sedentary endoparasites that survive as eggs (especially in temperate regions) and second-stage juveniles (mainly in rather warm areas) in the soil and in infested plant parts. In areas with mild winters, adult females may overwinter within the roots and other plant parts. Eggs are stimulated to hatch by adequate soil moisture and temperature and are not dependent upon root exudates of host plants. The second-stage juvenile emerging from the egg is the only infective stage and may persist in the soil as the active stage during cool periods; during hot and dry seasons their populations, as well as those of the eggs, decline rapidly (up to 80 percent in a month). Only second-stage juveniles of *Meloidogyne artiellia* are able to survive dry seasons, as they can enter an anydrobiotic survival stage. Under suit-

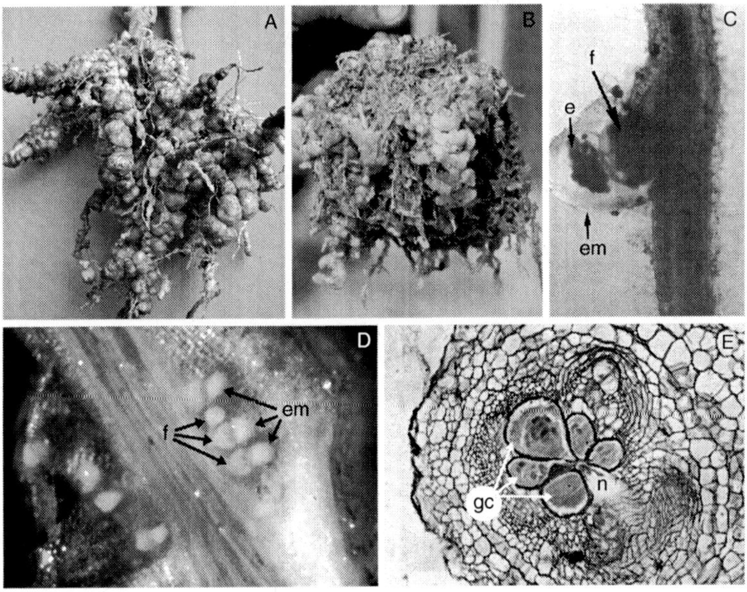

FIGURE 5.8. Root-knot nematodes, *Meloidogyne* spp. Roots of (A) tomato and (B) melon showing large galls induced by infection by *M. incognita* and *M. javanica*, respectively. (C) Female (f) of *M. artiellia* partly embedded in the root with external egg mass (em). (D) Longitudinal section of a root of banana showing pear-shaped females (f) of *M. incognita* bearing large egg masses (em), both completely embedded in the root tissues. (E) Transverse section of root infected by *M. incognita* and showing giant cells (gc) induced by the nematode feeding (photographs C and E courtesy of Dr. Nicola Vovlas; photograph D courtesy of Dr. Renato Crozzoli).

able environmental conditions, the second-stage juvenile penetrates the root tip and, after moving a few millimeters, selects a feeding site, where it induces the formation of giant cells (four-eight per nematode; Figure 5.8E) around its head, in the stelar region of the root. As with cyst nematodes, the formation of these giant ("nurse") cells is a prerequisite for the feeding and development of the nematode. Furthermore, the nematode stimulates the proliferation of many small normal cells surrounding its body, thus giving rise to the formation of the galls on the infected roots. The giant cells and the nematode body also cause disruption of the root vascular tissues, thus impairing root function. The life cycle proceeds as described for Heteroderidae, but the adult

FIGURE 5.9. Damage caused by *Meloidogyne* spp. on (A) tomato (arrow) and (D) cucumber in plastic-houses and on (B) kiwi, (C) tobacco, and (E) cauliflower, under field conditions.

nematode females become pear-shaped, remain embedded within the roots, and lay eggs in a gelatinous matrix or egg mass (Figure 5.8C and D), which can contain up to 1,500 eggs. Usually, the egg mass protrudes from the roots (Figure 5.8C), but in the case of large roots and in crop plants such as banana (Figure 5.8D), it can remain embedded in the root tissue. No cysts are formed, and generally the eggs hatch as soon as they complete embryonic development and environmental conditions are suitable. Root-knot nematodes may complete three to eight generations per year, depending on the nematode and host plant species and environmental conditions. The reproduction rate of these nematodes is very high (up to 1,000-fold the population at planting), as is their decline after harvest, especially under hot and dry conditions (about 80 percent in a month). The tolerance limit of the host plants is rather low (usually <1 egg/cm^3 soil) and sometimes is at an undetectable level (Table 5.2).

In temperate climates, root-knot nematode species of economic relevance on field crops are *M. artiellia* Franklin on cruciferous, leguminous, and cereal crops in Europe and Mediterranean countries; *Meloidogyne hapla* Chitwood on carrots, sugar beet, potatoes, brassicas, and many other field crops in several countries; *Meloidogyne naasi* Franklin on cereals and several other crops in Europe, the United States, New Zeland, and Chile; *Meloidogyne chitwoodi* Golden, O'Bannon, Santo, *et* Finley on potatoes in North America and some European countries; and *Meloidogyne fallax* Karssen on potatoes in the Netherlands and Belgium. The last two species are considered quarantine pests in Europe. Except for *M. hapla,* these species have host ranges much narrower than those of most of the tropical and subtropical species. Eggs of *M. naasi* hatch only after a chilling period of seven weeks at 5-10°C, and only one generation of the nematode per crop cycle develops. The crops affected are mainly field crops, which may not always provide high income; therefore, the use of both fumigant and nonfumigant nematicides is likely to be uneconomical. The only feasible and economical ways of controlling these root-knot nematode species are crop rotations and the use of healthy planting material. Sometimes, certain cultural practices, such as shifting planting times or introducing fallow intervals into the cropping system, may help to reduce damage.

Meloidogyne incognita (Kofoid *et* White) Chitwood, *Meloidogyne javanica* (Treub) Chitwood, *Meloidogyne arenaria* (Neal) Chitwood, and *M. hapla* must be regarded as very important pests of vegetable, fruit, and flower crops in regions with subtropical and Mediterranean climates. Moreover, they are highly pathogenic to several industrial crops, such as sugar beet, tobacco, and maize. However, their greatest effects are on vegetables and ornamentals, either outdoors or in protected culture in glass, plastic, or screen houses. These species infest and damage *Citrus* spp. but are unable to reproduce on these crops. Nevertheless, as many citrus groves contain many wild plants, the nematodes may reproduce on the wild plants and then infect citrus. Several *Meloidogyne* spp. that damage and reproduce on citrus are known from China (*Meloidogyne citri* Zhang, Gao, *et* Weng; *Meloidogyne donghaiensis* Zheng, Lin, *et* Zheng; *Meloidogyne fujianensis* Pan; *Meloidogyne jianyangenis* Yang, Hu, Chen, *et* Zhu; *M. mingnanica* Zhang; *Meloidogyne oteifae* Elmiligy) and India (*Meloidogyne indica* Whitehead). Moreover, a species not yet described but known as the Asiatic pyroid citrus nematode occurs in India and Taiwan.

Other tropical species of root-knot nematodes are severe pests of rice (Bridge, Luc and Plowright, 1990). Among them, *Meloidogyne graminicola* Golden *et* Birchfield occurs in southeast Asian countries and the United States and *Meloidogyne oryzae* occurs in central America, both in deep-water rice. They can persist in flooded soils for several months as eggs in egg masses or as second-stage juveniles and can be disseminated with irrigation and run-off water. *Meloidogyne salasi* Lopez-Chaves occurs in upland rice in Costa Rica and Panama and probably in other countries in Central and South America. In addition to rice, these root-knot nematodes reproduce on several Gramineae weeds that infest rice fields.

In the past, the control of root-knot nematodes (Sasser and Carter, 1985; Whitehead, 1998) was achieved by massive applications of synthetic nematicides, mainly fumigants containing 1,3-D, ethylene dibromide (EDB), 1,2-dibromo-3-chloropropane (DBCP), dazomet, and metham-sodium. However, because most of these chemicals possess almost exclusively nematicidal activity and nematodes often occur along with other soil-borne pathogens, farmers consider routine fumigation with methyl bromide to be more convenient. This fumigant has a very wide spectrum of action and efficiently controls various soil-borne pathogens and pests, including weeds. Its higher cost compared with individual nematicides, fungicides, insecticides, and herbicides is balanced by the need for only a single preplant application, often performed by companies with specialized personnel and machinery, which guarantees efficient distribution and in most cases very satisfactory control with few risks. In recent years, practically no crop grown in protected culture has not been preventively fumigated with methyl bromide.

However, the Montreal Protocol included methyl bromide among the stratospheric ozone-depleting compounds to be banned from routine agricultural uses as of January 1, 2005, in developed countries. Therefore, to reduce populations of soil-borne pathogens and pests to nonpathogenic levels, attention and efforts must be turned toward alternative chemicals or to alternative means of management.

With respect to nematicide treatments, tests were undertaken to choose the most effective chemicals for practical, economical, and environmentally safe methods of root-knot nematode control (Lamberti, Minuto and Filippini, 2003). Soil application of 1,3-D, either as a fumigant to be injected into the soil or as an emulsifiable concentrate to be distributed on the soil surface under plastic films in drip irrigation systems, is the most common practice. Chloropicrin, metham-sodium, and dazomet are also frequently applied. Among the nonvolatile compounds, the organophosphates fenamiphos, fosthiazate, ethoprophos, and cadusaphos and the carbamates carbofuran, oxamyl, and aldicarb are used with varying results. The systemic action of the organophosphates fenamiphos and fosthiazate gives them a more persistent effect. The carbamates carbofuran and oxamyl have less persistent systemic activity and, therefore, can be applied to established crops. Aldicarb has more restricted use because of its longer persistence.

Root-knot nematodes can also be controlled by steaming, a practice used mostly in glasshouses in the cultivation of flowers and seedlings owing to its high cost, or by soil solarization, again more effective and convenient in protected culture (Figure 5.14F) than outdoors (Gaur and Perry, 1991; Katan and De Vay, 1991; Cartia, Greco and Di Primo, 1998).

Biofumigation (Gamliel, 2000) is becoming more common for the control of root-knot nematodes. However, it is more feasible outdoors than in glasshouses, where intervals between successive crops are much reduced.

Crop rotation is neither common nor easily practicable to control most of the root-knot nematodes typical of subtropical and Mediterranean climates because of their wide host ranges.

Soil amendments with organic agricultural by-products can also give good results, but only from the second or third year of application (Akhtar and Malik, 2000; Bailey and Lazarovits, 2003).

Biological control (Kerry, 1987; Stirling, 1991; Chen and Dickson, 2004a, b) with antagonistic organisms such as *Paecilomyces lilacinus* (Samson) Thom., *Pochonia chlamydosporia* Zare et al., or *Pasteuria penetrans* (Thorne) Sayre *et* Starr is still in the experimental phase, although some commercial formulations are already available. Promising results are being obtained with plant extracts (Chitwood, 2002) such as those from *Azadirachta indica*.

The use of resistant cultivars (Sasser and Carter, 1985; Roberts, 1992, 2002; Starr and Roberts, 2004), when their produce is accepted by consumers, is the best solution for farmers. Cultivars resistant to root-knot nematodes are available mainly for tomato, soybean, French bean, and cowpea among annual crops and as resistant rootstocks for grapevine and peach in fruit crops. However, resistant tomato cultivars lose their resistance when soil temperature exceeds 28°C.

Planting material, such as tubers, bulbs, seedlings, and nursery stock, should always be free of parasitic nematodes. This can be achieved by producing them under strict hygiene conditions and, when necessary, by immersion in aqueous solutions or emulsions of non-phytotoxic nematicides.

In tropical climates, economically important *Meloidogyne* species are *M. arenaria, M. incognita,* and *M. javanica* as pests of vegetable crops, fruits such as banana, papaya, pineapple, and industrial crops such as sugarcane and soybean. *Meloidogyne exigua* Goeldi is a pest of coffee, *Meloidogyne brevicauda* Loos attacks tea, and *Meloidogyne mayaguensis* Kammah *et* Hirschmann has been described as a parasite of basil, French bean, cowpea, and vegetable crops as well as coffee and guava in Central and South America and in south and west African countries (Brito et al., 2003).

Control in the tropics is by nematicidal treatments for cash crops or by rotation in subsistence agriculture. Plant sanitation for crops such as banana, plantain, and pineapple is not frequently adopted but is strongly recommended.

Paratylenchidae

Nematodes of this family are ectoendoparasites with sedentary habits. The most important species are listed here.

- *Cacopaurus pestis* Thorne has been reported as a pest of walnut (Nyczepir and Halbrendt, 1993) in Italy and the United States but also occurs in France and Spain.
- *Paratylenchus peraticus* (Raski) Siddiqi *et* Goodey has been observed feeding on olive roots (Inserra and Vovlas, 1977).

A number of species of *Paratylenchus* have been found in association with declining herbaceous crops, such as *Paratylenchus projectus* Jenkins, *Paratylenchus dianthus* Jenkins *et* Taylor, and *Paratylenchus nanus* Cobb, or in the rhizosphere of fruit trees, such as *Paratylenchus hamatus* Thorne *et* Allen, *Paratylenchus curvitatus* Van der Linde, and *Paratylenchus neoamblycephalus* Geraert. They are weak parasites that cause crop losses only if present in very high population densities.

In annual crops they can be managed very well by rotations; fruit tree nursery stock should be free of nematodes and planted in non-infested soils.

Pratylenchidae

The Pratylenchidae family mostly comprises migratory endoparasitic nematodes of roots and subterranean plant organs, in which they cause cavities in the parenchyma (Figure 5.10B). They include several genera, of which only the most important species will be mentioned. Most of them have wide host ranges.

Many species of *Hirschmanniella* attack rice. The most common and widespread is *Hirschmanniella oryzae* (Van Breda de Haan) Luc *et* Goodey, but *Hirschmanniella spinicaudata* (Schuurmans Stekhoven) Luc *et* Goodey and *Hirschmanniella mucronata* (Das) Khan, Siddiqi, Khan, Husain, *et* Saxena also have economic importance. They are relatively long and thin nematodes (about 1.5 mm long and 24 μm wide in *H. oryzae*). *H. oryzae* completes only one or two generations per crop cycle and survives better in the roots of other hosts or ratooning rice roots and in decaying roots—for more than twelve months in nonwetted soil. Crop rotation with nonhost plants for more than one year and with appropriate weed control is effective in reducing nematode damage (Bridge, Luc and Plowright, 1990). *Sesbania rostrata,* a leguminous green manure crop, consistently increases the next rice yield. Nematode infestation very often starts in the seed beds. Treating the nursery soil with low doses of either fumigant or nonvolatile nematicides will produce healthy seedlings (Whitehead, 1998).

Nacobbus aberrans is a sedentary nematode that attacks several plant species. It is known as the false root-knot nematode as infested roots show galls similar to those caused by *Meloidogyne* spp. It is

FIGURE 5.10. Root-lesion nematodes. (A) A citrus grove in Florida (United States) with declining oranges (arrows) affected by *R. similis*. Note that in the front row plants are dead because the nematode attacks were replaced (arrows). Longitudinal sections showing (B) cavities (ca) containing a nematode specimen (n) and necrotic cells (nc) caused by *P. thornei* in roots of chickpea; (C) a specimen (n) and eggs (e) of *P. vulnus* in roots of olive.

a major pest of potatoes in South America and Mexico, sugarbeet in the United States, and tomato in Mexico and subtropical Argentina and Ecuador (Manzanilla-Lopez et al., 2002). In Bolivia the nematode is so common that it can easily be found on wild plants. Some Gramineae and Leguminosae are nonhosts for the nematode and can be used effectively in rotation schemes. However, a Mexican race is pathogenic to French bean and chilli pepper. The tolerance limit of sugar beet to the nematode was estimated as 0.54 second-stage juveniles/cm^3 soil. The nematode completes a life cycle in forty-eight days at the optimal temperature of 25°C and survives between crops in the soil or in crop residues, including tubers, and therefore is considered a quarantine pest. The use of potato tubers free of the nematode is a prerequisite to avoid spreading the nematode and ensure good crop performance. A minimum of three to four years of crop rotation along with weed control can give satisfactory control. There are wild *Solanum* species resistant to the potato race of *N. aberrans,* which might profitably be used in breeding programs for resistance. Some potato cultivars resistant to the nematode are also available. Nematicides are too expensive and may or may not give satisfactory results.

Pratylenchus spp., known as root-lesion nematodes, are migratory root endoparasites that cause histological, and thereby economic, damage to many herbaceous and tree crops (Figure 5.10B and C; Whitehead, 1998). *Pratylenchus brachyurus* (Godfrey) Filipjev *et* Schuurmans Stekhoven is a parasite of groundnut, maize, pineapple, and citrus. Control of this nematode is achieved by preplanting soil fumigation with 1,3-D or by nonvolatile nematicides incorporated into the soil or sprayed on the plants.

Pratylenchus coffeae (Zimmermann) Filipjev *et* Schuurmas Stekhoven is a pest of yams, banana, citrus, and coffee, among other crops. The control of this nematode on trees must be based on the use of healthy plants transplanted into fields free of the nematode. However, infested yam tubers can be freed of infection by immersion for thirty minutes in water at 54°C or in fensulphothion dips. In established plantations of citrus, coffee, and banana, results with nonvolatile nematicides are inconsistent even when systemic nematicides are used. Preplanting fumigation is necessary to obtain acceptable results, but such treatments must then be integrated with application to the growing crop of nonphytotoxic chemicals, granular or liquid in drip irrigation. For citrus and coffee, resistant rootstocks such as *Poncirus trifoliata* or *Coffea canephora* can be obtained (Villain, Anzueto and Sarah, 2004).

Pratylenchus goodeyi Sher *et* Allen causes severe damage to banana in every banana-growing area of east Africa (Gowen and Quénéhervé, 1990). Leaving fields fallow in the absence of banana volunteers and host weeds for ten to twelve months and paring diseased tissue from corms or immersing them in hot water can control this nematode.

Pratylenchus loosi Loof on tea may be controlled by preplant soil fumigation, which is very expensive, or by rotation or intercropping with grasses such as *Tripsacum laxum, Cymbopogon caufertiflorus,* and *Eragrostis curvula* (Campos, Sivapalan and Gnanapragasam, 1990).

Pratylenchus neglectus (Rensch) Filipjev *et* Schuurmans Stekhoven has worldwide distribution and damages mainly winter cereals, such as wheat. A four- to five-year crop rotation controls the nematode satisfactorily. Better results can be achieved by rotating winter cereals with summer crops in irrigated areas.

Pratylenchus penetrans (Cobb) Filipjev *et* Schuurmas Stekhoven is a serious pest of various fruit trees, strawberry, and roses (Whitehead, 1998). Preplant soil fumigation may be effective in its control.

Pratylenchus thornei Sher *et* Allen has a wide distribution and wide host range, attacking and damaging mainly wheat and chickpea, for which a tolerance limit of 0.03 nematodes/cm^3 of soil was estimated under field conditions (Table 5.2; Di Vito, Greco and Saxena, 1992). Crop rotation is the only way of controlling this nematode pest economically; however, soil solarization suppressed the nematode population to nonpathogenic levels (Di Vito, Greco and Saxena, 1991).

Pratylenchus vulnus Allen *et* Jensen occurs worldwide and is highly pathogenic on various fruit and nut trees in warm temperate and subtropical soils (Whitehead, 1998). It is a very difficult nematode to control and even large doses of fumigant nematicides may fail to eradicate it from nursery beds (Lamberti et al., 2001). Chinese wingnut, *Pterocarya stenoptera,* is a tolerant rootstock.

Radopholus similis (Cobb) Thorne is a very destructive pest on bananas, pepper, and coconut (Whitehead, 1998). Moreover, it attacks roots and rhizomes of several ornamental plants (Vovlas et al., 2003). A physiological race of *R. similis,* previously identified as *Radopholus citrophilus* but recently synonymized with *R. similis* (Valette et al., 1998), is the causal agent of the "spreading decline" of citrus on the sandy soils of the central ridge area of Florida (Figure 5.10A).

In bananas and plantains, control of the nematode can be achieved by sanitation (hot-water treatment, paring, or immersion in nematicidal solution dips of infested material, to be planted in nematode-free soil). Micropropagating material from tissue culture is today readily available in many countries. Preplant soil fumigation with large doses or incorporation of systemic nematicides such as fenamiphos, ethoprophos, aldicarb, or carbofuran at the planting sites gives good control of the nematode. Diploid clones of *Musa* spp. possess good resistance to *R. similis* but yield poorly. However, a few tetraploid banana cultivars obtained in India showed resistance to *R. similis,* with some also giving acceptable yield performance (Krishnamoorthy, Kumar and Poornima, 2004). In citrus plantations, keeping the infested soil free of host roots for twelve to twenty-four months eradicates the nematode. A few *Citrus* species and selections are tolerant or resistant to the nematode (Whitehead, 1998).

Rotylenchulidae

The reniform nematode, *R. reniformis,* is widespread in the tropics and in the subtropics, where it is a major pest of various vegetable crops, chickpea, pigeon pea (Figure 5.11), banana, coffee, papaya, pineapple, cotton, soybean, and so on (Robinson et al., 1997; Whitehead, 1998). Five races have been reported, each reproducing on a different range of host plants (Rao and Ganguly, 1996). The tolerance

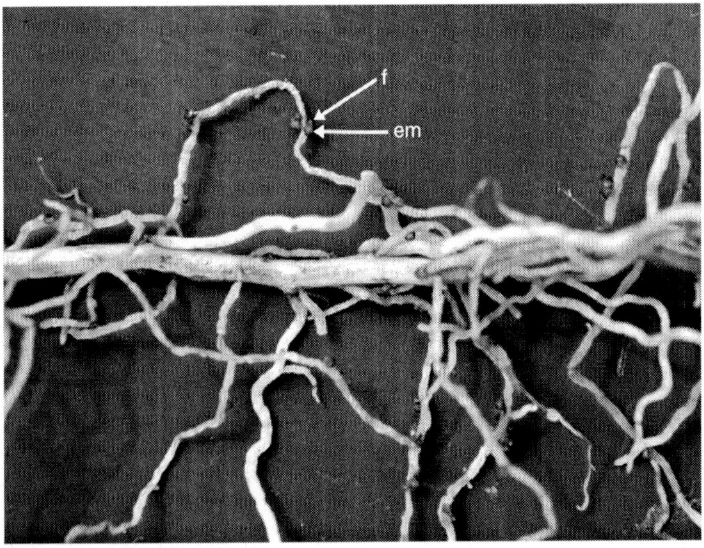

FIGURE 5.11. Root of pigeon pea from India showing several females (f) of the reniform nematode, *R. reniformis,* bearing egg masses (em) (photograph courtesy of Dr. Shashi B. Sharma).

limit to the nematode is less than 1 egg/cm^3 and yield losses of chickpea of 80 percent were observed at a density of 10 nematodes/cm^3 soil.

Fumigant and nonvolatile nematicides are effective in the control of the nematode, as are management techniques such as rotation, fallowing, and soil solarization. Cultivars resistant to *R. reniformis* are available for chickpea, cotton, cowpea, papaya, potato, soybean, and tomato.

Telotylenchidae

Several *Tylenchorhynchus* species are migratory ectoparasites of various crops. Among them, *Tylenchorhynchus dubius* is the most widespread. It causes damage to cereals and leguminous plants. The nematode can be controlled by nematicides, which are expensive, or by crop rotation (Whitehead, 1998).

Tylenchulidae

The citrus nematode, *Tylenchulus semipenetrans* Cobb, is a rather warm-season nematode that occurs in all regions where citrus fruits are cultivated, causing more or less severe damage depending on the pedo-climatic conditions (Duncan and Cohn, 1990). The nematode has semiendoparasitic habits (Figure 5.12). Three biotypes have been identified: citrus, Mediterranean, and poncirus. All can infest citrus and grapes, but the Mediterranean biotype reproduces poorly on *P. trifoliata* (L.) Raf. and derived rootstocks, such as Swingle citrumelo, and not at all on olive, which is infested by the citrus and poncirus biotypes. *P. trifoliata* is infested only by the poncirus biotype (Inserra et al., 1994).

Tylenchulus semipenetrans has a very limited number of host species; therefore replanting problems can be solved with 1-2 years of rotation or fallow, provided that all the volunteer citrus are eliminated. To eradicate the nematode from infested soils, large doses of fumigants, mainly 1,3-D, can be useful. Much care should be taken in the selection of planting material, which should be nematode free. When only infested material is available, its immersion in nematicidal solutions, such as fenamiphos, aldicarb, or carbofuran, can suppress the nematode. Infestations in established orchards can be controlled by the application of systemic nematicides, aldicarb, or fenamiphos in deep irrigation, but control may routinely require two or three ap-

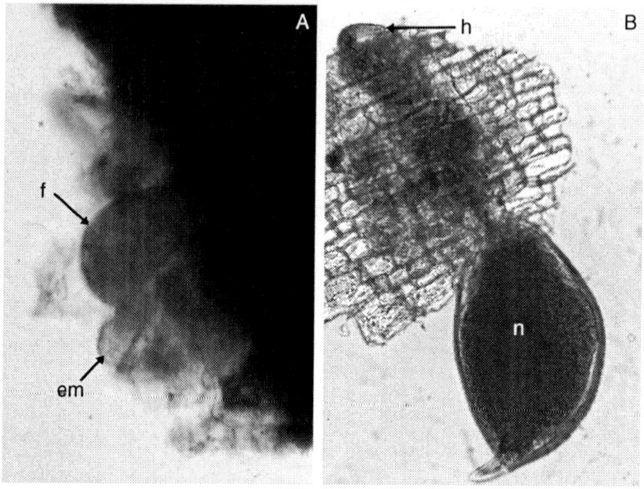

FIGURE 5.12. Swollen females (f) of the citrus nematode, *T. semipenetrans*: (A) protruding from a citrus root and bearing egg masses (em); (B) anterior region crossing several cell layers (n = nematode, h = head region) (photograph B courtesy of Dr. Renato Crozzoli).

plications per growing season. In many cases, the damage caused by *T. semipenetrans* is aggravated by other stress factors. Therefore, improvement of cultural practices could minimize the effects or adversely affect the nematode and improve plant growth and yield.

T. semipenetrans is a nematode pest well managed by resistant or tolerant rootstocks, provided its host race is known. *P. trifoliata,* Troyer citrange, and Swingle citrumelo are also resistant to other diseases (Whitehead, 1998).

FACTORS AFFECTING SELECTION OF CONTROL METHODS

Strategies for nematode control and management include the identification of the nematode problem, the evaluation of the pedo-climatic situation, and the appreciation of the economics of the enterprise (Lamberti, Greco and Basile, 1986; Brown and Kerry, 1987).

Nematode Species

The parasitic habit and the biology of the nematode species causing the phytopathological problem must be known. Control of free-living ectoparasitic nematodes, which lay eggs free in the soil, is much easier and will require lower doses of nematicides than control of nematodes whose eggs are protected by a gelatinous matrix or contained in cysts and/or that spend all their life cycle as endoparasites. Nematodes that complete a single or a few generations during a growing season will be controlled more easily than nematodes with several generations. In fact, once the initial nematode population is reduced to nonpathogenic levels, plants will have a good start and the population increases occurring after the crop growth phase is completed will have only a modest effect on the yield. This situation is more complex when preplant treatments are performed for perennial crops. In this case, greater levels of control are required to greater soil depths to avoid rapid recolonization of the soil. When virus vector nematodes must be controlled, almost complete eradication must be achieved, as a few surviving virus-infected nematodes can quickly spread the infection to healthy replants (Thomason and McKenry, 1975).

Determination of Population Density

When the nematode species has been identified, the population density must also be determined. Control and management strategies depend on the risk that the parasite brings with its presence. It might be a well established pest for which reducing population levels below the economic threshold will be sufficient. However, the nematode may be a recent introduction or may have been endemic in the past and liable to become a severe pest on new crop species never cultivated before in the region. For an annual crop, the soil may be heavily treated at the beginning of each growing cycle. With perennial crops, perphaps with the presence of a virus vector, populations should be more or less completely suppressed before planting and then maintained at low levels for the remainder of the crop cycle.

Determination of population densities is not easy. Which stage of the nematode species should be extracted and quantified to evaluate the situation properly: eggs, juvenile stages, adults, cysts? Where are these stages to be found at the most significant level: in soil, root tissues, stem tissues, leaf tissues, trunks (wood), cysts, egg masses? Various methods are available for the extraction. Cobb's sieving and Baermann's funnel methods and their modifications can be very useful, and centrifugation methods, such as that of Coolen (1979), work with a wide range of nematode species. Several other methods of extracting and processing nematodes have been discussed by Hooper (1990) and Hooper and Evans (1993). The best season in which to sample is also important, especially with respect to the periods in which peak population densities occur.

One of the major considerations in planning a control strategy is to judge whether the crop to be protected is a low-value or a high-value crop.

Generally, "low value" is defined as an extensive culture cropped over large areas, with high levels of mechanization and large yields, such as cereals, legumes, and many industrial plants. Conversely, "high-value" crops are flower, vegetable, and fruit crops intensively cultivated on smaller areas, which require careful handling and large amounts of manual labor to be introduced into the food chain or market. All plants cultivated in glass or plastic houses are, by definition, considered of high value. Brown (1987) suggests that a low-value crop might be defined as "a crop on which disease control by conventional use of a nematicide cannot be economically justified."

Economic Threshold

The economic threshold of a crop plant to a nematode can be defined as the initial population of the nematode causing a yield loss whose value is equal to the cost of the management required to avoid that yield loss. Therefore, it is not a constant measure of the nematode population density but, as it depends on the market prices of both the crop product and the selected management measure, may vary from year to year. Moreover, the estimation of the economic threshold requires information on the relationship between a range of nematode densities and the yield of the host plant and on the relationship between the rate of a treatment (kilograms of nematicides or soil amendment per hectare, years of rotation, exposure time to a given temperature) and the rate of nematode survival/mortality (Ferris and Greco, 1992).

Usually, for the same host plant–nematode combination, the economic threshold is larger than the tolerance limit. Seinhorst (1965, 1979) demonstrated that the relationship between initial population densities of nematodes and crop production (Figure 5.13) can be expressed by the equation $y = m + (1 - m)z^{P-T}$, where y is the ratio of the yield at density P to the yield at density $P \leqslant T$, T is the tolerance limit (the nematode soil population density at planting up to which no yield loss occurs), m is the minimum relative yield (the yield at very large initial nematode densities), P is the initial population density, and z is a constant $<T$ with z^{-T} equal to 1.05.

Applying this equation, various authors have calculated, from experiments performed under controlled or semicontrolled conditions, the tolerance limits of some crops to major nematode pests (Table 5.2; Lamberti and Greco, 1989; Sasanelli, 1994, 1996).

If, in this equation, one substitutes for P the actual value of the initial population density, it is possible to estimate the crop loss expected

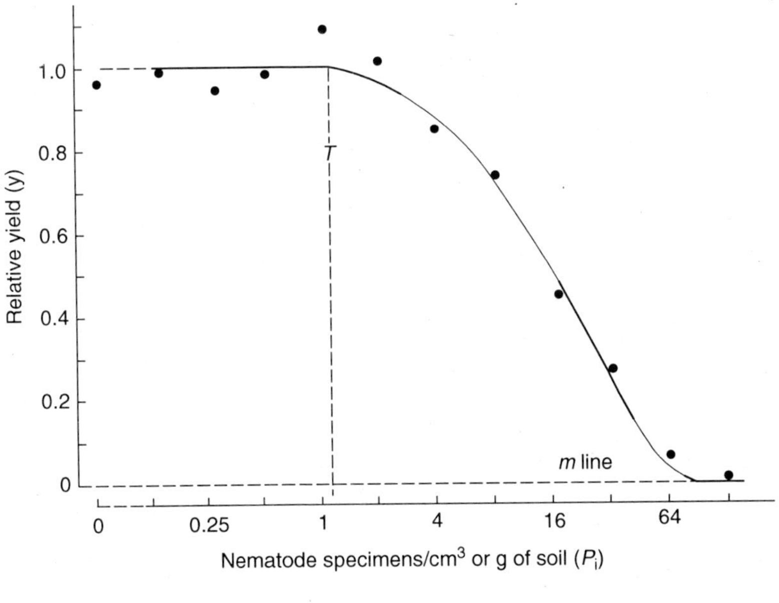

FIGURE 5.13. Curve according to Seinhorst's equation $y = m + (1 - m)z^{P-T}$ to relate nematode population density at sowing or planting with the relative yield of the host crop. Points are experimental data.

as a result of the presence of a nematode infestation (Sasanelli, 1994, 1996). However, Seinhorst's model applies only to short-cycle plants, such as annuals, and is better for nematodes that pass only one generation per crop cycle.

Very little information exists on the rate of nematode mortality as affected by increasing the rate of a nematicide. Moreover, most of the relevant investigations were conducted under controlled conditions so that, in most cases, only information on the suggested minimum and maximum rates of application is available. In addition, it must be taken into account that at the suggested rates, although fumigant nematicides can kill all nematode stages in the soil (including eggs contained within cysts and egg masses), nonfumigant nematicides can kill only free-living nematode stages and, if they have systemic activity, also endoparasitic stages—but not eggs. However, nonfumigant nematicides may delay hatching of eggs, thus preventing nematode infection at an early plant growth stage. Finally, as the extent of nematode damage and efficacy of control measures may vary according to the crop involved and pedo-climatic conditions, the economic thresholds will vary accordingly.

Adoption of Strategies

The economic threshold and the expected crop losses depend on the nematode species, the crop involved, and the pedo-climatic conditions in which the plant is cultivated. Consequently, control strategies will be adopted on the basis of the costs/benefits in each particular situation.

The rate of nematode mortality will suggest the crop sequence and the duration of rotations. The availability of tolerant or resistant cultivars or both for the same crop will affect cultivar choice in terms of future prospects. Modification of some cultural practices will suppress multiplication and restrict movements of nematodes. Soil amendments may enhance the establishment of antagonistic organisms. Introduction into the environment of fungal or bacterial parasites of nematodes will allow a less drastic treatment of the soil for nematode control.

The choice must be particularly careful where synthetic nematicides are involved. These compounds are used because their application results in good and economic control of the nematodes along with satisfactory yields from both quantitative (Figures 5.1H and 5.14A-D) and

FIGURE 5.14. Control by nematicides on (A) *G. rostochiensis* on potato and (B) *H. carotae* on carrot; nematicide combined with solarization for (C) 7 weeks against *D. dipsaci* and (D) 8 weeks against *H. carotae* on carrots; soil solarization for (E) 8 weeks against *H. ciceri* on chickpea and (F) 45 days against *M. incognita* and basal necrosis of pepper in a plastic house. (Controls are identified by arrows in A and "nt" in B-F.)

qualitative points of view. They generally affect nematodes by inhibiting or interfering with their vital processes. Nematicides must be injected or incorporated into the soil or sprayed on the crop. However, the risks associated with their application must be considered seriously. Today, the application of many nematicides, either fumigants or nonvolatile compounds, is possible via drip irrigation. This is an application method that is receiving priority because it is safer for both the applicator and the environment. Nematicide dispersal in the water phase is much less than in the air phase, thus limiting pollution, and the method is less costly because, generally, reduced rates of the chemical and less labor for its application are required.

Soils exhibit intrinsic characteristics, such as texture and chemical composition, and external characteristics, such as temperature, humidity, and organic matter content, that affect the behavior of pesticides. When nematicides are applied to the soil, conditions may influence the efficiency of nematode control. Percolation, diffusion, evaporation, chemical and biological degradation, photodecomposition, adsorption to soil particles, and uptake by plants are all factors that affect movement and persistence of nematicides. Percolation and diffusion constitute the driving forces that move chemicals through the soil; evaporation is responsible for losses from the soil to the atmosphere. The rate of chemical transport through the soil depends on the fraction of the soluble and the reversibly adsorbed product and the temperature. Thus, the higher the solubility of the chemical, the more it is transported by water flow through the soil. The dispersal of a chemical from the soil surface is not constant during the year, and its downward distribution is a direct function of its rate of application (Lamberti, Greco and Basile, 1986).

Side Effects

Control practices can also cause undesirable side effects. This occurs mainly with chemical nematicides, which, if used indiscriminately and improperly, can reduce or suppress the benefits of their application. Improper application of pesticides will include excessive or too low a dose, the incorrect interval between application and planting, intolerance of the nitrification process, environmental pollution, and residues in plant products. Low doses may not be effective at all, and high doses may kill the organisms responsible for biodegradation, thus increasing nematicide persistence and perhaps even causing phytotoxicity. Populations of endomycorrhizae may be suppressed by a drastic soil treatment. Altered fertility of the soil can cause nutritional problems. Often, after treatment, nematode populations increase rapidly owing to the lack of competitors or natural enemies.

The time of application of a nematicide is important. Temperature and moisture content of the soil can affect nematicide persistence and may result in phytotoxicity if residues of fumigants, such as 1,3-D or dazomet, are still present in the soil when a crop is planted. Generally, appropriate time intervals between nematicide application and planting vary with the season. Nonvolatile contact or systemic nematicides need some time to disperse in the soil and to act as toxicants on nematodes. They are partially adsorbed onto organic matter or clay particles present in the soil, thus reducing their effective concentration and their dispersal or penetration into the soil (Loffredo et al., 1991). Alkalinity of the soil is sometimes the cause of unsatisfactory results in nematode control as it accelerates the inactivation or degradation of many nematicides.

Nematicides should be applied long enough before planting to avoid phytotoxicity and to protect the plants for as long as possible. Therefore, as all fumigant nematicides are phytotoxic and kill all nematode stages, they should be applied three to five weeks before planting to kill the nematodes and then disperse from the soil by planting time. Nonfumigant nematicides generally are not phytotoxic at the recommended rates, so can be applied seven to ten days before planting to kill most of the free-living nematode stages by planting and remain at effective concentrations in the soil after transplanting/planting, thus protecting the crop during early stages of plant development. Application of nonfumigant nematicides earlier would reduce their efficacy after planting as they would have largely degraded by planting time.

FUTURE TRENDS

The extent of damage caused by nematodes, the increased awareness of the impact that these parasites have on agricultural crops, the ban on methyl bromide as a soil fumigant for use in developed countries (which is baffling farmers), the general concern about environmental pollution, and the concern of consumers about the residues of pesticides in edible plant products are all challenging nematologists to search for control methods that are at once effective, safe, and nonpolluting. Although rather difficult, because the task is also ambitious, a number of investigations have been undertaken that aim at reducing the amount of chemicals used in agriculture to meet these needs.

Several chemicals that have no or poor ozone depletion potential are being screened as substitutes for methyl bromide (Mirusso, Chellemi and Nance, 2002). Some are as effective or even more effective than methyl bromide and, although so far not permitted for use, at least a few should become available in the near future. Moreover, some nonfumigant nematicides that have become available recently are effective at much lower rates than earlier products, thus limiting their negative impact on the environment. Reductions in the amount of nematicides used can also be achieved by banded application or by placing the chemical at sowing or planting sites, and by the use of liquid formulations applied via drip irrigation. Combining soil solarization with reduced rates of nematicides is also promising and may help reduce the amount of chemicals used for nematode control (Figure 5.14C and D). This issue is of particular interest, and the role that precision agriculture can play to control nematodes has been discussed in terms of using nematicides only as much as and where necessary (Evans and Barker, 2004; Lawrence et al., 2004).

Thus, it is not surprising that much effort is also being put into a search for chemicals of natural origin (plant extracts) and resistant or nematicidal plants to be used as intercrops or to be incorporated into the soil as biofumigants (e.g., several crucifers; Gamliel, 2000). A number of microorganisms, such as *P. penetrans, P. chlamydosporia,* and *P. lilacinus,* are being tested for their antagonism to nematodes (Kerry, 1987; Stirling, 1991; Chen and Dickson, 2004a, b), and some commercial formulations of them are becoming available. These formulations may not be as effective as chemicals, but if they are included in an integrated control program they may play an important role in some circumstances. Other microorganisms, such as rhizobacteria, are being investigated for their antagonistic effects on nematodes or for their ability to induce resistance to nematodes in plants. All this is shortly expected to provide new information and possible solutions based on means of control that are environmentally friendly.

The number of cultivars resistant to nematodes is continuously increasing, especially for potato, tomato, soybean, French bean, and cowpea, and certainly these will play a key role in the control of nematodes in the future (Roberts, 1992, 2002; Peng and Moens, 2003; Starr and Roberts, 2004).

Crop management based on the exclusion or at least reduction of chemicals is expected to increase. There has been an increase in consumer demand for food products free of pesticide residues, through consumers' organizations. In turn, this is resulting in direct agreements between farmers or their organizations on one side and the major food distributors on the other that require farmers to follow agreed production protocols aiming at limiting the risk of pesticide residues in food products. This, combined with the need to record (trace) all steps of food production along the distribution chain (from producer to consumer), enforced by law in Europe, will certainly result in a further reduction in the use of chemicals in agriculture. Moreover, organic farm produce, grown according to established crop protocols that permit the use only of non- or less polluting means of production, excludes the use of most nematicides. So far, approximately 1,000,000 ha are cultivated under organic conditions in Europe, and the area is expected to increase after support from governments and consumer demand. A final point is that limitation of the use of chemicals for soil disinfestations, already enforced in several countries, is also fostering the use of soilless crop production. If they are managed properly, soilless crops may contribute strongly to the exclusion of nematodes and other soil-borne pathogens and, therefore, of the use of nematicides. However, the adoption of a single method may not be sufficient to control nematodes. Agriculturalists and farmers should not feel very satisfied when they succeed in obtaining good yields in heavily infested soil only by using high rates of a nematicide. Rather, they should manage to keep soil population densities of nematodes at as low a level as possible so that the use of polluting means of control can be minimized. This goal can be achieved by rotating crops or by combining the different methods of control. In many instances, combining resistant cultivars and soil solarization, solarization alone (Figure 5.14E and F) or with low rates of chemicals (Figure 5.14C and D), crop rotations (Figure 5.15), weed control, anticipating or delaying sowing time, early harvesting, or working and leaving the field uncropped for a few months in summer in dry and hot areas may greatly contribute to maintaining nematode soil populations at nondamaging or at least very low levels. However, as nematodes often may not be the only soil-borne pathogen present in a field, the choice of the control method must be made only after a complete consideration of the problems likely to be faced by the crop.

Although in developing countries the problem is one of how to increase food production, many developed countries have food surpluses. Therefore, governments are supporting, encouraging, and al-

FIGURE 5.15. Control by crop rotation. (A) *H. avenae, M. artiellia,* and *P. neglectus* affecting hard wheat in Italy. Note the poor growth of wheat in a plot cultivated for four consecutive years with wheat (foreground) and the good crop stand in the plot (background) in which the cultivation of wheat had been suspended for three consecutive years. (B) *H. ciceri* on chickpea in Syria: cultivation of chickpea for three consecutive years (white arrows) and every other year (black arrows).

locating funds for agricultural production and management systems and research projects that aim to improve food quality rather than food quantity. This means that, in the future, agriculture will mostly rely, and nematology research will mainly focus, on ways of controlling nematodes that are alternative to chemicals.

REFERENCES

Abawi, G.S. and J. Chen. (1998). Concomitant pathogen and pest interactions. In *Plant and Nematode Interactions,* K.R. Barker, G.A. Pederson and G.L. Windham, eds. American Society of Agronomy, Crop Science Society of America, Soil Science Society of America, Madison, pp. 135-158.

Akhtar, M. and A. Malik. (2000). Roles of organic amendments and soil organisms in the biological control of plant-parasitic nematodes: a review. *Bioresource Technology* 74: 35-47.

Bailey, K.L. and G. Lazarovits. (2003). Suppressing soil-borne diseases with residue management and organic amendments. *Soil and Tillage Research* 72: 169-180.

Baldwin, J.G. and M. Mundo-Ocampo. (1991). Heteroderinae, cyst- and non-cyst-forming nematodes. In *Manual of Agricultural Nematology,* W.R. Nickle, ed. Marcel Dekker Inc., New York, pp. 275-362.

Bridge, J., M. Luc and R.A. Plowright. (1990). Nematode parasites of rice. In *Plant Parasitic Nematodes in Subtropical and Tropical Agriculture,* M. Luc, R.A. Sikora and J. Bridge, eds. CAB International, Wallingford, UK, pp. 69-108.

Brito, J., R. Inserra, P. Lehman and W. Dickson. (2003). The root knot nematode *Meloidogyne mayaguensis* Rammah *et* Hirshmann, 1988 (Nematoda:

Tylenchida). In *Electronic publication, Pest Alert,* Web site of Florida Department of Agriculture and Consumer Service, Division of Plant Industry, Gainesville, Florida (http://www.doacs.state.fl.us/∼pi/enpp/nema/m-ayaguensis.html).
Brodie, B.B., K. Evans and J. Franco. (1993). Nematode parasites of potato. In *Plant Parasitic Nematodes in Temperate Agriculture,* K. Evans, D.L. Trudgill and J.M. Webster, eds. CAB International, Wallingford, UK, pp. 87-132.
Brown, D.J.F. and D.L. Trudgill. (1997). Longidorid nematodes and their associated viruses. In *An Introduction to Virus Vector Nematodes and Their Associated Viruses,* M.S.N. de A. Santos, I.M. de O. Abrantes, D.J.F. Brown and R.M. Lemos, eds. Instituto Do Ambiente e Vida, Coimbra, Portugal, pp. 1-40.
Brown, D.J.F., J. Zheng and X. Zhou. (2004). Virus vectors. In *Nematology, Advances and Perspectives. Vol. 2 Nematode Management and Utilization,* Z.X. Chen, S.Y. Chen and D.W. Dickson, eds. CABI Publishing, Wallingford, UK, pp. 717-770.
Brown, R.H. (1987). Control strategies in low-value crops. In *Principles and Practice of Nematode Control,* R.H. Brown and B.R. Kerry, eds. Academic Press Australia, Marrickville, Australia, pp. 351-387.
Brown, R.H. and B.R. Kerry. (1987). *Principles and Practice of Nematode Control.* Academic Press Australia, Marrickville, Australia, p. 447.
Campos, V.P., P. Sivapalan and N.C. Gnanapragasam. (1990). Nematode parasites of coffee, cocoa and tea. In *Plant Parasitic Nematodes in Subtropical and Tropical Agriculture,* M. Luc, R.A. Sikora and J. Bridge, eds. CAB International, Wallingford, UK, pp. 387-430.
Cartia, G., N. Greco and P. Di Primo. (1998). Experience acquired in southern Italy in controlling soilborne pathogens by soil solarization and chemicals. In *Soil Solarization and Integrated Management of Soilborne Pests,* J.J. Stapleton, J.E. DeVay and C.L. Elmore, eds. FAO Plant Production and Protection Paper, 147, Rome, Italy, pp. 351-366.
Castillo, P. and N. Vovlas. (2005). *Bionomics and Identification of Rotylenchus Species. (Nematology Monographs and Perspectives, Vol. 3).* Brill Academic Publishers, Leiden, The Netherlands, p. 316
Chen, Z.X. and D.W. Dickson. (2004a). Biological control of nematodes by fungal antagonists. In *Nematology, Advances and Perspectives. Vol. 2 Nematode Management and Utilization,* Z.X. Chen, S.Y. Chen and D.W. Dickson, eds. CABI Publishing, Wallingford, UK, pp. 979-1039.
Chen, Z.X. and D.W. Dickson. (2004b). Biological control of nematodes by bacterial antagonists. In *Nematology, Advances and Perspectives. Vol. 2 Nematode Management and Utilization,* Z.X. Chen, S.Y. Chen and D.W. Dickson, eds. CABI Publishing, Wallingford, UK, pp. 1041-1082.
Chitwood, D.J. (2002). Phytochemical based strategies for nematode control. *Annual Review of Phytopathology* 40: 221-249.
Cohn, E. and L.W. Duncan. (1990). Nematode parasites of subtropical and tropical fruit trees. In *Plant Parasitic Nematodes in Subtropical and Tropical Agriculture,* M. Luc, R.A. Sikora and J. Bridge, eds. CAB International, Wallingford, UK, pp. 347-362.

Coiro, M.I., C.E. Taylor and F. Lamberti. (1987). Population changes of *Xiphinema index* in relation to host plant, soil type and temperature in southern Italy. *Nematologia Mediterranea* 15: 173-181.

Cooke, D. (1993). Nematode parasites of sugarbeet. In *Plant Parasitic Nematodes in Temperate Agriculture,* K. Evans, D.L. Trudgill and J.M. Webster, eds. CAB International, Wallingford, UK, pp. 133-169.

Coolen, W.A. (1979). Methods for the extraction of *Meloidogyne* spp. and other nematodes from roots and soil. In *Root-Knot Nematodes (Meloidogyne species) Systematics, Biology and Control,* eds. F. Lamberti and C.E. Taylor. Academic Press, New York, pp. 317-329.

Decramer, W. (1995). *The Family Tricodoridae: Stubby Root and Virus Vector Nematodes.* Kluwer Academic Publishers, Dordrecht, The Netherlands, p. 360.

Di Vito, M. and N. Greco. (1986). The pea cyst nematode. In *Cyst Nematodes,* F. Lamberti and C.E. Taylor, eds. Plenum Press, New York, pp. 321-332.

Di Vito, M., N. Greco and M.C. Saxena. (1991). Effectiveness of soil solarization for control of *Heterodera ciceri* and *Pratylenchus thornei* on chickpea in Syria. *Nematologia Mediterranea* 19: 109-111.

Di Vito, M., N. Greco and M.C. Saxena. (1992). Pathogenicity of *Pratylenchus thornei* on chickpea in Syria. *Nematologia Mediterranea* 20: 71-73.

Duncan, L.W. and E. Cohn. (1990). Nematodes parasites of citrus. In *Plant Parasitic Nematodes in Subtropical and Tropical Agriculture,* M. Luc, R.A. Sikora and J. Bridge, eds. CAB International, Wallingford, UK, pp. 321-346.

Eisenback, J.D. and H.H. Triantaphillou. (1991). Root-knot nematodes: *Meloidogyne* species and races. In *Manual of Agricultural Nematology,* W.R. Nickle, ed. Marcel Dekker Inc., New York, pp. 191-274.

Esquibet, M., S. Bekal, P. Castagnone-Sereno, J.P. Gauthier, R. Rivoal and G. Caubel. (1998). Differentiation of normal and giant *Vicia faba* populations of the stem nematode *Ditylenchus dipsaci*: agreement between RAPD and phenotypic characteristics. *Heredity* 81: 291-298.

Evans, K. and A.D.P. Barker. (2004). Economies in nematode management from precision agriculture—Limitations and possibilities. In *Proceedings of the Fourth International Congress of Nematology,* R. Cook and D.J. Hunt, eds. June 8-13, 2002, Tenerife, Spain. Nematology Monographs and Perspective 2, Koninklijke Brill NV, Leiden, The Netherlands, pp. 23-32.

Evans, K. and J.A. Rowe. (1998). Distribution and economic importance. In *The Cyst Nematodes,* S.B. Sharma, ed. Kluwer Academic Publishers, Dordrecht, The Netherlands, pp. 1-30.

Ferris, H. and N. Greco. (1992). Management strategies for *Heterodera goettingiana* in a vegetable cropping system in Italy. *Fundamental and Applied Nematology* 15: 25-33.

Fleming, C.C. and T.O. Powers. (1998). Potato cyst nematode: Species, pathotypes and virulence. In *Potato Cyst Nematodes: Biology, Distribution and Control,* R.J. Marks and B.B. Brodie, eds. CAB International, Wallingford, UK, pp. 51-57.

Gamliel, A. (2000). Soil amendments: a non chemical approach to the management of soilborne pest. In *Proceedings of the Fifth International Symposium on Chemical and Non-Chemical Soil and Substrate Disinfestation,* M.L. Gullino, J. Katan and A. Matta, eds. *Acta Horticulturae* 532: 39-47.

Gaur, H.S. and R.N. Perry. (1991). The use of soil solarization for control of plant parasitic nematodes. *Nematological Abstracts* 60: 153-167.

Gowen, S. and P. Quénéhervé. (1990). Nematode parasites of bananas, plantains and abaca. In *Plant Parasitic Nematodes in Subtropical and Tropical Agriculture,* M. Luc, R.A. Sikora and J. Bridge, eds. CAB International, Wallingford, UK, pp. 431-460.

Greco, N. (1986). The carrot cyst nematode. In *Cyst Nematodes,* eds F. Lamberti and C.E. Taylor. Plenum Press, New York, pp. 333-346.

Greco, N. and A. Brandonisio. (1987). Investigation on *Heterodera avenae* in Italy. *Nematologia Mediterranea* 15: 225-234.

Greco, N., A. Brandonisio and F. Elia. (1985). Control of *Ditylenchus dipsaci, Heteodera carotae* and *Meloidogyne javanica* by soil solarization. *Nematologia Mediterranea* 13: 191-197.

Greco, N., M. Di Vito, A. Brandonisio, I. Giordano and G. De Marinis. (1982). The effect of *Globodera pallida* and *G. rostochiensis* on potato yield. *Nematologica* 28: 379-386.

Greco, N., M. Di Vito, M.C. Saxena and M.V. Reddy. (1988). Investigation on the root lesion nematode *Pratylenchus thornei,* in Syria. *Nematologia Mediterranea* 16: 101-105.

Greco, N. and D. Esmenjaud. (2004). Management strategies for nematode control in Europe. In *Proceedings of the Fourth International Congress of Nematology,* R. Cook and D.J. Hunt, eds. June 8-13, 2002, Tenerife, Spain. Nematology Monographs and Perspectives 2, Koninklijke Brill NV, Leiden, The Netherlands, pp. 33-43.

Greco, N., H. Ferris and A. Brandonisio. (1991). Effect of *Heterodera goettingiana* population densities on the yield of pea, broad bean and vetch. *Revue de Nématologie* 14: 619-624.

Greco, N. and I.L. Moreno. (1992). Influence of *Globodera rostochiensis* on yield of summer, winter and spring sown potato in Chile. *Nematropica* 22: 165-173.

Griffith, R. and P.K. Koshy. (1990). Nematode parasites of coconut and other palms. In *Plant Parasitic Nematodes in Subtropical and Tropical Agriculture,* M. Luc, R.A. Sikora and J. Bridge, eds. CAB International, Wallingford, UK, pp. 363-386.

Hesling, J.J. (1974). *Ditylenchus myceliophagus.* CIH Description of Plant-Parasitic Nematodes. Set 3, No. 36. Commonwealth Institute of Helminthology, St Albans, UK, p. 4.

Hooper, D.J. (1990). Extraction and processing of plant and soil nematodes. In *Plant Parasitic Nematodes in Subtropical and Tropical Agriculture,* M. Luc, R.A. Sikora and J. Bridge, eds. CAB International, Wallingford, UK, pp. 45-68.

Hooper, D.J. and K. Evans. (1993). Extraction, identification and control of plant parasitic nematodes. In *Plant Parasitic Nematodes in Temperate Agriculture,* K. Evans, D.L. Trudgill and J.M. Webster, eds. CAB International, Wallingford, UK, pp. 1-59.

Hunt, D.J. (1993). *Aphelenchidae, Longidoridae and Trichodoridae: Their Systematics and Bionomics.* CAB International, Wallingford, UK, p. 352.

Inserra, R.N., L.W. Duncan, J.H. O'Bannon and S.A. Fuller. (1994). Citrus biotypes and resistant root-stocks in Florida. Nematology Circular No. 205, Florida Department of Agriculture and Consumer Services, Division of Plant Industry, Gainesville, Florida, p. 4.

Inserra, R.N. and N. Vovlas. (1977). Parasitic habits of *Gracilacus peratica* on olive feeder roots. *Nematologia Mediterranea* 5: 345-348.

Jatala, P. and J. Bridge. (1990). Nematode parasites of root and tuber crops. In *Plant Parasitic Nematodes in Subtropical and Tropical Agriculture,* M. Luc, R.A. Sikora and J. Bridge, eds. CAB International, Wallingford, UK, pp. 137-180.

Katan, J. and J.E. De Vay. (1991). *Soil Solarization.* CRC Press Inc., Boca Raton, FL, p. 267.

Kerry, B.R. (1987). Biological control. In *Principles and Practice of Nematode Control,* R.H. Brown and B.R. Kerry, eds. Academic Press Australia, Marrickville, Australia, pp. 233-263.

Khan, M.W. (1993). *Nematode Interactions.* Chapman & Hall, London, UK, p. 377.

Krishnamoorthy, V., N. Kumar and K. Poornima. (2004). Reaction of some tetraploid banana hybrids to *Radopholus similis. Nematologia Mediterranea* 32: 175-179.

Lamberti, F., T. D'Addabbo, N. Sasanelli and A. Carella. (2001). Control of *Pratylenchus vulnus* in stone fruit nurseries. *Mededelingen Faculteit LandbouwkundigeUniversiteit Gent* 66/2b: 629-632.

Lamberti, F. and N. Greco. (1989). Perdite di produzione causate da nematodi fitoparassiti in Italia. *Informatore Fitopatologico* 29: 35-39.

Lamberti, F., N. Greco and M. Basile. (1986). Treatments of soil—Nematolgical aspects. *EPPO Bulletin* 16: 327-333.

Lamberti, F., A. Minuto and L. Filippini. (2003). I fumiganti per la disinfestazione del terreno. *Informatore Fitopatologico* 53: 38-43.

Lamberti, F. and C.E. Taylor. (1979). *Root-Knot Nematodes (Meloidogyne species) Systematics, Biology and Control.* Academic Press, New York, p. 477.

Lawrence, G.W., A.T. Kelley, R.L. King, J. Vickery, H.K. Lee and K.S. McLean. (2004). Remote sensing and precision nematicide applications for *Rotylenchulus reniformis* management in cotton. In *Proceedings of the Fourth International Congress of Nematology,* R. Cook and D.J. Hunt, eds. June 8-13, 2002, Tenerife, Spain. Nematology Monographs and Perspective 2, Koninklijke Brill NV, Leiden, The Netherlands, pp. 13-21.

Loffredo, E., N. Senesi, V.A. Melillo and F. Lamberti. (1991). Influence of irrigation and time of leaching of fenamiphos in soil columns and its uptake and accumulation by tomato plants. *Nematologia Mediterranea* 19: 335-340.

Luc, M. (1986). Cyst nematodes in equatorial and hot tropical regions. In *Cyst Nematodes,* F. Lamberti and C.E. Taylor, eds. Plenum Press, New York, pp. 355-372.

Maggenti, A.R. (1991). Nemata: Higher classification. In *Manual of Agricultural Nematology,* W.R. Nickle, ed. Marcel Dekker Inc., New York, USA, pp. 147-187.

Mamiya, Y. (1984). The pine wood nematode. In *Plant and Insect Nematodes,* W.R. Nickle, ed. Marcel Dekker Inc., New York, pp. 589-626.

Manzanilla-Lopez, R.H., M.A. Costilla, M. Doucet, J. Franco, R.N. Inserrra, P.S. Lehmann, I. Cid del Prado-Vera, R.M. Souza and K. Evans. (2002). The genus

Nacobbus Thorne & Allen, 1944 (Nematoda: Pratylenchidae): systematics, distribution, biology and management. *Nematropica* 32: 149-227.

Marks, R.J. and B.B. Brodie. (1998). Introduction: potato cyst nematodes—an international pest complex. In *Potato Cyst Nematodes: Biology, Distribution and Control,* R.J. Marks and B.B. Brodie, eds. CAB International, Wallingford, UK, pp. 1-4.

Meagher, J.W. and R.H. Brown. (1974). Microplots experiments on the effect of plant hosts on populations of the cereal cyst nematode (*Heterodera avenae*) and on subsequent yield of wheat. *Nematologica* 20: 337-346.

Minton, N.A. and P. Baujard. (1990). Nematode parasites of peanut. In *Plant Parasitic Nematodes in Subtropical and Tropical Agriculture,* M. Luc, R.A. Sikora and J. Bridge, eds. CAB International, Wallingford, UK, pp. 285-320.

Mirusso, J., D. Chellemi and J. Nance. (2002). Field evaluation of methyl bromide alternatives. Proceedings of *Annual International Research Conference on Methyl Bromide Alternatives and Emission Reduction,* 6-8 November 2002, Orlando, Florida, p. 18.

Mota, M.M., H. Braach, M.A. Bravo, A.C. Penas, W. Burgermeister, K. Metge and E. Sousa. (1999). First report of *Bursaphelenchus xylophilus* in Portugal and Europe. *Nematology* 1: 727-734.

Nyczepir, A.P. and J.M. Halbrendt. (1993). Nematode pests of deciduous fruit and nut trees. In *Plant Parasitic Nematodes in Temperate Agriculture,* K. Evans, D.L. Trudgill and J.M. Webster, eds. CAB International, Wallingford, UK, pp. 381-425.

Parveen, R., A.A. Khan, M. Imran and A.A. Ansari. (2003). Response of wheat varieties to the seed gall nematode, *Anguina tritici. Nematologia Mediterranea* 31: 103-104.

Peng, Y. and M. Moens. (2003). Host resistance and tolerance to migratory plant-parasitic nematodes. *Nematology* 5: 145-177.

Ploeg, A.T. and D.J.F. Brown. (1997). Trichodorid nematodes and their associated viruses. In *An Introduction to Virus Vector Nematodes and their Associated Viruses,* M.S.N. de A. Santos, I.M. de O. Abrantes, D.J.F. Brown and R.M. Lemos, eds. Instituto Do Ambiente e Vida, Coimbra, Portugal, pp. 41-68.

Poinar, G.O. Jr. (1983). *The Natural History of Nematodes.* Prentice-Hall, Englewood Cliffs, NJ, p. 323.

Powell, D.F. (1974). Fumigation of field bean against *Ditylenchus dipsaci. Plant Pathology* 23: 110-113.

Rao, G.M.V.P. and S. Ganguly. (1996). Host preference of 6 geographical isolates of reniform nematode, *Rotylenchulus reniformis. Indian Journal of Nematology* 26: 19-22.

Raski, D.J., W.B. Hewitt, A.C. Goheen, C.E. Taylor and R.H. Taylor. (1965). Survival of *Xiphinema index* and reservoirs of fan leaf virus in fallowed vineyard soil. *Nematologica* 11: 349-352.

Richardson, P.N. and P.S. Grewal. (1993). Nematode pests of glasshouse crops and mushrooms. In *Plant Parasitic Nematodes in Temperate Agriculture,* K. Evans, D.L. Trudgill and J.M. Webster, eds. CAB International, Wallingford, UK, pp. 501-544.

Riggs, R.D. and T.L. Niblack. (1993). Nematode pests of oilseed crops and grain legumes. In *Plant Parasitic Nematodes in Temperate Agriculture,* K. Evans, D.L. Trudgill and J.M. Webster, eds. CAB International, Wallingford, UK, pp. 209-258.

Rivoal, R. and R. Cook. (1993). Nematode pests of cereals. In *Plant Parasitic Nematodes in Temperate Agriculture,* K. Evans, D.L. Trudgill and J.M. Webster, eds. CAB International, Wallingford, UK, pp. 259-303.

Roberts, P.A. (1992). Current status, development and use of host plant resistance to nematodes. *Journal of Nematology* 24: 213-227.

Roberts, P.A. (2002). Concept and consequence of resistance. In *Plant Resistance to Parasitic Nematodes,* J.L. Starr, J. Bridge and R. Cook, eds. CAB International, Wallingford, UK, pp. 23-41.

Roberts, P.A. and W.C. Matthews. (1995). Disinfection alternatives for control of *Ditylenchus dipsaci* in garlic seed cloves. *Journal of Nematology* 27: 448-456.

Robinson, A.F., R.N. Inserra, E.P. Caswell-Chen, N. Vovlas and A. Troccoli. (1997). *Rotylenchulus* species: Identification, distribution, host ranges and crop plant resistance. *Nematropica* 27: 27-180.

Santos, M.S.N. de A., I.M. de O. Abrantes, D.J.F. Brown and R.M. Lemos. (1997). *An Introduction to Virus Vector Nematodes and Their Associated Viruses.* Instituto do Ambiente e Vida, Coimbra, Portugal, p. 535.

Sasanelli, N. (1994). Tables of nematode-pathogenicity. *Nematologia Mediterranea* 22: 153-157.

Sasanelli, N. (1996). Economic importance of plant parasitic nematodes and crop loss assessment. In *Proceedings of the V Congresso della Società Italiana di Nematologia,* F. Lamberti and N. Greco, eds. October 19-21, 1995, Martina Franca (TA), Italy. *Nematologia Mediterranea* 23 (Supplement): 5-13.

Sasser, J.N. (1979). Pathogenicity, host range and variability in *Meloidogyne* species. In *Root-knot Nematodes* (*Meloidogyne* species) *Systematics, Biology and Control,* F. Lamberti and C.E. Taylor, eds. Academic Press, New York, pp. 257-268.

Sasser, J.N. and C.C. Carter. (1985). *An Advanced Treatise on Meloidogyne. Volume I. Biology and Control.* North Carolina State University Graphics, Raleigh, NC, p. 422.

Schmitt, D.P., J.A. Wrather and R.D. Riggs. (2004). *Biology and Management of Soybean Cyst Nematode* (second edition). Walsworth Publishing Company, Marceline, MO, p. 262.

Seinhorst, J.W. (1979). Nematodes and growth of plants: Formalization of the nematode-plant system. In *Root-knot Nematodes* (*Meloidogyne* species) *Systematics, Biology and Control,* F. Lamberti and C.E. Taylor, eds. Academic Press, New York, pp. 231-256.

Seinhorst, J.W. (1965). The relation between nematode density and damage to plants. *Nematologica* 11: 137-154.

Seinhorst, J.W. (1982). The relationship in field experiments between population density of *Globodera rostochiensis* before planting potatoes and yield of potato tubers. *Nematologica* 28: 277-284.

Sharma, S.B., Y.L. Nene, M.V. Reddy and D. McDonald. (1993). Effect of *Heterodera cajani* on biomass and grain yield of pigeon pea on vertisol in pot and field experiments. *Plant Pathology* 42: 163-167.

Siddiqi, M.R. (2000). *Tylenchida Parasites of Plants and Insects*. CABI Publishing, Wallingford, UK, p. 833.

Sikora, R.A., N. Greco and J.F.V. Silva. (2005). Nematode parasites of food legumes. In *Plant Parasitic Nematodes in Subtropical and Tropical Agriculture* (second edition), M. Luc, R.A. Sikora and J. Bridge, eds. CABI Publishing, Wallingford, UK, pp. 259-318.

Simon, A. (1980). A plant assay of soil to assess potential damage to wheat by *Heterodera avenae*. *Plant Disease* 64: 917-919.

Siti, E., E. Cohn, J. Katan and M. Mordechai. (1982). Control of *Ditylenchus dipsaci* in garlic by bulb and soil treatments. *Phytoparasitica* 10: 93-100.

Smart, G.C. and K.B. Nguyen. (1991). Sting and awl nematodes: *Belonolaimus* spp. and *Dolichodorus* spp. In *Manual of Agricultural Nematology,* W.R. Nickle, ed. Marcel Dekker Inc., New York, pp. 627-667.

Starr, J.L. and P.A. Roberts. (2004). Resistance to plant-parasitic nematodes. In *Nematology, Advances and Perspectives. Vol. 2 Nematode Management and Utilization,* Z.X. Chen, S.Y. Chen and D.W. Dickson, eds. CABI Publishing, Wallingford, UK, pp. 879-907.

Stirling, G.R. (1991). *Biological Control of Plant Parasitic Nematodes*. CAB International, Wallingford, UK, p. 282.

Sturhan, D. and M.W. Brzeski. (1991). Stem and bulb nematodes, *Ditylenchus* spp. In *Manual of Agricultural Nematology,* W.R. Nickle, ed. Marcel Dekker Inc., New York, pp. 423-464.

Sutherland, J.R. and J.M. Webster. (1993). Nematode pests of forest trees. In *Plant Parasitic Nematodes in Temperate Agriculture,* K. Evans, D.L. Trudgill and J.M. Webster, eds. CAB International, Wallingford, UK, pp. 351-380.

Swarup, G. and C. Sosa-Moss. (1990). Nematode parasites of cereals. In *Plant Parasitic Nematodes in Subtropical and Tropical Agriculture,* M. Luc, R.A. Sikora and J. Bridge, eds. CAB International, Wallingford, UK, pp. 109-136.

Tacconi, R. (1989). Produzione di "Aglio bianco piacentino" esente dal nematode dello stelo e dei bulbi. *Informatore Agrario* 45: 63-64.

Taylor, C.E. and D.J.F. Brown. (1997). *Nematode Vectors of Plant Viruses*. CAB International, Wallingford, UK, p. 296.

Thomason, I.J. and M. McKenry. (1975). Chemical control of nematode vectors of plant viruses. In *Nematode Vectors of Plant Viruses,* F. Lamberti, C.E. Taylor and J.W. Seinhorst, eds. Plenum Press, London, UK, pp. 423-439.

Valette, C., D. Mounport, M. Nicole, J.L. Sarah and P. Baujard. (1998). Scanning electron microscopy study of two African populations of *Radopholus similis* (Nematoda: Pratylenchidae) and proposal of *R. citrophilus* as a junior synonym of *R. similis*. *Fundamental and Applied Nematology* 21: 139-146.

Venter, C., Van G. Aswegen, A.J. Meyer and D. De Waele. (1995). Histological studies of *Ditylenchus africanus* within peanut pods. *Journal of Nematology* 27: 284-291.

Villain, L., F. Anzueto and J.L. Sarah. (2004). Resistance to root-lesion nematodes on *Coffea canephora*. In *Proceedings of the Fourth International Congress of Nematology*, R. Cook and D.J. Hunt, eds. June 8-13, 2002, Tenerife, Spain. Nematology Monographs and Perspectives 2, Koninklijke Brill NV, Leiden, The Netherlands, pp. 289-302.

Vovlas, N., P. Castillo and A. Troccoli. (1998). Histology of nodular tissue of three leguminous hosts infected by three root-knot nematode species. *International Journal of Nematology* 8: 105-110.

Vovlas, N., A. Troccoli, M. Pestana, I.M. de O. Abrantes and M.S.N. de A. Santos. (2003). Parasitization of vascular bundles of Anthurium rhizomes by *Radopholus similes*. *Nematropica* 33: 209-213.

Whitehead, A.G. (1998). *Plant Nematode Control*. CAB International, Wallingford, UK, p. 384.

Whitehead, A.G. and S.J. Turner. (1998). Management and regulatory control strategies for potato cyst nematodes (*Globodera rostochiensis* and *Globodera pallida*). In *Potato Cyst nematodes: Biology, Distribution and Control,* R.J. Marks and B.B. Brodie, eds. CAB International, Wallingford, UK, pp. 135-152.

Chapter 6

Nematophagous Fungi As Biocontrol Agents of Plant-Parasitic Nematodes

Geeta Saxena

INTRODUCTION

Plant-parasitic nematodes are microscopic eelworms that live in the soil and attack the roots of plants. Crop production problems induced by nematodes, therefore, generally occur as a result of root dysfunction and reduced rooting volume and efficiency of water and nutrient foraging and utilization. Many different genera and species of nematodes are important to crop production in the world. In addition to the direct crop damage caused by nematodes, many of the nematode species predispose plants to infection by fungal or bacterial pathogens or transmit viral disease that leads to additional yield reduction.

The use of biological control agents in the nematode management system has received enthusiastic attention in recent years. Biological control agents are natural enemies such as predators, parasites, and competitive antagonists of crop pests that are either introduced into an area or augmented in their natural surroundings to enhance their establishment and control potential. Health hazards and environmental pollution problems necessitate the use of nonchemical means of nematode control. Plant-parasitic nematodes, especially root-knot

Management of Nematode and Insect-Borne Plant Diseases

doi:10.1300/5754_06

nematodes, cause enormous damage to all major field crops, vegetable crops, certain cash crops, ornamentals, and grass plants.

Biological control will play an increasing role in practical nematode control in future. This chapter reviews the use of nematophagous fungi along with organic soil amendments to develop management strategies for nematode pests. Several reports have indicated natural control of specific nematode pests where nematophagous fungi have limited nematode multiplication (Jatala, 1986; Saxena and Mukerji, 1988; Saxena et al., 1991; Sikora, 1992; Kerry, 1993; Mittal et al., 1996; Kerry and Jaffee, 1997; Kerry and Hominick, 2000; Saxena, 2004).

NEMATOPHAGOUS FUNGI

Nematophagous-, or nematode-destroying, fungi comprise those fungi that attack living nematodes or their eggs and cysts and utilize them as a source of nutrients. These fungi are common and abundant in various soils (Saxena and Mittal, 1997), and they play a major role in maintaining the balance of microbial life by recycling carbon, nitrogen, and other important elements from the biomass of nematodes in the soil. Nematophagous fungi have been categorized into three groups, depending upon the mode of attack: predatory, endoparasitic, and parasites of root-knot and cyst nematodes.

Predatory Fungi

Predatory fungi are also known as nematode-trapping fungi as they kill nematodes by producing various trapping devices at intervals along their hyphae. The predacious fungi have extensive hyphal development on the substratum. After capture, an appressorium-like outgrowth develops at the point of contact with the nematode. The captured nematode, after a prolonged struggle, becomes exhausted and is penetrated by the fungus. After penetration, elongated, unbranched absorptive hyphae grow in both directions along the length of the body. After complete exploitation of the contents, protoplasm is migrated back from the absorption hyphae and a cross-wall is formed. Trapping organs may be adhesive or nonadhesive.

Adhesive Trapping Organs

Adhesive organs of capture have been categorized into hyphae, branches, knobs, and nets. Adhesive hyphae are characteristic of zygomycetes. Nematodes are captured by means of adhesive produced directly on hyphae. Hyphae either are coated with adhesive along their entire length or produce adhesive at any point in response to nematodes. A common example of such type of trapping device is found in *Stylopage hadra* (Barron, 1977). Adhesive branches are found in Deuteromycotina. They arise as short laterals that grow erect from the prostrate hyphae. They are mostly a few cells in height. In a few nematophagous fungi, such as *Monacrosporium cionopagum* and *Monacrosporium gephyropagum,* the organs of capture are adhesive branches, but even in these species there is a tendency to form simple two-dimensional nets (Figure 6.1). Adhesive knobs are found in Basidiomycotina and Deuteromycotina. They are either sessile or produced at the apex of a slender nonadhesive stalk composed of one to three cells. The knob is separated by a septum from the support stalk. Nematodes adhere to a knob and eventually are caught by several knobs, with subsequent penetration and destruction. Adhesive nets are ubiquitous and the most common organ form. They vary from simple loops, as in *Arthrobotrys musiformis,* to complex three-dimensional

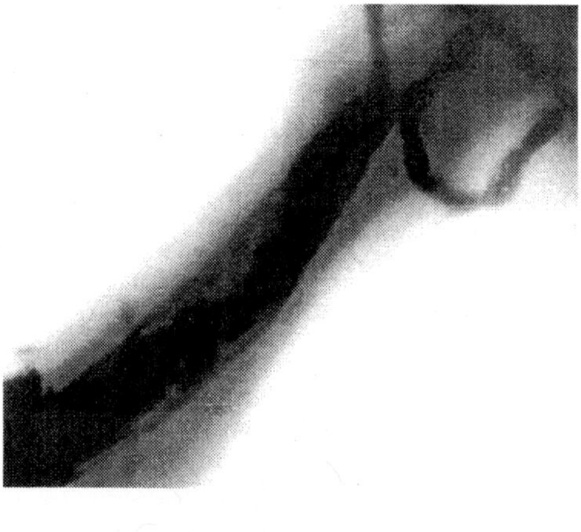

FIGURE 6.1. A nematode trapped in adhesive branches of *Monacrosporium gephyropagum* that have united to form scalariform traps. Assimilative hyphae inside the nematode are also seen.

networks, as in *Arthrobotrys oligospora*. The adhesive trapping is highly effective, and by struggling the nematode becomes attached at several points in the net system (Figure 6.2).

Nonadhesive Trapping Organs

Nonadhesive trapping organs are either nonconstricting or constricting rings. Members of the Deuteromycotina produce rings on lateral branches arising from the prostrate hyphae. On a slender stalk, three-celled nonconstricting rings are produced. These are passive in their action.The nematode enters the ring at speed, and its forward motion causes its body to become wedged inside the ring sufficiently tightly to cause a constriction of the cuticle. Production of nonconstricting rings is often associated with adhesive knobs, as in *Dactylaria candida* and *Dactylaria lysipaga*. Constricting rings are produced on a shorter stalk. They capture nematodes by a "garroting" action; the ring cells swell to grasp the nematodes in a strangle hold, with no chance of escape (Figures 6.3 and 6.4). They are found, for example, in *Dactylaria brochopaga, Dactylella doedycoides,* and *Arthrobotrys anchonia* (Drechsler, 1937, 1954).

Endoparasitic Fungi

In the soil, endoparasites exist mainly as conidia. In Deuteromycotina, conidia either adhere to the nematode cuticle or are ingested

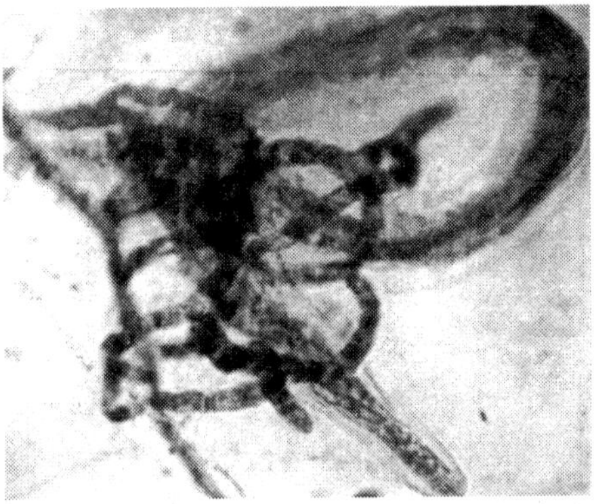

FIGURE 6.2. A nematode captured in adhesive three-dimensional networks of *Arthrobotrys oligospora*.

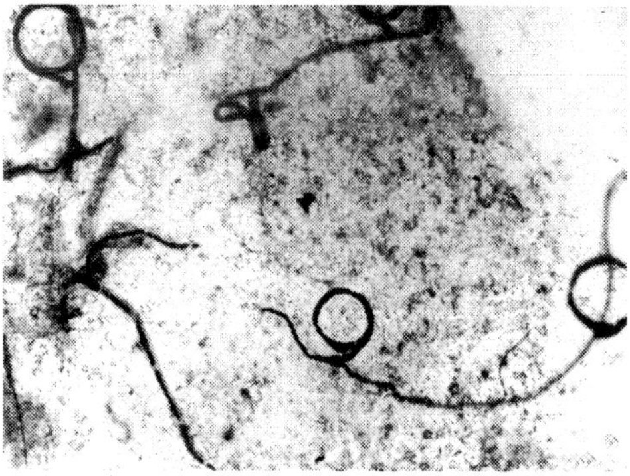

FIGURE 6.3. Open constricting rings produced on mycelial hyphae of *Arthrobotrys dactyloides.*

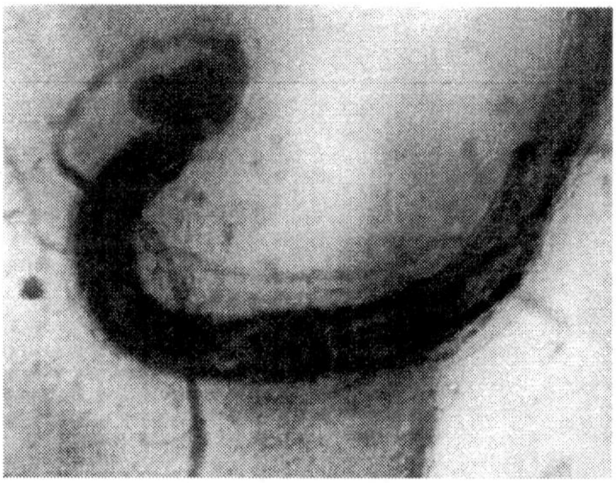

FIGURE 6.4. A nematode captured in a closed three-celled constricting ring of *Arthrobotrys brochopaga.*

by them. The attached spore germinates, penetrates directly through the cuticle, and forms an infection hypha in the nematode body cavity. Hyphae grow and proliferate by drawing on the body contents, and within a few days the body of the nematode is filled with the hyphae of the parasite. External development is restricted to the development of conidiophores and conidia. In the case of chytridiomycetes and oomycetes, flagellate spores are produced that either directly adhere to the cuticle or produce adhesive buds. Evacuation tubes are produced from the sporangia for the release of zoospores.

Catenaria anguillulae shows consistent orientation (polarity) of zoospore encystment. Zoospores are attracted to and encysted near the mouth, excretory pore, and anus of nematodes and eventually become encysted on nematode surfaces. Cysts germinate within 20-60 minutes via a narrow germ tube. The germ tube penetrates the nematode and forms a the vesicle inside the host (Deacon and Saxena, 1997).

One exceptional genus, *Nematoctonus,* a member of the Basidiomycotina, includes species that have been reported both as predatory and as endoparasitic. *Harposporium anguillulae, Drechmeria coniospora, Hirsutella rhossiliensis,* and *Verticillium balanoides* are the most studied endoparasites (Figures 6.5 and 6.6).

Parasites of Cyst and Root-Knot Nematodes

The third group of nematophagous fungi comprises parasites of cyst and root-knot nematodes that attack eggs or females of these nematodes by ingrowth of vegetative hyphae (Figure 6.5). During parasitization, an appressorium-like swelling may be involved. The cyst nematode parasites mainly infect larvae, eggs, or females. Fungi in this group have attracted much attention as biological control agents. Parasites of the nematode genera Heterodera, Globodera, Meloidogyne, Rotylenchus, Tylenchorhynchus, Pratylenchus, Ditylenchus,

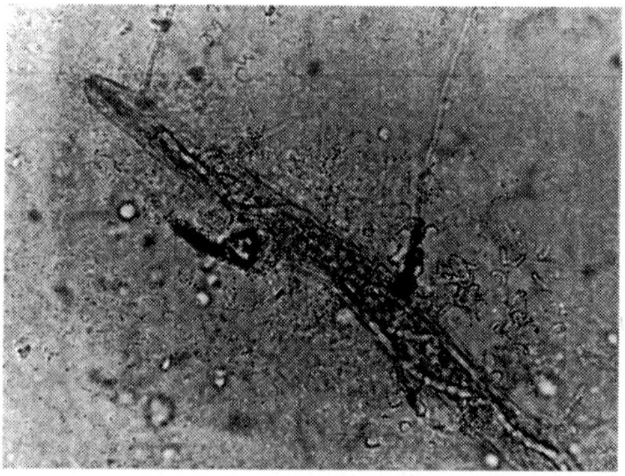

FIGURE 6.5. Conidiophores of the endoparasite *Harposporium anguillulae* emerging from an infected nematode. Crescent-shaped conidia are seen on the surface.

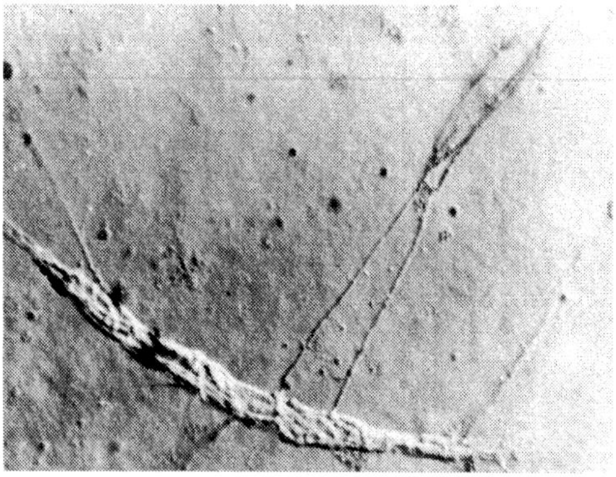

FIGURE 6.6. A nematode filled with hyphae of the endoparasite *Verticillium balanoides.* Conidiophores can be seen breaking out of the nematode body.

and *Helicotylenchus* have received attention as biocontrol agents. In general, fungi colonizing eggs and cysts are more numerous than those parasitizing females (Gintis, Morgan-Jones and Rodriguez-Kabana, 1983).

PREDATORY FUNGI AS BIOCONTROL AGENTS

Linford and Yap (1938, 1939) added five species of nematode-trapping fungi *(A. musiformis, A. oligospora, D. candida, Dactylaria thaumasia,* and *Dactylella ellipsospora)* to soil in which pineapple plants infested with larvae of *Heterodera marioni* were grown. These species yielded only a slight beneficial effect. Olthof and Estey (1966) found that the predacious activity of *A. oligospora* increased when it was grown on a medium supplemented with dextrose and ammonium nitrate. Similarly, a reduction in root-knot of tomato caused by *Meloidogyne hapla* was observed in sterilized soil amended with dextrose and ammonium nitrate. Mankau and McKenry (1976) observed thirteen species of nematode-trapping fungi associated with *Meloidogyne incognita* in a twenty-year-old peach orchard. These fungi appeared to account for considerable natural control of the root-knot nematode population. *Dactylella oviparasitica* was found to be a successful biocontrol agent of *Meloidogyne* (Stirling, McKenry and

Mankau, 1979). Mitsui (1985) added an inoculum of nematode-trapping fungi cultured with 100 g of barley grains to volcanic ash soil infested with *M. incognita. Arthrobotrys haptotyla* decreased the nematode population most effectively among the six fungus species tested. Quinn (1987) concluded that competitive stress imposed by common soil saprophytes may cause an increase in predation by *A. oligospora* and *M. cionopagum* on *Acrobeloides buetschlii.* Mankau and Wu (1985) studied the effect of six isolates of *Monacrosporium ellipsosporum* on *M. incognita* population in the field. The most aggressive isolate was tested for protection of tomato from *M. incognita.* The fungus effectively controlled plant damage at 5 and 10 g levels and reduced galling by 42 and 49 percent respectively, in pot trials. In field tests, reduction of *M. incognita* and improved plant growth were directly correlated with the amount of fungus used.

Some commercial preparations of nematode-trapping fungi have been marketed. Cayrol et al. (1978) developed a commercially prepared isolate of *Arthrobotrys* to protect mushrooms from attack by *Ditylenchus myceliophagus*. They chose *Arthrobotrys robusta* var. *antipolis* for development as a biological control agent (Royal 300). Field trials with the fungus in mushroom compost containing *Agaricus bisporus* showed harvests of mushroom increased by more than 28 percent, and the *D. myceliophagus* population in the compost was reduced by approximately 40 percent. Cayrol and Frankowski (1979) developed another isolate, *Arthrobotrys irregularis,* commercially (Royal 350) and tried it in the fields of several vegetable growers at 140 g/m^2. This rate resulted in good protection of tomatoes against *Meloidogyne*.

Nemastin, a bionematicide has been introduced recently. It is a wettable powder containing the nematophagous fungi *Arthrobotrys conoides, A. oligospora, Paecilomyces lilacinus, Paecilomyces fumo soroseus,* and *Verticillium chlamydosporium.* Target nematodes controlled are *Meloidogyne, Heterodera, Globodera, Pratylenchus, Tylenchulus, Trichodorus, Xiphinema* and other tylenchid nematodes parasitizing food, fiber and ornamental crops. All stages of target nematodes, including cysts, larvae, and adults are susceptible. Nemastin can be applied at any stage of crop development, including like preseeding, vegetative growth, and fruiting. The product should be used at monthly intervals.

ENDOPARASITES AS BIOCONTROL AGENTS

Catenaria anguillulae is a widespread endoparasite. Pathogenicity studies by Esser and Ridings (1973) added thirteen genera and nine species of nematodes to the host range of the fungus. Sayre and Keeley (1969) found that infection by the zoospore of *C. anguillulae* was reduced in acid conditions and that it was more active at 24-28°C. The antagonistic potential of different isolates of *C. anguillulae* vary against *Heterodera schachtii* (Voss, Utermohl and Wyss, 1992).

Jaffee and Zehr (1985) found that *H. rhossiliensis* produces spores that adhere to and penetrate the nematode cuticle and assimilate the body contents before sporulation. It has a broad nematode host range that includes plant-parasitic nematodes: *Heterodera, Globodera, Meloidogyne, Pratylenchus,* and *Ditylenchus* (Timper, Kaya and Jaffee, 1991; Lackey, Muldoon and Jaffee, 1993; Tedford, Jaffee and Muldoon, 1993; Timper and Brodie, 1993). The life history of this fungus is simple (Jaffee et al., 1990). The conidium, surrounded by an adhesin and borne on a phialide, attaches to the nematode. The germtube penetrates the cuticle directly beneath the conidium after twelve hours. After penetration, it forms an infection bulb, from which assimilative hyphae grow and ramify through the nematode body. Hyphae completely fill the nematode body within three days. In suitable conditions, the fungus grows from the nematode and produces new conidia on phialides. The spores do not attach to nematodes if they are detached from the conidiophore (McInnis and Jaffee, 1989); hence, the spread of infection is limited to hosts that pass close to colonized cadavers.

Shufen and Senyu (2005) reported the biocontrol potential of *H. rhossiliensis* and *Hirsutella minnesotensis* against *Heterodera glycines*. In this study the potential of liquid cultures of the two fungi was evaluated. Liquid cultures at 0.2, 0.4, and 0.8 g of fresh mycelium per 300 cm^3 soil (per pot) and solid culture at one percent (corn grits: soil, w/w) reduced nematode egg population densities in both autoclaved and unheated soils compared with a soil-only control or a corn grits control. However, the liquid culture at 0.2-0.8 g of mycelium/pot appeared to be more effective in reducing the nematode population than the solid culture. *H. rhossiliensis* resulted in lower nematode population density than *H. minnesotensis* only in unheated soil.

D. coniospora was used in biocontrol experiments because of its ability to attract nematodes and its known specificity of conidial adhesion to nematodes (Jansson and Nordbring-Hertz, 1984). *D. coniospora* significantly reduced root-knot nematode *Meloidogyne* spp. galling on tomatoes in greenhouse pot trials (Jansson, Jeyaprakash and Zuckerman, 1985).

PARASITES OF CYST NEMATODES AS BIOCONTROL AGENTS

Parasites of Females

Interest in the pathology of cyst-forming nematodes has been growing. Parasitic fungi that invade eggs or females have attracted much attention as biocontrol agents. In general, fungi colonizing eggs and cysts are more numerous than those parasitizing females (Gintis, Morgan-Jones and Rodriguez-Kabana, 1983).

Catenaria auxiliaris parasitizes the females of *Globodera rostochiensis, Heterodera avenae,* and *H. glycines* (Kerry and Crump, 1977; Kerry and Mullen, 1981; Crump, Sayre and Young, 1983; Kerry, 1987). Kerry (1975) suggested that nematode populations decline because females on the root system either fail to form cysts or produce fewer eggs than expected. Kerry (1980) discussed the potential role of *C. auxiliaris* as a biological control agent. Crump, Sayre and Young (1983) surveyed several soybean growing areas of the United States and collected females of *H. glycines* infected with *C. auxiliaris.*

Kerry (1975) surveyed cereal fields for parasites of *H. avenae.* He observed four species of nematode parasitic fungi: *Cylindrocarpon destructans, Tarichium auxiliare, V. chlamydosporium,* and an *Entomophthora*-like fungus. Kerry and Crump (1977) found the *Entomophthora*-like fungus to be capable of infecting *Heterodera carotae, Heterodera cruciferae, H. avenae, Heterodera goettingiana, H. schachtii* and *Heterodera trifolii.*

Nematophthora gynophila is an oomycetous fungus with biflagellate spores. Kerry (1980) found that *N. gynophila* is widespread in Great Britain and reduced the population of *H. avenae* to nondamaging levels. The fungus killed the females in less than seven days, and the resting spores survived in soil for at least two years. The fungus completes its life cycle in the cereal cyst nematode within five days at 13°C (Kerry and Crump, 1980). This fungus was active at all the three sites surveyed by Kerry, Crump and Mullen (1982), and the populations of *H. avenae* failed to increase. The activity of *N. gynophila* is affected by soil moisture and density of females as well as density of spores in soil. The thick-walled spores of this fungus appear to have a longevity of at least five years in the absence of nematodes (Kerry, 1984).

Parasites of Eggs

Egg parasites are important biological control agents of nematodes. Nematode eggs of the group Heteroderidae are vulnerable to attack by parasites. In contact with the cyst or egg masses, parasitic fungi grow rapidly and eventually parasitize all the eggs that are in an early embryonic stage of development.

Verticillium chlamydosporium, a member of the Deuteromycotina, is widespread and has been isolated from cyst and root-knot nematodes around the world (Kerry, 1988). Infection occurs when a hypha comes into contact with a nematode egg or sedentary female in the rhizosphere. Penetration of the egg shell by *V. chlamydosporium* is an enzymatic and physical phenomenon (Morgan-Jones, White and Rodriguez-Kabana, 1983; Segers et al., 1994). Subtilisin serine proteases are thought to be key enzymes in the infection of nematode eggs by *V. chlamydosporium* (Segers et al., 1994, 1995, 1996). It takes several days to destroy a nematode egg; eventually the mycelium lyses, leaving an empty, shrivelled egg shell. The fungus produces thick-walled dictyochlamydospores on short pedicels and conidia on simple phialides (Barron and Onions, 1966). Infection of cyst nematode females within two weeks of their emergence on the root surface results in premature maturation to a small cyst, reduced fecundity, and extensive parasitism of eggs (Kerry, 1990). Hyphae, chlamydospores, and conidia have been observed within infected cyst nematodes.

Crump, Sayre and Young (1983) observed *V. chlamydosporium* to be the most frequent parasite of the eggs and larvae of *H. schachtii*. Kerry (1981) recorded that *V. chlamydosporium* attacked the eggs of *H. avenae* and reduced their populations to nondamaging levels in a wide range of soils in Europe. Kerry et al. (1986) tested six strains of this fungus isolated from *H. avenae* eggs and found that they varied significantly in their pathogenicity to *H. avenae* eggs. Similar results were found in *H. schachtii* eggs using six other strains of this fungus (Irving and Kerry, 1986). Strains differed in their pathogenicity, but all were parasitic and capable of colonizing viable eggs, including those containing second-stage juveniles. Crump and Irving (1992) studied the efficacy of *V. chlamydosporium* as a biocontrol agent of beet and potato cyst nematodes. The most effective isolates of the fungus tested gave 75 and 76 percent control of first-generation of eggs of *H. schachtii* and *Globodera pallida,* respectively.

Morgan-Jones and Rodriguez-Kabana (1986) found *C. destructans* to be the most frequently encountered species associated with *G. pallida,* the potato cyst nematode. Crump (1987) isolated *C. destructans* from infected females of *H. avenae* and *H. schachtii* on sugar beet roots in two soils.

PARASITES OF ROOT-KNOT NEMATODES AS BIOCONTROL AGENTS

Most work on the regulation of nematode populations through the application of *V. chlamydosporium* has concerned the control of root-knot nematodes. In general, the control of root-knot nematodes depends greatly on factors that affect the growth of the fungus during its saprophytic phase and colonization of the rhizosphere by the fungus. In general, the more abundant the fungus is in the rhizosphere, the greater the extent of parasitism and the control of nematode multiplication. Proliferation of the fungus after application to soil is less in mineral than organic soils (De Leij, Kerry and Dennehy, 1993) and is affected by temperature (De Leij, Dennehy and Kerry, 1992). Chlamydospores are the most convenient form of inoculum, as shown by De Leij, Kerry and Dennehy (1993) in small plot trial. In this trial, a broadcast application of 5,000 chlamydospores per gram of soil incorporated at planting established the fungus throughout the growing season and controlled a small population of *Meloidogyne hapla* on a tomato crop grown in microplots.

Morgan-Jones, White and Rodriguez-Kabana (1983) evaluated an isolate of *V. chlamydosporium* from females of *Meloidogyne arenaria in vitro* for its ability to parasitize eggs of the nematode.

It prevented egg hatching and colonized eggs by hyphal penetration. Both the egg shell and the larval cuticle were disrupted, and hyphae readily proliferated endogenously within the eggs and larvae. Some ultrastructural disorganization of the chitin and lipid layers of the shell and the basal layer of the larval cuticle was evident. Bourne, Kerry and de Leij (1996) studied the effect of the host plant on the efficacy of *V. chlamydosporium* as a biological control agent for root-knot nematodes in four experiments. The growth of the fungus in the rhizosphere differed significantly with different plant species. Cabbage and kale supported the most extensive colonization of the fungus. Rao, Reddy and Nagesh (1998) evaluated the effect of *V. chlamydosporium* and *Pasteuria penetrans* either singly or in combination for the management of *M. incognita* infecting tomato. The fungus alone or in combination with *P. penetrans* was effective in increasing plant growth parameters of tomato seedlings. In the field integration of both the biogents was most effective in reducing root galling, number of eggs per egg mass, and nematode population in roots and soil and in increasing egg parasitization with the fungus. When Stirling and Smith (1998) applied *V. chlamydosporium* in granulated form in the field to tomato, the population density of the fungus increased to approximately 10^4 colony-forming units in soil after seven to fourteen weeks. It was found that 37 and 82 percent of the first-generation egg masses produced by *Meloidogyne javanica* contained parasitized eggs.

Dactylella oviparasitica was found to keep *Meloidogyne* populations at levels that had no economic impact on a crop (Stirling and Mankau, 1978). The fungus actively parasitized *Meloidogyne* eggs, which were more vulnerable to attack than larvae. Stirling and Mankau (1979) found that the fungus was able to survive saprophytically on dead roots or by parasitizing eggs of other nematodes.

Paecilomyces lilacinus has potential as a biocontrol agent of *M. incognita.* Jatala, Kaltenbach and Bocangel (1979) reported parasitism of *M. incognita* acrita egg mass *by P. lilacinus* in Peru. The fungus consistently infected eggs and occasionally infected females. Some 80-90 percent of the egg masses of the nematode were found to be destroyed by the fungus. Jatala et al. (1980) reported that inoculation by *P. lilacinus* parasitized the egg masses of *M. incognita* acrita, concluding that the fungus had the potential to control *M. incognita* on potato under field conditions. Godoy et al. (1983) found *P. lilacinus* to be the most frequently occurring egg parasite of *M. arenaria.* Noe and Sasser (1984) noted significant control of *M. incognita* infected with *P. lilacinus* in tomato and okra.

Morgan-Jones, White and Rodriguez-Kabana (1984) found that *P. lilacinus* occurred in significant numbers in eggs of *M. arenaria. P. lilacinus* is typically a soil-borne fungus and appears to be relatively common and ubiquitous in the tropics and subtropics (Domsch, Gams and Anderson, 1980). The capability of this species to degrade chitin has been reported by Okafer (1967). The efficiency and adaptability of *P. lilacinus* in controlling various pathogenic nematodes has been studied under different climatic and soil environmental conditions all over the world. As a potential biocontrol agent, *P. lilacinus* appears to have a number of advantages. It is a good competitor in most tropical and subtropical soils as its optimum growth temperature is 26-30°C. It acts as a chitin degrader and has proteolytic properties (Endreeva, Ushakova and Egorov, 1972). Owing to its qualities, this organism has been applied as a biocontrol agent in several recent investigations.

In a three-year trial with *P. lilacinus,* which was added only in the first year, crop damage by *M. incognita* was reduced each year. The root galling index on the original crop was severe, but on successive crops, it was reduced to moderate levels (Jatala et al., 1981). Cabanillas, Barker and Daykin (1988) observed the absence of root galling and giant cell formation in tomato roots inoculated with nematode eggs infected with *P. lilacinus*. Microplot experiments were conducted to evaluate the effects of inoculum level of this fungus on protection of tomato plants against *M. incognita*. The efficacy of *P. lilacinus* alone and in combination with chitin in controlling *M. incognita* was studied by Mittal, Saxena and Mukerji (1992, 1995, 1999). Growth parameters were assessed in terms of shoot/root length, fresh weight and dry weight, and number of galls per gram fresh root weight. The results showed that plants treated with *P. lilacinus* and chitin had improved growth and the lowest galling. Khan and Saxena (1997) reported successful control of *M. javanica* in tomato using organic materials and *P. lilacinus.* Rao, Reddy and Nagesh (1997) controlled *M. incognita* in okra by combining *P. lilacinus* with castor cake suspension. Qualitative analysis of microflora present in the rhizosphere of okra during all stages of plant growth showed the presence of egg parasitizing and antagonistic fungi such as *P. lilacinus, Paecilomyces variotii,* and *Paecilomyces fusisporus* etc. (Rawat et al., 1999). Several reports have been made on the use of *P. lilacinus* in the control of root-knot nematodes of economically important crop plants (Candamedo-Lay et al., 1982; Morgan-Jones, White and Rodriguez-Kabana, 1984; Roman and Rodriguez-Marcano, 1985; Khan and Hussain, 1986; Cabanillas and Barker, 1989; Saxena, 2004). The extent of its activity was found to be better than that of any of the commonly used nematicides (Jatala, 1983). Several experiments on *P. lilacinus* infestations in fields in Peru, Panama, Puerto Rico, the Philippines, Malaysia, and the United States indicate significant potential of the fungus as a potential biocontrol agent for *M. incognita.*

Atkins et al. (2003) used *Pochonia chlamydosporia,* a fungal parasite of root-knot nematodes, for the management of *M. incognita.* The fungus significantly reduced nematode infestations in soil after a tomato crop in a strategy that combined the use of the fungus with crop rotation. The survival of the fungus in soil was also examined in controlled conditions in which it remained in soil in densities significantly greater than its original application rate for at least five months. The long-term efficacy of *P. chlamydosporia* was tested in two cropping systems, one consisting of three consecutive lettuce crops and another consisting of one tomato crop followed by two lettuce crops (Veerie, Annemie and Nicole, 2005). A one-time application of *P. chlamydosporia* was able to slow down the build up of the *M. javanica* population for at least five to seven months.

INTEGRATED NEMATODE MANAGEMENT

The most practical way of achieving biological control of nematodes may be to adopt practices that enhance the activity of resident anatagonists present in most cultivated soil. Incorporation of crop residues and organic wastes into soil often favors antagonists, improves plant growth, and reduces nematode populations. Linford, Yap and Oliveira (1938) suggested that organic amendments enhanced the activity of resident antagonists of nematodes. They believed that the increase in the number of soil microorganisms that occurs in amended soil caused populations of free-living nematodes to increase. The presence of these nematodes then stimulated resident nematode-trapping fungi into predacious activity, and thus plant-parasitic nematodes were destroyed. This led to acceptance of the hypothesis that organic amendments stimulate the activity of nematode-trapping fungi. Decomposition of organic materials in soil is a complex microbiological process, in which marked changes take place in the population of fungi and bacteria in soil. An initial microbial explosion as soluble and readily available energy sources are consumed, is followed by a series of changes in the composition of the microflora as a succession of different microorganisms utilize the remaining chemical constituents. After the addition of organic amendments, fungi such as *Phoma, Chaetomium, Fusarium oxysporum,* and *C. destructans,* which are parasitic in the females and eggs of cyst and root-knot nematodes, become active and reduce the nematode population. A wide range of chemicals are also produced during the surge in microbial activity after the addition of organic matter to the soil. Ammonia, nitrites, tanins, phenols, fatty acids, volatile compounds and other organic acids are released during the decomposition process. All these substances either are nematicidal, affect egg hatching and the motility of juveniles, or increase the resistance of roots to nematodes (Badra, Saleh and Oteifa, 1979; Alam, Khan and Saxena, 1979; Mian and Rodriguez-Kabana, 1982).

Organic Manure

Population levels of plant-parasitic nematodes may decrease after organic manures are added to the soil. It is thought that manure causes an increase in the natural enemies of nematodes such as bacteria, fungi, and predacious nematodes (Stirling, 1991). Crop yield responses in these amended soils, particularly for amendments high in nitrogen, are often attributed to improved soil fertility and water-holding capacity as well as suppression of plant-parasitic nematodes. Although the mode of action has not been demonstrated in many studies, most explanations suggest that amendments or their degradation products either are directly toxic to nematodes or serve to enhance the proliferation of soil-borne antagonists to nematodes (Lazzeri and Mancini, 2000).

Stockli (1952) reported the effects of a wide variety of organic additives on total soil nematode numbers. Johnson (1959) reported the reduction of severity of root-knot disease on tomato by eleven crop residues incorporated into soil. Mankau and Das (1969) studied quantitative effects of five organic additives and plant roots on nematodes of an agricultural soil lightly infested with a plant-parasitic species. Materials used in the respective treatments were commercially packaged dung, chopped green alfalfa plants, tomato seedlings (plant roots), dried rooted wood shavings, chopped oat hay, and screened dry chicken manure. The results indicated that dung and green manure (alfalfa) supported a greater variety of predecious fungi such as *Arthrobotrys, D. brochopaga, Dactylella* sp., *S. hadra,* and the endoparasite *H. anguillulae.* These treatments also resulted in a sharp decline in total nematode numbers.

Mojtahedi et al. (1991) reported the suppression of root-knot nematode populations using selected *Brassica* cultivars as green manure. Devi and Gupta (1995) tested the utility of four plants (mung, cowpea, sunhemp, and dhaincha [*Sesbania aculeata*]) as green manure on pigeon pea plants infected with *Heterodera cajani,* where sunhemp gave best control. Crow, Guertal and Rodriguez-Kabana (1996) tested velvet bean along with rapeseed and supplemental urea to control *M. arenaria* and *M. incognita.* The biofumigation potential of *Brassica* and other members has been investigated by several workers (Sarwar and Kirkegaard, 1998; Sarwar et al., 1998). Al-Rehiayani and Hafez (1998) amended soil with green residues of horsebean, velvet bean, castorbean, sudangrass, rapeseed, corn, or oil radish. They effectively reduced *M. chitwoodi* populations by 79-94 percent. Chavarria-Carvajal and Rodriguez-Kabana (1998) evaluated four organic amendments (velvet bean, kudzu, pinebark, and urea N) for the management of the root-knot nematode. Most organic amendments were effective in reducing the nematode population and increasing the population of nonparasitic nematodes. Catalase and esterase were sharply increased by velvet bean, kudzu, and pine bark. Their results suggest that suppression of root-knot nematodes is due to complex modes of action in amended soils.

Various organic amendments, such as crop residue, hay, coffee grinds, poultry manure, and oil seed cakes, etc., are effective in controlling plant-parasitic nematodes. The nitrogen content of an amendment is particularly important in determining its effectiveness against nematodes, and ammonia and possibly nitrite are the compounds thought to be responsible for this effect (Mian et al., 1982; Mian and Rodriguez-Kabana, 1982). The use of oil radish as a green manure has dramatically reduced stubby root nematode *(Trichodorus)* and root-lesion nematode *(Pratylenchus)* in Idaho potato fields (Anon., 2001). For hundreds of years, Indian farmers have used the neem tree *(Azadirachta indica)* for its pesticidal, antifungal, and antifeedant properties. In research trials, potting soil amended with plant parts from the neem tree and Chinaberry tree *(Melia azedarach)* inhibited root-knot nematode development on tomatoes. Reddy, Nagesh and Denappa (1997) evaluated the effect of neem cake at 1 kg/m^2, *P. penetrans,* and *P. lilacinus* either singly or in combination for the management of *M. incognita* infecting tomato under nursery and field conditions. Parasitism of *M. incognita* females was maximal when neem cake was integrated with *P. penetrans,* and egg parasitism was highest when neem cake was integrated with *P. lilacinus.* Neem cake, made from crushed neem seeds, provides nitrogen in slow release form in addition to protecting plants against parasitic nematodes. Neem cake is toxic to plant-parasitic nematodes but not as detrimental to beneficial free-living soil organisms (Riga and Lazarovits, 2001).

Wang, Sipes and Schmitt (2001, 2002, 2003) managed the population of *Rotylenchus reniformis* in pineapple soils by adding *Crotalaria juncea, Brassica napus, Tagetes erecta,* and intercycle cover crops. Jaffee (2002) used soil cages to quantify how nematode-trapping fungi responded to grape leaf amendment in two vineyards. In one vineyard, grape leaf amendment stimulated *Dactylellina haptotyla* population density and trapping in one year but not in another. In the second vineyard, grape leaf amendment stimulated *A. oligospora* population density in one year but not in another. Jaffee (2004) found in a vineyard study that *D. haptotyla* population density and trapping were most enhanced by a smaller quantity of alfalfa amendment, whereas *A. oligospora* population density was most enhanced by a larger quantity of alfalfa amendment. Organic matter can be utilized in large quantities in high-value crops or where it is readily available.

Chitin

Chitin and inorganic fertilizers that release ammoniac nitrogen into soil have been shown to suppress nematode populations owing to plasmolysing effects and to selective proliferation of microbial antagonists (Rodriguez-Kabana, 1986). Chitin soil amendments have been used to increase chitinolytic activity within the soil microflora (Spiegel, Cohn and Chet, 1986; Spiegel, Chet and Chon, 1987). Chitin amendments to soil are effective for the control of nematodes (Mankau and Das, 1969; Mian et al., 1982; Godoy et al., 1983; Rodriguez-Kabana, Morgan-Jones and Gintis, 1984; Mittal, Saxena and Mukerji, 1992, 1995, 1999). One possible amendment is to use nematode components, for example cuticle egg shell or gelatinous matrix. Addition of these specific compounds to soil is expected to stimulate development of microbial species capable of degrading similar compounds present in live nematodes. The effect of chitin amendments on nematodes may last for several months.

Godoy et al. (1983) studied the effect of chitin added to soil for the control of *M. arenaria* and its effect on the soil microflora. They found that the addition of chitin to soil reduced nematode populations. Rodriguez-Kabana, Morgan-Jones and Gintis (1984) added crustacean chitin to soil to control *H. glycines* in the roots of soybean plants. The total number of chitinolytic fungi and actinomycetes in the soil increased after eight weeks of amendment. Fungal species isolated from chitin-treated soils are known parasites of eggs of *Globodera, Heterodera,* and *Meloidogyne* spp. The numbers of chitinolytic microorganisms, especially actinomycetes and bacteria, were higher in chitin-amended soil than in control soil, and the amended soil had significantly reduced populations of root-knot nematodes (Rodriguez-Kabana and Morgan-Jones, 1987; Spiegel, Chet and Cohn, 1987). Culbreath, Rodriguez-Kabana and Morgan-Jones (1985) demonstrated that the addition of lignohemicellulosic materials to soil amended with chitin can increase the effectiveness of chitin against nematodes. Mittal, Saxena and Mukerji (1995) found that chitin- and fungus-treated tomato plants showed improved plant growth and the lowest galling among all the treatments tried.

Various experiments suggest that chitin or other appropriate materials serve as the substrate for the selective development of biological control agents in soil. An increase in the number and activity of a specialized mycoflora is likely responsible for the extended control of plant-parasitic nematodes observed in soil amended with chitin.

MOLECULAR APPROACHES

The potential of molecular techniques in the development of biocontrol agents of nematodes appears promising, but it is largely unexploited. The rapidly expanding fields of biotechnology and genetic engineering are likely to have an impact on biological control, although the use of new technologies is at an early stage. Development of selected fungal isolates to be used in biocontrol is dependent on the development of suitable molecular markers and DNA probes. Different *V. chlamydosporium* isolates grown in pure culture can be discriminated using polymerase chain reaction (PCR)-based DNA fingerprinting (Arora, Hirsch and Kerry, 1996). Hirsch et al. (2000) developed PCR-based methods to detect *V. chlamylosporium* on infected plant root. Arbitrary ERIC primers and those based on rRNA genes, to identify fungi grown in pure culture, were unsuitable for DNA extracted from nematode-infested roots because of interference by plant and nematode DNA. A novel method utilizing specific primers designed from an amplified and cloned fragment of the *V. chlamydosporium* P-tubulin gene was developed. Hirsch et al. (2001) compared several methods for studying *V. chlamydosporium* in soil and root environments. These included a semiselective medium for the fungus, PCR primers specific for the fungal P-tubulin gene, and monoclonal antibodies. The P-tubulin gene was implicated in resistance to the fungicides used in the semiselective medium, and the genetic basis for this was investigated. Culture and PCR-based methods were used to screen for the presence of *V. chlamydosporium* in field soils known to have been suppressive to cereal cyst nematodes.

Ahman et al. (2002) reported that a subtilisin-like extracellular serine protease, designated PII, is an importrant pathogeneicity factor in *A. oligospora*. The transcript of PII was not detected during the early stages of infection, but high levels were expressed during the killing and colonization of the nematode. Disruption of the PII gene by homologous recombination had a limited effect on the pathogenicity of the fungus. However, mutants containing additional copies of the PII gene developed a higher number of infection structures and had an increased speed of killing than the wild type. This is the first report showing that genetic engineering can be used to improve the pathogenicity of a nematode-trapping fungus.

Mauchline, Kerry and Hirsch (2002) quantified *V. chlamydosporium* in soil by a competitive PCR (cPCR) assay. Ar-irradiated soil was seeded with different numbers of chlamydospores from the fungal isolate, and ten samples were obtained at time intervals up to eight weeks. Samples were analyzed by cPCR and by plating on to a semiselective medium. The results suggested that saprophytic *V. chlamydosporium* growth did not occur in soil and that the two methods detected different phases of growth. The first stage of growth, DNA replication, was demonstrated by a rapid increase in cPCR estimates, and the presumed carrying capacity (PCC) of the soil was reached only after one week of incubation. The second stage was indicated by a less rapid increase in colony-forming units, and to reach the PCC three weeks were required. Mauchline, Kerry and Hirsch (2004) developed a random amplification of polymorphic DNA-PCR (RAPD-PCR) assay to test for competitive variability in growth of the fungus *P. chlamydosporia*. Saprophytic competence in soil with or without tomato plants was examined in three isolates of the fungus: RES 280(J), RES 200(I), and RES 279(S). Viable counts taken at seventy days indicated that RES 200(I) was the best saprophyte, followed by isolate S, with J the poorest. RAPD-PCR analysis of colonies from mixed treatments revealed that there was a cumulative effect of adding isolates to the system.

Jinkui et al. (2005) cloned the gene encoding a cuticle-degrading serine protease from three isolates of *Lecanicillium psalliotae* (syn. *Verticillium psalliotae*) using the 3′ and 5′ rapid amplification of cDNA ends method. The gene encodes for 382 amino acids, and the protein shares conserved motifs with subtisilin N and peptidase S8. Comparison of translated cDNA sequences of three isolates revealed one amino acid polymorphism at position 230. The deduced protease sequence had a high degree of similarity to other cuticle-degrading proteases from other nematophagous fungi.

CONCLUSION

In the past, the image of biological control has suffered because instant results were expected by the addition of antagonists to the soil. However, it is much more likely that biological control methods will be used as an adjunct to other methods of control. The future lies in the encouragement of resident antagonists or the introduction of antagonists from elsewhere. Efforts should be directed to endowing organisms with characteristics likely to favor their survival in a hostile environment, or to developing techniques of inoculum preparation and environment modification that aid their establishment. New biotypes of antagonists with superior properties to those of wild-type strains should be developed through genetic engineering.

Treating the soil with various organic amendments, such as green manure and chitin, stimulates microbial activity, which in turn increases predators and parasites of nematodes. The use of organic amendments with biocontrol is a promising aspect of the development of new nematode management practices. Most nematode control tactics must be implemented before or after planting. Once the crop is planted, farmers have few management alternatives for that growing season. Adoption of integrated nematode management practices along with cultural practices can serve to further reduce nematode populations, thus reducing the use of nematicides.

REFERENCES

Ahman, J., T. Johansson, M. Olsson, P.J. Punt, C.A.M.J.J. van den Hondel and A. Tunlid. (2002). Improving the pathogenicity of a nematode-trapping fungus by genetic engineering of a subtilisin with nematotoxic activity. *Applied and Environmental Microbiogy* 68: 3408-3415.

Alam, M.M., A.M. Khan and S.K. Saxena. (1979). Mechanism of control of plant-parasitic nematodes as a result of the application of organic amendments to the soil V. Role of phenolic compounds. *Indian Journal of Nematology* 9: 136-142.

Al-Rehiayani, S. and S. Hafez. (1998). Host status and green manure effect of selected crops on *Meloidogyne chitwoodi* race 2 and *Pratylenchus neglectus*. *Nematropica* 28: 213-230.

Anonymous. (2001). Oil radish green manure continues promise against nematodes. *The Grower* June-July 7.

Arora, D.K., P.R. Hirsch and B.R. Kerry. (1996). PCR based discrimination of *Verticillium chlamydosporium* isolates. *Mycological Research* 100: 801-809.

Atkins, S.D., L. Hidalgo-Diaz, H. Kalisz, T.H. Mauchline, P.R. Hirsch and B.R. Kerry. (2003). Development of a new management strategy for the control of root-knot nematodes (*Meloidogyne* spp) in organic vegetable production. *Pest Management Service* 59: 183-189.

Badra, T., M.A. Saleh and B.A. Oteifa. (1979). Nematicidal activity and composition of some organic fertilizers and amendments. *Review of Nematology* 2: 29-36.

Barron, G.L. (1977). *The Nematode-Destroying Fungi: Topics in Mycobiology 1.* Canadian Biological Publications, Guelph, Canada, p. 140.

Barron, G.L. and A.H.S. Onions. (1966). *Verticillium chlamydosporium* and its relationships to *Diheterospora, Stemphyliopsis* and *Paecilomyces. Canadian Journal of Botany* 44: 861-869.

Bourne, J.M., B.R. Kerry and F.A.A.M. de Leij. (1996). The importance of the host plant on the interaction between root-knot nematodes (*Meloidogyne* spp.) and the nematophagous fungus *Verticillium chlamydosporium* Goddard. *Biocontrol Science and Technology* 6: 539-548.

Cabanillas, E. and K.R. Barker. (1989). Impact of *Paecilomyces lilacinus* inoculum level and application time on control of *Meloidogyne incognita* on tomato. *Journal of Nematology* 21: 115-120.

Cabanillas, E., K.R. Barker and M.E. Daykin. (1988). Histopathology of the interactions of *Paecilomyces lilacinus* with *Meloidogyne incognita* on tomato. *Journal of Nematology* 20: 362-364.

Candamedo-Lay, E., J. Lara, P. Jatala and F. Gonzalez. (1982). Evaluacion preliminer del comportamiento de *Paecilomyces lilacinus* como controlado biologico del nematode, nodulador *Meloidogyne incognita* en tomato industrial. *Nematropica* 12: 154.

Cayrol, J.C. and J.P. Frankowski. (1979). Une methode de lutte biologique contre les nematodes a galles des racines appartenant an genere *Meloidogyne. Revue Horticole* 193: 15-23.

Cayrol, J.C., J.P. Frankowski, A. Laniece, G. d'Hardmare and J.P. Talon. (1978). Contre les nematodes en champignonniere, Mise. au point dune methods, de lutte biologique a l'aide d' un hyphomycete predateur *Arthrobotrys robusta* souche *antipolis* (Royal 300). *Revue Horticole* 184: 23-30.

Chavarria-Carvajal, J.A. and R. Rodriguez-Kabana. (1998). Changes in soil enzymatic activity and control of *Meloidogyne incognita* using four organic amendments. *Nematropica* 28: 7-18.

Crow, W.T., E.A. Guertal and R. Rodriguez-Kabana. (1996). Responses of *Meloidogyne arenaria* and *M. incognita* to green manures and supplemental urea in glasshouse culture. *Supplement to Journal of Nematology* 28: 648-654.

Crump, D.H. (1987). A method for assessing the natural control of cyst nematode populations. *Nematologica* 33: 232-243.

Crump, D.H. and F. Irving. (1992). Selection of isolates and methods of culturing *Verticillium chlamydosporium* and its efficacy as a biological control agent of beet potato cyst nematodes. *Nematologica* 38: 367-374.

Crump, D.H., R.M. Sayre and L.D. Young. (1983). Occurrence of nematophagous fungi in cyst nematode population. *Plant Disease* 67: 63-64.

Culbreath, A.K., R. Rodriguez-Kabana and G. Morgan-Jones. (1985). The use of hemicellulosic waste matter for reduction of the phytotoxic effects of chitin and control of root-knot nematodes. *Nematropica* 15: 49-75.

Deacon, J.W. and G. Saxena. (1997). Orientated zoospore attachment and cyst germination in *Catenaria anguillulae,* a facultative endoparasite of nematodes. *Mycological Research* 101: 513-522.

De Leij, F.A.A.M., J.A. Dennehy and B.R. Kerry. (1992). The effect of temperature and nematode species on interaction between the nematophagous fungus *Verticillium chlamydosporium* and root-knot nematodes (*Meloidogyne* spp.). *Nematologica* 38: 65-79.

De Leij, F.A.A.M., B.R. Kerry and J.A. Dennehy. (1993). *Verticillium chlamydosporium* as a biological control agent of *Meloidogyne incognita* and *M. hapla* in pot and microplot tests. *Nematologica* 39: 115-126.

Devi, S. and P. Gupta. (1995). Effect of four green manures against *Heterodera cajani* on pigeonpea sown with or without *Rhizobium* seed treatment. *Indian Journal of Plant Pathology* 25: 254-256.

Domsch, K.H., W. Gams and T.H. Anderson. (1980). *Compendium of Soil Fungi.* Vol. I, Academic Press, New York.

Drechsler, C. (1937). Some hyphomycetes that prey on free living terricolous nematodes. *Mycologia* 29: 447-552.

Drechsler, C. (1954). Some hyphomycetes that capture eelworms in southern states. *Mycologia* 46: 762-782.

Endreeva, N.A., V.I. Ushakova and N.S. Egorov. (1972). Study of proteolytic enzymes of different strains of *Paecilomyces lilacinus* Thom. in connection with their fibrinolytic activity. *Mikrobiologiva* 41: 364-368.

Esser, R.P. and W.H. Ridings. (1973). Pathogenicity of selected nematodes by *Catenaria anguillulae. Proceedings of Soil and Crop Science Society of Florida, USA* 33: 60-64.

Gintis, B.O., G. Morgan-Jones and R. Rodriguez-Kabana. (1983). Fungi associated with several developmental stages of *Heterodera glycines* from an Alabama soil. *Nematropica* 13: 181-200.

Godoy, G., R. Rodriguez-Kabana, A. Shelby and G. Morgan-Jones. (1983). Chitin amendments for control of *Meloidogyne arenaria* in infested soil II. Effects on microbial population. *Nematropica* 13: 63-74.

Hirsch, P.R., S.D. Atkins, T.H. Mauchline, C.O. Morton, K.G. Davies and B.R. Kerry. (2001). Methods for studying the nematophagous fungus *Verticillium chlamydosporium* in the root environment. *Plant and Soil* 232: 21-30.

Hirsch, P.R., T.H. Mauchline, T.A. Mendum and B.R. Kerry. (2000). Detection of the nematophagous fungus *Verticillium chlamydosporium* in nematode infested plant roots using PCR. *Mycological Research* 104: 435-439.

Irving, F. and B.R. Kerry. (1986). Variation between strains of the nematophagous fungus *Verticillium chlamydosporium* Goddard II, Factors affecting parasitism of cyst nematode eggs. *Nematologica* 32: 474-485.

Jaffee, B.A. (2002). Soil cages for studying how organic amendments affect nematode-trapping fungi. *Applied Soil Ecology* 21: 1-9.

Jaffee, B.A. (2004). Do organic amendments enhance the nematode-trapping fungi *Dactylellina haptotyla* and *Arthrobotrys oligospora? Journal of Nematology* 36: 267-275.

Jaffee, B.A., A.E. Muldoon, R. Philips and M. Mangel. (1990). Rates of spore transmission, mortality and production for the nematophagous fungus *Hirsutella rhossiliensis. Phytopathology* 80: 1083-1088.

Jaffee, B.A. and E.I. Zehr. (1985). Parasitic and saprophytic abilities of the nematodes attacking fungus *Hirsutella rhossiliensis. Journal of Nematology* 17: 341-345.

Jansson, H.B., A. Jeyaprakash and B.M. Zuckerman. (1985). Control of root-knot nematodes on tomato by the endoparasitic fungus *Meria coniospora. Journal of Nematology* 17: 327-329.

Jansson, H.B. and B. Nordbring-Hertz. (1984). Involvement of sialic acid in nematode chemotaxis and infection by an endoparasitic nematophagous fungus. *Journal of General Microbiology* 130: 39-43.

Jatala, P. (1983). Biological control with the fungus *Paecilomyces lilacinus*. In *International Meloidogyne Project. Proceedings of the Third Research* and *Planning Conference on Root-Knot Nematodes Meloidogyne* spp. Coimbra, Portugal, pp. 183-187.

Jatala, P. (1986). Biological control of plant parasitic nematodes. *Annual Review of Phytopathology* 24: 453-489.

Jatala, P., R. Kaltenbach and M. Bocangel. (1979). Biological control of *Meloidogyne incognita acrita* and *Globodera pallida* on potatoes. *Jorunal of Nematology* 11: 303.

Jatala, P., R. Kaltenbach, M. Bocangel, A.J. Devax and R. Campos. (1980). Field application of *Paecilomyces lilacinus* for controlling *Meloidogyne incognita* on potatoes. *Journal of Nematology* 12: 226-227.

Jatala, P., R. Salas, R. Kaltenbach and M. Bocangel. (1981). Multiple application and long term effect on *Paecilomyces lilacinus* in controlling *Meloidogyne incognita* under field applications. *Journal of Nematology* 13: 445.

Jinkui, Y., H. Xiaowei, T. Baoyu, S. Hui, D. Junxin, W. Wenping and Z. Keqin. (2005). Characterization of an extracellular serine protease gene from the nematophagous fungus *Lecanicillium psalliotae. Biotechnology Letters* 27: 1329-1334.

Johnson, L.F. (1959). Effect of the addition of organic amendments to soil on root-knot of tomatoes I. *Plant Disease Reporter* 43: 1059-1062.

Kerry, B.R. (1975). Fungi and the decrease of cereal cyst nematode population in cereal monoculture. *EPPO Bulletin* 5: 353-361.

Kerry, B.R. (1980). Biocontrol: Fungal parasites of female cyst nematodes. *Journal of Nematology* 12: 253-259.

Kerry, B.R. (1981). Progress in the use of biological agents for control of nematodes. In *Biological Control in Crop Production, BARC Symposium No.5,* G.C. Papavizas. Allenheld and Osmum, ed. Totowa, NJ, pp. 79-90.

Kerry, B.R. (1984). Nematophagous fungi and the regulation of nematode population in soil. *Helminthological Abstract Series B* 53: 1-14.

Kerry, B.R. (1987). Biological control. In *Principles and Practice of Nematode Control in Crops,* R.H. Brown and B.R. Kerry, eds. Academic Press, New York, pp. 233-263.

Kerry, B.R. (1988). Fungal parasites of cyst nematodes. *Agriculture, Ecosystems and Environment* 24: 293-305.

Kerry, B.R. (1990). Selection of exploitable biological control agents for plant parasitic nematodes. *Aspect of Applied Biology* 24: 1-9.

Kerry, B.R. (1993). The use of microbial agents for the biological control of plant parasitic nematodes. In *Exploitation of Microorganisms,* D. Gareth Jones, ed. Chapman and Hall, London, pp. 81-104.

Kerry, B.R. and D.H. Crump. (1977). Observation of fungal parasites of females and eggs of the cereal cyst nematode *Heterodera avenae* and other cyst nematodes. *Nematologica* 23: 193-201.

Kerry, B.R. and D.H. Crump. (1980). Two fungi parasitic on females of cyst nematodes (*Heterodera* spp). *Transactions of the British Mycological Society* 74: 119-125.

Kerry, B.R., D.H. Crump and L.A. Mullen. (1982). Natural control of the cereal cyst nematode *Heterodera avenae* Woll by soil fungi at three sites. *Crop Protection* 1: 99-109.

Kerry, B.R. and W.M. Hominick. (2000). Biological control. In *Biology of Nematodes,* D.L. Lee, ed. Harwood Academic, Reading, UK, pp. 483-509.

Kerry, B.R., F. Irving and J.C. Hornsey. (1986). Variation between strains of the nematophagous fungus *Verticillium chlamydosporium* Goddard I. Factors affecting growth *in vitro. Nematologica* 32: 461-473.

Kerry, B.R. and B.A. Jaffee. (1997). Fungi as biological control agents for plant parasitic nematodes. In *The Mycota IV: Environmental and Microbial Relationships,* D.T. Wicklow and B. Soderstrom, eds. Springer Verlag, Berlin, pp. 203-218.

Kerry, B.R. and L.A. Mullen. (1981). Fungal parasites of some plant parasitic nematodes. *Nematropica* 11: 187-190.

Khan, T.A. and S.I. Hussain. (1986). Biological control of reniform nematode disease of cowpea by the application of *Paecilomyces lilacinus. Proceedings of XVIII International Nematology Symposium,* Antibes, France.

Khan, T.A. and S.K. Saxena. (1997). Integrated management of root-knot nematode *Meloidogyne javanica* infecting tomato using organic materials and *Paecilomyces lilacinus. Bioresource Technology* 61: 247-250.

Lackey, B.A., A.E. Muldoon and B.A. Jaffee. (1993). Alginate pellet formulation of *Hirsutella rhossiliensis* for biological control of plant parasitic nematodes. *Biological Control* 3: 155-160.

Lazzeri, L. and L.M. Mancini. (2000). The glucosinolate-myrosinase system, a natural and practical tool for biofumigation. In *Proceedings of the International Symposium on Chemicals and Non-Chemicals Soil and Substrate Disinfestation,* J. Kattan, A. Matta, A. Arzone, D.O. Chellemi and F. Lamberti, eds. *Acta Horticulture* 532: 89-95.

Linford, M.B. and F. Yap. (1938). Root-knot injury restricted by a nematode-trapping fungus. *Phytopathology* 28: 14-15.

Linford, M.B. and F. Yap. (1939). Root-knot nematode injury restricted by a fungus. *Phytopathology* 29: 596-609.

Linford, M.B., F. Yap and J.M. Oliveira. (1938). Reduction of soil populations of the root-knot nematode during decomposition of organic matter. *Soil Science* 45: 127-141.

Mankau, R. and S. Das. (1969). The influence of chitin amendments on *Meloidogyne incognita. Journal of Nematology* 1: 15-16.

Mankau, R. and M.V. McKenry. (1976). Spatial distribution of nematophagous fungi associated with *Meloidogyne incognita* on peach. *Journal of Nematology* 8: 294-295.

Mankau, R. and X. Wu. (1985). Effects of nematode-trapping fungus *Monacrosporium ellipsosporum* on *Meloidogyne incognita* population in field soil. *Revue de Nematologie* 8: 147-153.

Mauchline, T.H., B.R. Kerry and P.R. Hirsch. (2002). Quantification in soil and the rhizosphere of the nematophagous fungus *Verticillium chlamydosporium* by competitive PCR and comparison with selective plating. *Applied and Environmental Microbiology* 68: 1846-1853.

Mauchline, T.H., B.R. Kerry and P.R. Hirsch. (2004). The biocontrol fungus *Pochonia chlamydosporia* shows nematode host preference at the infraspecific level. *Mycological Research* 108: 161-169.

McInnis, T.M. and B.A. Jaffee. (1989). An assay for *Hirsutella rhossiliensis* spores and the importance of phialides for nematode inoculation. *Journal of Nematology* 21: 229-234.

Mian, I.H., G. Godoy, R.A. Shelby, R. Rodriguez-Kabana and G. Morgan-Jones. (1982). Chitin amendments for control of *Meloidogyne arenaria* in infested soil. *Nematropica* 12: 71-84.

Mian, I.H. and R. Rodriguez-Kabana. (1982). Soil amendments with oil cakes and chicken litter for control of *Meloidogyne arenaria. Nematropica* 12: 205-220.

Mittal, N., G. Saxena and K.G. Mukerji. (1992). Biological control of *Meloidogyne incognita* by *Paecilomyces lilacinus.* In *Root Ecology and Its Practical Application* 3 ISSR Symposium, Klagenfurt, Austria, L. Kutschera, E. Hubl, E. Lichtenegger, H. Persson and M. Sobolik, eds. Worzelforschung, Klagenfurt, Wien, Austria, pp. 547-550.

Mittal, N., G. Saxena and K.G. Mukerji. (1995). Integrated control of root-knot disease in three crop plants using chitin and *Paecilomyces lilacinus. Crop Protection* 14: 647-651.

Mittal, N., G. Saxena and K.G. Mukerji. (1999). Biological control of root-knot nematode by nematode-destroying fungi. In *From Ethnomycology to Fungal Biotechnology—Exploiting Fungi from Natural Resources for Novel Products,* J. Singh and K.R. Aneja, eds. Kluwer Academic/Plenum Publishers, New York, pp. 163-171.

Mittal, N., G. Saxena, K.G. Mukerji and R.K. Upadhyay. (1996). Biocontrol of nematodes by nematode-destroying fungi. In *IPM System in Agriculture: Biocontrol in Emerging Biotechnology.* Vol. 2, R.K. Upadhyay, K.G. Mukerji and R.L. Rajak, eds. Aditya Books Pvt. Ltd., New Delhi, pp. 453-502.

Mitsui, Y. (1985). Distribution and ecology of nematode-trapping fungi in Japan. *Japan Agriculural Research Quartely* 18: 182-193.

Mojtahedi, H., G.S. Santo, A.N. Hang and J.H. Wilson. (1991). Suppression of root-knot nematode populations with selected rapeseed cultivated as green manure. *Journal of Nematology* 23: 170-174.

Morgan-Jones, G. and R. Rodriguez-Kabana. (1986). Fungi associated with cysts of potato cyst nematodes in Peru. *Nematropica* 16: 21-31.

Morgan-Jones, G., J.F. White and R. Rodriguez-Kabana. (1983). Phytonematode pathology: Ultrastructural studies I. Parasitism of *Meloidogyne arenaria* eggs by *Verticillium chlamydosporium. Nematropica* 13: 245-260.

Morgan-Jones, G., J.F. White and R. Rodriguez-Kabana. (1984). Phytonematode pathology: Ultrastructural studies II. Parasitism of *Meloidogyne arenaria* eggs and larvae by *Paecilomyces lilacinus. Nematropica* 14: 57-71.

Noe, J.P. and J.N. Sasser. (1984). Efficacy of *Paecilomyces lilacinus* in reducing yield losses due to *Meloidogyne incognita. Proceedings of the First International Congress of Nematology,* Canada (Abstr.), p. 116.

Okafer, N. (1967). Decomposition of chitin by microorganisms isolated from a temperate and a tropical soil. *Nova Hedwigia* 13: 209-226.

Olthof, T.H.A. and R.H. Estey. (1966). Carbon and nitrogen levels of a medium in relation to growth and nematophagous activity of *Arthrobotrys oligospora* Fresenius. *Nature* 209: 1158.

Quinn, M.A. (1987). The influence of saprophytic competition on nematode predation by nematode-trapping fungi. *Journal of Invertebratre Pathology* 49: 170-174.

Rao, M.S., P.P. Reddy and M. Nagesh. (1997). Integrated management of *Meloidogyne incognita* on okra by castor cake suspension and *Paecilomyces lilacinus. Nematologia Mediterranea* 25: 17-19.

Rao, M.S., P.P. Reddy and M. Nagesh. (1998). Use of neem based formulation of *Paecilomyces lilacinus* for the effective management of *Meloidogyne incognita* infecting egg plant. In *Proceedings of the Third International Symposium of Afro-Asian Society of Nematologists* Coimbatore, (Abstr.) p. 73.

Rawat, R., A. Pandey, G. Saxena and K.G. Mukerji. (1999). Rhizosphere biology of root-knot diseased *Abelmoschus esculentus* in relation to its biocontrol. In *From Ethnomycology to Fungal Biotechnology. Exploiting fungi from Natural Resources for Novel Products,* J. Singh and K.R. Aneja, eds. Kluwer Academic/ Plenum Publishers, New York, pp. 173-183.

Reddy, P.P., M. Nagesh and V. Denappa. (1997). Effect of integration of *Pasteuria penetrans, Paecilomyces lilacinus* and neem cake for the management of root-knot nematodes infecting tomato. *Pest Management Horticulture Ecosystem* 3: 100-104.

Riga, E. and G. Lazarovits. (2001). Development of an organic pesticide based on neem tree products. American Phytopathological society/Mycological society of America/Society of Nematology Joint Meeting Abstracts of Presentations Salt Lake City, Utah. *Phytopathology* 91: S141.

Rodriguez-Kabana, R. (1986). Organic and inorganic nitrogen amendments to soil as nematode suppressants. *Journal of Nematology* 18: 129-135.

Rodriguez-Kabana, R. and G. Morgan-Jones. (1987). Biological control of nematodes: Soil amendments and microbial antagonists. *Plant and Soil* 100: 237-248.

Rodriguez-Kabana, R., G. Morgan-Jones and B.O. Gintis. (1984). Effects of chitin amendments to soil on *Heterodera glycines,* microbial populations and colonization of cysts by fungi. *Nematropica* 14: 10-25.

Roman, J. and A. Rodriguez-Marcano. (1985). Effect of the fungus *Paecilomyces lilacinus* on the larval population and root-knot formation of *Meloidogyne incognita* in tomato. *Journal of Agriculture University Puerto Rico* 69: 159-167.

Sarwar, M. and J.A. Kirkegaard. (1998). Biofumigation potential of brassicas II. Effect of environment and ontogeny on glucosinolate production and implications for screening. *Plant and Soil* 201: 91-101.

Sarwar, M., J.A. Kirkegaard, P.T.W. Wong and J.M. Desmarchelier. (1998). Biofumigation potential of brassicas III. In vitro toxicity of isothiocyanates to soil-borne fungal pathogens. *Plant and Soil* 201: 103-112.

Saxena, G. (2004). Biocontrol of Nematode-borne diseases in vegetable crops. In *Disease Management of Fruits and Vegetables: Fruit and Vegetable Diseases.* Vol. I., K.G. Mukerji, ed. Kluwer Academic Publishers, Dordecht, The Netherlands, pp. 397-450.

Saxena, G. and N. Mittal. (1997). Ecology of nematode-destroying fungi. In *New Approaches in Microbial Ecology,* J.P. Tiwari, G. Saxena, N. Mittal, I. Tewari and B.P. Chamola, eds. Aditya Books Pvt. Ltd., New Delhi, pp. 39-85.

Saxena, G., N. Mittal, K.G. Mukerji and D.K. Arora. (1991). Nematophagous fungi in the biological control of nematodes. In *Handbook of Applied Mycology—Humans, Animals and Insects.* Vol. 2, D.K. Arora, L. Ajello and K.G. Mukerji, eds. Marcel Dekker Inc., New York, pp. 707-733.

Saxena, G. and K.G. Mukerji. (1988). Biological control of nematodes. In *Biocontrol of Plant Diseases.* Vol. I, K.G. Mukerji and K.L. Garg, eds. CRC Press, Boca Raton, FL, pp. 113-127.

Sayre, R.M. and L.S. Keeley. (1969). Factors influencing *Catenaria anguillulae* infections in a free living and a plant parasitic nematode. *Nematologica* 15: 492-502.

Segers, R., T.M. Butt, J.F. Keen, B.R. Kerry and J.F. Peberdy. (1995). The subtilisins of the invertebrate mycopathogens *Verticillium chlamydosporium* and *Metarhizium anisopliae* are serologically and functionally related. *FEMS Microbiology Letters* 126: 227-232.

Segers, R., T.M. Butt, B.R. Kerry, A. Beckett and J.F. Peberdy. (1996). The role of the proteinase VCP1 produced by the nematophagous *Verticillium chlamydosporium* on the infection process of nematode eggs. *Mycological Research* 100: 421-428.

Segers, R., T.M. Butt, B.R. Kerry and J.F. Peberdy. (1994). The nematophagous fungus *Verticillium chlamydosporium* produces a chymoelastase like protease that hydrolyses host nematode proteins in situ. *Microbiology* 140: 2715-2723.

Shufen, L. and C. Senyu. (2005). Efficacy of the fungi *Hirsutella minnesotensis* and *H. rhossiliensis* from liquid culture for control of the soybean cyst nematode *Heterodera glycines. Nematology* 7: 149-157.

Sikora, R.A. (1992). Management of the antagonistic potential in agricultural ecosystems for the biological control of plant parasitic nematodes. *Annual Review of Phytopathology* 30: 245-270.

Spiegel, Y., E. Cohn and I. Chet. (1986). Use of chitin for controlling plant parasitic nematodes. Direct effects on nematode reproduction and plant performance. *Plant and Soil* 95: 87-95.

Spiegel, Y., I. Chet and E. Cohn. (1987). Use of chitin for controlling plant-parasitic nematodes II. Mode of action. *Plant and Soil* 98: 337-345.

Stirling, G.R. (1991). *Biological Control of Plant Parasitic Nematodes: Progress, Problems and Prospects.* CAB International, UK.

Stirling, G.R. and R. Mankau. (1978). *Dactylella oviparasitica* a new fungal parasite of *Meloidogyne* eggs. *Mycologia* 70: 774-783.
Stirling, G.R. and R. Mankau. (1979). Mode of parasitism of *Meloidogyne* and other nematode eggs by *Dactylella oviparasitica. Journal of Nematology* 11: 282-288.
Stirling, G.R., M.V. McKenry and R. Mankau. (1979). Biological control of root-knot nematodes *Meloidogyne* spp. on peach. *Phytopathology* 69: 806-809.
Stirling, G.R. and L.J. Smith. (1998). Field tests of formulated products containing either *Verticillium chlamydosporium* or *Arthrobotrys dactyloides* for biological control of root-knot nematodes. *Biological Control* 11: 231-239.
Stockli, A. (1952). Studien Uber Bodennematoden mit besonderer berucksichtigung des Nematodengehalters von Walot, Grunland und ackerbalich genutzten Boden. *Zeitschrift für Pflanzen Ernabr. Dung* 59: 97-139.
Tedford, E.C., B.A. Jaffee and A.E. Muldoon. (1993). Effect of temperature on infection of the cyst nematode *Heterodera schachtii* by the nematophagous fungus *Hirsutella rhossiliensis. Journal of Invertebrate Pathology* 66: 6-10.
Timper, P. and B.B. Brodie. (1993). Infection of *Pratylenchus penetrans* by nematode-pathogenic fungi. *Journal of Nematology* 25: 297-302.
Timper, P., H.K. Kaya and B.A. Jaffee. (1991). Survival of entomogenous nematodes in soil infested with the nematode-parasitic fungus *Hirsutella rhossiliensis* (Deuteromycotina: Hyphomycetes). *Biological Control* 1: 42-50.
Veerie, V.D., H. Annemie and V. Nicole. (2005). Long term efficacy of *Pochonia chlamydosporia* for management of *Meloidogyne javanica* in glasshouse crops. *Nematology* 7: 727-736.
Voss, B., P. Utermohl and U. Wyss. (1992). Variation between strains of the nematophagous endoparasitic fungus *Catenaria anguillulae* Sorokin II. Attempts to achieve parasitism of *Heterodera schachtii* Schmidt in pot trials. *Zeitschrift für Pflanzenkrankenheit und Pflanzenschutz* 99: 311-318.
Wang, K.H., B.S. Sipes and D.P. Schmitt. (2001). Suppression of *Rotylenchus reniformis* by *Crotalaria juncea, Brassica napus* and *Tagetes erecta. Nematropica* 31: 235-249.
Wang, K.H., B.S. Sipes and D.P. Schmitt. (2002). Management of *Rotylenchus reniformis* in pineapple, *Ananas comosus* by intercycle cover crops. *Journal of Nematology* 34: 106-114.
Wang, K.H., B.S. Sipes and D.P. Schmitt. (2003). Enhancement of *Rotylenchus reniformis* suppressiveness by *Crotalaria juncea* amendment in pineapple soils. *Agriculture Ecosystem Environment* 94: 197-203.

Chapter 7

Biofumigation to Manage Plant-Parasitic Nematodes

Antoon Ploeg
José-Antonio López-Pérez
Antonio Bello

INTRODUCTION

Annual yield losses worldwide attributed to plant-parasitic nematodes are estimated to range between 5 and 12 percent (Sasser and Freckman, 1987). Depending on climate, crops grown, nematode species and levels present, and economic factors, a number of tactics can be employed to minimize nematode damage. Preventing the introduction of nematodes with planting material, seeds, or soil, including nonhost crops in rotation schemes, using nematode-resistant varieties or rootstocks, and lowering nematode populations through nematicides are among the most frequently used strategies.

However, concern about the negative impact of synthetic nematicides on the environment and on general public health has led to a reevaluation of these methods. For example, high use of the soil fumigant methyl bromide and the resultant contamination of ground, surface, and drinking water in the Netherlands led to a ban on its use in the 1980s (Mus and Huygen, 1992). Later, methyl bromide was listed as an ozone-depleting compound at the fourth meeting of the Montreal Protocol in Copenhagen in 1992, and, in accordance with the U.S.

Management of Nematode and Insect-Borne Plant Diseases

doi:10.1300/5754_07

Clean Air Act, its use as a fumigant is now banned in developed nations (United Nations Environment Programme, 1992). Methyl bromide was previously used as a preplanting broad-spectrum soil fumigant to control soil-borne diseases, nematodes, insects, and weeds in high-value crops such as tomato, strawberry, cucurbits, nursery crops, and flowers and to avoid replanting problems in vineyards and orchards (Rodríguez-Kábana, 1997).

With the disappearance of nematicides or restrictions on their allowed use, interest in the development of safe, sustainable, and economically viable nematode management strategies has increased. One such strategy that potentially fulfills these requirements is biofumigation. This method was included as a nonchemical alternative to methyl bromide by the Methyl Bromide Technical Options Committee (MBTOC, 1997).

MECHANISM

Kirkegaard, Angus et al. (1993) described a process of "biological fumigation" that was later called "biofumigation" (Kirkegaard, Gardner et al., 1993; Matthiessen and Kirkegaard, 1993). The first article on biofumigation in a refereed international journal dealt with the inhibition of the fungus *Gaeumannomyces graminis* by root tissue of *Brassica* species (Angus et al., 1994). In that paper, biofumigation is referred to as the release of volatile breakdown products, mainly isothiocyanates (ITCs), from *Brassica* roots. Initially, the term was limited to "the suppression of soil-borne pests and pathogens by biocidal compounds released . . . when glucosinolates in *Brassica* . . . crops are hydrolized" (Kirkegaard and Sarwar, 1998).

The mechanism responsible for the biocidal effect of decomposing brassica crops is thought to be a chain of chemical reactions resulting in the formation of biologically active products (Underhill, 1980). Brassica crops contain glucosinolates located in the cell vacuoles. Glucosinolates are sulfur-containing stable and nontoxic compounds, but upon tissue damage they come into contact with myrosinase (=thioglucosidase), an enzyme endogenously present in brassica tissues but stored in the cell walls or the cytoplasm, away from the glucosinolates (Poulton and Moller, 1993). The enzymatic hydrolysis of glucosinolates produces volatile ITCs, nitriles, and thiocyanates (Cole, 1976; Fenwick, Heaney and Mullin, 1983). The ITCs in particular have general biocidal properties (Kirkegaard and Sarwar, 1998). ITCs also form the active ingredient of some synthetic nematicides (methyl isothiocyanate releasers).

There are more than 100 different glucosinolates (Manici et al., 2000; Underhill, 1980). A single *Brassica* species can contain several different types of glucosinolates (Sang et al., 1984), and the types and quantities of glucosinolates are highly variable between species and even varieties of brassica (Rosa et al., 1997). As a result, the quantities and types of biocidal ITCs resulting from the breakdown of glucosinolates are highly variable. Furthermore, the concentration of ITCs produced is also influenced by soil texture, moisture, temperature, microbial community, and pH (Bending and Lincoln, 1999; Price, 1999; Morra and Kirkegaard, 2002).

To increase the efficiency of biofumigation using *Brassica* species, initial research focused on the selection and characterization of varieties with high glucosinolate content (Kirkegaard and Sarwar, 1998, 1999; Potter, Davies and Rathjen, 1998; Potter et al., 2000). However, results from a number of studies indicated that when brassica material was added to soil, the actual conversion of glucosinolates into ITCs was very low and that the glucosinolate content in the biofumigant material does not predict the biocidal activity (Angus et al., 1994; Bending and Lincoln, 1999; Charron and Sams, 1999; McLeod and Steel, 1999; Lazzeri and Manici, 2001; Harvey, Hannahan and Sams, 2002; Morra and Kirkegaard, 2002). Compounds other than ITCs, such as alkenals or alkenols (Potter, Davies and Rathjen, 1998) or sulfur-containing compounds such as dimethyl-disulphide (Bending and Lincoln, 1999), may also play an important role. Finally, Morra and Kirkegaard (2002) reported that rather than the glucosinolate content, the most important factor limiting the efficacy of brassica biofumigants was the efficient disruption of the plant material.

The initial definition of "biofumigation" as the breakdown of brassica tissue was expanded by Halbrendt (1996) and Bello et al. (2004) to describe the process of biological decomposition of plant or animal by-products leading to the production of volatile compounds with disease and pest suppressive properties.

STUDIES INVOLVING NEMATODES

The negative impact of brassica tissues on soil-borne pathogens and parasites has repeatedly been demonstrated and was reviewed by Brown and Morra (1997). Mojtahedi et al. (1991, 1993) were among the first to study in detail the effects of amending soil with brassica tissue on the plant-parasitic nematode *Meloidogyne chitwoodi*. They reported that incorporation of *Brassica napus* shoots into *M. chitwoodi*-infested soil reduced nematode levels to very low levels and that amendment with *B. napus* was more effective than amendment with wheat or corn. The effect of the amendment, however, did not extend into the soil layers below the zone of incorporation, and the amendment was more effective against second-stage juveniles than against egg masses. The amendment protected host plant roots growing in the zone of *B. napus* incorporation from nematode infestation for up to six weeks. Soil incorporation rates of 4 percent (w/w) killed nearly all second-stage juveniles, whereas rates of 6 percent were required to prevent hatching of juveniles from egg masses (Mojtahedi et al., 1991, 1993). Stapleton and Duncan (1998) used biofumigation with brassica tissue to control *Meloidogyne incognita* and reported that the efficacy was much higher when soils were heated to sublethal temperatures. Similar results were obtained by Ploeg and Stapleton (2001), who reported that biofumigation with broccoli failed to control *M. incognita* or *Meloidogyne javanica* at 20°C but resulted in near-complete control at temperatures of 30 and 35°C. In addition, they found that the time required to achieve control was shorter as soil temperatures increased and concluded that biofumigation to control *M. incognita* or *M. javanica* should be done when soil temperatures of approximately 25°C can be achieved. Bello et al. (2004) also recommended using biofumigation when soil temperatures are above 20°C. In contrast, Mojtahedi et al. (1993) reported that *M. chitwoodi* was controlled by brassica amendments at average soil temperatures of 19°C and speculated that the slower decomposition of the plant tissues under cool conditions resulted in an extended period of nematode control. However, the fact that *M. incognita* and *M. javanica* are nematodes typical of (sub)tropical climates, whereas *M. chitwoodi* generally occurs in temperate climates, might be partly responsible for the differences in soil temperature requirements. The temperature at which development starts is much lower for *M. chitwoodi* than for *M. javanica* or *M. incognita* (Ploeg and Maris, 1999). Thus, at soil temperatures of 20°C, it is likely that more *M. chitwoodi* than *M. incognita* or *M. javanica* second-stage juveniles would have hatched from egg masses. As second-stage juveniles are more susceptible to biofumigation than egg masses (Mojtahedi et al., 1993), this could in part explain the success of biofumigation against *M. chitwoodi* at relatively low soil temperatures where biofumigation against *M. incognita* or *M. javanica* was not effective. McLeod and Steel (1999) significantly reduced *M. javanica* soil levels by incorporation of plant tissue from a range of *Brassica* species at rates of 1-2 percent (w/w) and produced evidence from in vitro experiments that glucosinolate-derived volatiles played an important role. However, they also concluded that in a soil environment other mechanisms, possibly the stimulation of nematode-antagonistic organisms, played an important role in the observed nematicidal effects (McLeod and Steel, 1999).

Thus, the cultivation of brassicas as green manure crops and their subsequent soil incorporation as a biofumigant appears to be an attractive option to control root-knot nematode species. However, as most brassicas are hosts to root-knot nematodes (McSorley and Frederick, 1995; McLeod and Steel, 1999; McLeod, Kirkegaard and Steel, 2001), there is a risk of nematode increase rather than decrease (McLeod and Warren, 1993; McLeod and Steel, 1999; Stirling and Stirling, 2003). To avoid nematode build-up on brassicas, it is advisable to grow crops during cool seasons, when nematode multiplication is slow, and to focus research on identifying brassicas with high glucosinolate content and resistance to root-knot nematodes (Kirkegaard and Sarwar, 1998; McLeod and Steel, 1999; Stirling and Stirling, 2003).

Reports on the management of other plant-parasitic nematodes using soil incorporation of brassica tissue include that by Halbrendt (1996), who reported a lowering of *Xiphinema americanum* population levels after incorporation of a rapeseed green manure crop into infested orchard soil. Several reports by Potter, Davies and Rathjen (1998) and Potter et al. (1999, 2000) discuss the control of *Pratylenchus neglectus* using canola *(B. napus)* and other *Brassica* species.

The management of soil-borne pathogens and pests, including plant-parasitic nematodes, by amending soil with organic material is a well-known and long-practiced strategy that has been reviewed by Bridge (1996), Stirling (1991), Hoitink (1988), and Lazarovitis et al. (2001). According to Bello, López-Pérez and Díaz-Viruliche (2000) and Bello et al. (2000, 2004), biofumigation is not limited to using brassica residues; all organic materials can be used as biofumigants. As biofumigation relies on the production of volatile substances during the decomposition process, organic material used for biofumigation should not be (fully) decomposed before use and should preferably have a C:N ratio between 8 and 20. In addition, after incorporation of organic matter, soils should be watered to field capacity and sealed with plastic to increase soil temperatures and trap developing gases (Rodríguez-Kábana, 1997; Bello et al., 2004). A general recommended dose for incorporation is 50 tonnes/ha, although the efficacy of materials can vary depending on the biochemical and biological properties and the method of application (Bello et al., 2004). For example, Bello et al. (2004) tested the biofumigant effect of a range of agroindustrial by-products and livestock manures in different doses and combinations on levels of *M. incognita* control and concluded that the majority of materials could effectively be used. In commercial greenhouse trials in Spain, with an integrated management system including biofumigation with sheep manure and mushroom residue and the cultivation of short-cycle vegetables acting as trap crops, initially very high levels of *M. incognita* were reduced to close to zero in the main cucumber and tomato crops (Bello, 1998). Similarly, biofumigation combined with soil solarization in a greenhouse in Spain provided levels of *M. incognita* control in bell pepper that were similar to levels achieved using methyl bromide (Bello et al., 2004). In potato field trials in Idaho and Washington, cropping sequences including the cultivation and incorporation of sudangrass, a crop also known to release nematicidal compounds, dramatically reduced *M. chitwoodi* infestation levels (Riga et al., 2004). Bello et al. (2004), from results obtained in a vineyard with a high incidence of grapevine fanleaf virus and its vector nematode species *Xiphinema index,* suggested that biofumigation may also be useful to reduce the fallow period necessary to eliminate *X. index* virus-vector nematodes in uprooted vineyards. Bello et al. (2004) also obtained promising levels of *M. incognita* and *M. javanica* control in banana plantations in the Canary Islands, of *M. incognita* in peach orchards, and of the citrus nematode *Tylenchulus semipenetrans* in an uprooted orange orchard in Spain with biofumigation using urban waste and urea, or manure and banana residue.

CONCLUSION AND OUTLOOK

Biofumigation using brassica tissue or other sources of organic material appears to be a promising strategy for the management of soil-borne diseases, pests, and weeds. The general benefits of amending soil with organic matter are well known and include improvements in the soil nutrient status and water-holding capacity and an increase in the presence and activity of beneficial soil organisms, including those that are antagonistic to plant-parasitic nematodes. In addition, it may provide a use for agroindustrial and municipal "waste" products (Stirling, 1991; Bridge, 1996; Lazarovits, Tenuta and Conn, 2001). The mechanisms of pest and pathogen control by biofumigation are still largely unknown and, although the production of biocidal gases is undoubtedly important, several researchers have indicated that other mechanisms also likely play an important role (Potter, Davies and Rathjen, 1998; Bending and Lincoln, 1999). In fact, a few studies have compared the efficacy of biofumigation under plastic to trap gases and biofumigation without plastic. In one study, biofumigation with manure under plastic to control *M. incognita* in tomato gave only a slight reduction in root-galling indices compared with biofumigation without plastic, and *M. incognita* soil populations were controlled to very similar levels by both methods (Bello, 1998). It is unlikely that biofumigation alone will provide sufficient levels of nematode control over multiple seasons, but advantages are that this method is also useful in the management of other soil-borne problems and that it can easily be combined with other strategies such as soil solarization and the use of resistant varieties.

REFERENCES

Angus, J.F., P.A. Gardner, J.A. Kirkegaard and J.M. Desmarchelier. (1994). Biofumigation: Isothiocyanates released from *Brassica* roots inhibit growth of the take-all fungus. *Plant and Soil* 162: 107-112.

Bello, A. (1998). Biofumigation and integrated crop management. In *Alternatives to Methyl Bromide for the Southern European Countries.* A. Bello, J.A. González,

M. Arias and R. Rodríguez-Kábana, eds. Valencia, Spain, Phytoma-España, DG XI EU, CSIC, pp. 99-126.

Bello, A., J.A. López-Pérez and L. Díaz-Viruliche. (2000). Biofumigación y solarización como alternativas al bromuro de metilo. In *Memorias del Simposium Internacional de la Fresa.* J.Z. Castellanos and F. Guerra O'Hart, eds. Zamora, México, pp. 24-50.

Bello, A., J.A. López-Pérez, A. García-Álvarez, R. Sanz and A. Lacasa. (2004). Biofumigation and nematode control in the Mediterranean region. In *Proceedings of the Fourth International Congress of Nematology.* R.C. Cook and D.J. Hunt, eds. June 8-13, 2002, Tenerife, Spain. Nematology monographs and perspectives, Vol. 2. Brill, Leiden and Boston, pp. 133-149.

Bello, A., J.A. López-Pérez, R. Sanz, M. Escuer and J. Herrero. (2000). Biofumigation and organic amendments. Regional workshop on methyl bromide alternatives for North Africa and Southern European countries. Paris, France, United Nations Environment Programme (UNEP), pp. 113-141.

Bending, G.D. and S.D. Lincoln. (1999). Characterisation of volatile sulphur-containing compounds produced during decomposition of *Brassica juncea* tissues in soil. *Soil Biology and Biochemistry* 31: 695-703.

Bridge, J. (1996). Nematode management in sustainable and subsistence agriculture. *Annual Review of Phytopathology* 34: 201-225.

Brown, P.D. and M.J. Morra. (1997). Control of soil-borne plant pests using glucosinolate-containing plants. *Advances in Agronomy* 61: 167-231.

Charron, C.S. and C.E. Sams. (1999). Inhibition of *Pythium ultimum* and *Rhizoctonia solani* by shredded leaves of *Brassica* species. *Journal of the American Society of Horticultural Sciences* 124: 462-467.

Cole, R.A. (1976). Isothiocyanates, nitriles and thiocyanates as products of autolysis of glucosinolates in Cruciferae. *Phytochemistry* 15: 759-762.

Fenwick, G.R., R.K. Heaney and W.J. Mullin. (1983). Glucosinolates and their breakdown products in food and food plants. In *Critical Reviews in Food Science and Nutrition.* T.E. Furia, ed. CRC Press, Boca Raton, pp. 123-201.

Halbrendt, J.M. (1996). Allelopathy in the management of plant-parasitic nematodes. *Journal of Nematology* 28: 8-14.

Harvey, S.G., H. Hannahan and C.E. Sams. (2002). Indian mustard and allyl isothiocyanate inhibit *Sclerotium rolfsii. Journal of the American Society of Horticultural Sciences* 127: 27-31.

Hoitink, H.A. (1988). Basis for the control of soilborne plant pathogens with composts. *Annual Review of Phytopathology* 24: 93-114.

Kirkegaard, J.A., J.F. Angus, P.A. Gardner and H.P. Cresswell. (1993). Benefits of brassica break crops in the Southeast wheat belt. *Proceedings of the Seventh Australian Agronomy Conference,* Adelaide, Australia, pp. 282-285.

Kirkegaard, J.A., P.A. Gardner, J.M. Desmarchelier and J.F. Angus. (1993). Biofumigation—Using *Brassica* species to control pests and diseases in horticulture and agriculture. In *Proceedings of the Ninth Australian Assembly on Brassicas.* N. Wratten and R. Mailer. Wagga Wagga, eds. 5-7 October, 1993, pp. 77-82.

Kirkegaard, J.A. and M. Sarwar. (1998). Biofumigation potential of brassicas. *Plant and Soil* 201: 71-89.

Kirkegaard, J.A. and M. Sarwar. (1999). Glucosinolate profiles of Australian canola (*Brassica napus annua* L.) and Indian mustard (*Brassica juncea* L.) cultivars: Implications for biofumigation. *Australian Journal of Agricultural Research* 50: 315-324.

Lazarovits, G., M. Tenuta and K.L. Conn. (2001). Organic amendments as a disease control strategy for soilborne diseases of high-value agricultural crops. *Australasian Plant Pathology* 30: 111-117.

Lazzeri, L. and L.M. Manici. (2001). Allelopathic effect of glucosinolate-containing plant green manure on *Pythium* sp. and total fungal population in soil. *HortScience* 36: 1283-1289.

Manici, L.A., L. Lazzeri, G. Baruzzi, O. Leoni, S. Galletti and S. Palmieri. (2000). Suppressive activity of some glucosinolate enzyme degradation products on *Pythium irregulare* and *Rhizoctonio solani* in sterile soil. *Pest Management Science* 56: 921-926.

Matthiessen, J.N. and J.A. Kirkegaard. (1993). Biofumigation, a new concept for "clean and green" pest and disease control. *Western Australian Potato Grower,* October:14-15.

MBTOC. (1997). Report of the Methy Bromide Technical Options Committee. Nairobi, Kenya, UNEP, 221p.

McLeod, R.W., J.A. Kirkegaard and C.C. Steel. (2001). Invasion, development, growth and egg laying by *Meloidogyne javanica* in Brassicaceae crops. *Nematology* 3: 463-472.

McLeod, R.W. and C.C. Steel. (1999). Effects of brassica-leaf green manures and crops on activity and reproduction of *Meloidogyne javanica. Nematology* 1: 613-624.

Mcleod, R. and M. Warren. (1993). Effects of cover crops on inter-row nematode infestation in vineyards. 1. Relative increase of root knot nematode *Meloidogyne incognita* and *M. javanica* on legume, cereal and brassica crops. *The Australian Grapegrower and Winemaker* 357: 28-30.

McSorley, R. and J.J. Frederick. (1995). Responses of some common Cruciferae to root-knot nematodes. *Journal of Nematology* 27: 550-554.

Mojtahedi, H., G.S. Santo, A.N. Hang and J.H. Wilson. (1991). Suppression of root-knot nematode populations with selected rapeseed cultivars as green manure. *Journal of Nematology* 23: 170-170.

Mojtahedi, H., G.S. Santo, J.H. Wilson and A.N. Hang. (1993). Managing *Meloidogyne chitwoodi* on potato with rapeseed as green manure. *Plant Disease* 77: 42-46.

Morra, M.J. and J.A. Kirkegaard. (2002). Isothiocyanate release from soil-incorporated *Brassica* tissues. *Soil Biology and Biochemistry* 34: 1683-1690.

Mus, A. and C. Huygen. (1992). Methyl bromide. The Dutch environmental situation and policy. TNO. Institute of Environmental Sciences. Order No. 50554, 13p.

Ploeg, A.T. and P.C. Maris. (1999). Effects of temperature on the duration of the life cycle of a *Meloidogyne incognita* population. *Nematology* 1: 389-393.

Ploeg, A.T. and J.J. Stapleton. (2001). Glasshouse studies on the effects of time, temperature and amendment of soil with broccoli plant residues on the infestation of melon plants by *Meloidogyne incognita* and *M. javanica. Nematology* 3: 855-861.

Potter, M.J., K. Davies and A.J. Rathjen. (1998). Suppressive impact of glucosinolates in *Brassica* vegetative tissues on root lesion nematode *Pratylenchus neglectus*. *Journal of Chemical Ecology* 24: 67-80.

Potter, M.J., V. Vanstone, K. Davies, J. Kirkegaard and A.J. Rathjen. (1999). Reduced susceptibility of *Brassica napus* to *Pratylenchus neglectus* in plants with elevated root concentrations of 2-phenylethyl glucosinolate. *Journal of Nematology* 31: 291-298.

Potter, M.J., V.A. Vanstone, K.A. Davies and A.J. Rathjen. (2000). Breeding to increase the concentration of 2-phenylethyl glucosinolate in the roots of *Brassica napus*. *Journal of Chemical Ecology* 26: 1811-1820.

Poulton, J.E. and B.L. Moller. (1993). Glucosinolates. In *Methods in Plant Biochemistry*. P.J. Lea, ed. Vol 9. PJ Academic press, London, pp. 209-237.

Price, A. (1999). *Quantification of Volatile Compounds Produced During Simulated Biofumigation Utilizing Indian Mustard Degrading in Soil Under Different Environmental Conditions*. MS Thesis. University of Tennessee, Knoxville.

Riga, E., H. Mojtahedi, R. Ingham and A. McGuire. (2004). Green manure amendments and management of root-knot nematodes on potato in the Pacific North West of USA. In *Proceedings of the Fourth International Congress of Nematology*. R.C. Cook and D.J. Hunt, eds. 8-13 June, 2002, Tenerife, Spain. Nematology monographs and perspectives, Vol. 2. Brill, Leiden and Boston, pp. 151-158.

Rodríguez-Kábana, R. (1997). Alternatives to methyl bromide (MB) soil fumigation. In *Alternatives to Methyl Bromide for the Southern European Countries*. A. Bello, J.A. González, M. Arias and R. Rodríguez-Kábana, eds. Valencia, Spain, pp. 17-33.

Rosa, E.A.S., R.K. Heaney, G.R. Fenwich and C.A.M. Portas. (1997). Glucosinolates in crop plants. *Horticultural Reviews* 19: 99-215.

Sang, J.P., I.R. Minchinton, P.K. Johnstone and R.J.W. Truscott. (1984). Glucosinolate profiles in the seed, root and leaf tissue of cabbage, mustard, rapeseed, radish and swede. *Canadian Journal of Plant Science* 64: 77-93.

Sasser, J.N. and D.W. Freckman. (1987). A world perspective of nematology: The role of the Society. In *Vistas on Nematology*. J.A. Veech and D.W. Dickson, eds. Society of Nematology, Hyattsville, MD, pp. 7-14.

Stapleton, J.J. and R.A. Duncan. (1998). Soil disinfestation with cruciferous amendments and sublethal heating: Effects on *Meloidogyne incognita, Sclerotium rolfsii* and *Pythium ultimum*. *Plant Pathology* 47: 737-742.

Stirling, G.R. (1991). *Biological Control of Plant-Parasitic Nematodes: Progress, Problems and Prospects*. CAB International, Wallingford, UK, 282p.

Stirling, G.R. and A.M. Stirling. (2003). The potential of *Brassica* green manure crops for controlling root-knot nematode (*Meloidogyne javanica*) on horticultural crops in a subtropical environment. *Australian Journal of Experimental Agriculture* 43: 623-630.

Underhill, E.W. (1980). Glucosinolates. In *Secondary Plant Products. Encyclopedia of Plant Physiology*. E.A. Bell and B.V. Charlwood, eds. New Series, Vol. 8. Springer-Verlag, Berlin, pp. 493-511.

Chapter 8

Potato Early Dying Complex: Key Factors and Management Practices

Nandita Selvanathan

INTRODUCTION

The potato (*Solanum tuberosum* L.) is an annual, dicotyledonous plant and is the fourth-largest food crop in the world, exceeded only by wheat, maize, and rice (Stevenson et al., 2001). Potato is subject to several diseases caused by fungi, bacteria, viruses, and nematodes. In the arid and semiarid regions of the world, Potato Early Dying, also known as Early Dying Complex or *Verticillium* Wilt, is a common limiting factor in potato production (Rowe et al., 1987) and is endemic in many potato production areas of North America.

When Potato Early Dying develops throughout a field midway through the growing season and then becomes severe during the period of maximum tuber bulking, a significant reduction in tuber size and total marketable yield can result (Rowe and Powelson, 2002). In North America, yield reduction in moderately diseased fields can easily be 10-15 percent and in severly diseased fields it can be as high as 30-50 percent (Powelson and Rowe, 1993). The economic impact of Potato Early Dying is significant because of the direct losses from low yields (Rowe et al., 1987). This chapter discusses the symptoms of the disease, causal agents, and management practices helpful in controlling this economically important disease complex.

Management of Nematode and Insect-Borne Plant Diseases

doi:10.1300/5754_08

SYMPTOMS

Symptoms of Potato Early Dying are highly variable and may be associated with other diseases or physiological problems. A typical visual symptom of Potato Early Dying is loss of plant vigor during mid- to late summer followed by senescence and death of the crop before normal maturity. The symptoms are difficult to distinguish from normal senescence and may initially involve reduced growth. Initially the disease causes uneven chlorosis of the lower leaves on a few plants, but these early stages of Potato Early Dying may be confused with normal senescence of the lower leaves of healthy plants. Areas between the leaf veins turn yellow and later brown. Later wilting of the lower leaflets of the plants may occur, and infection proceeds up the infected stems, which often remain erect. The disease causes vascular discoloration of stems at their tuber ends and plants may die without reaching maturity. It is not uncommon for one or two stems of the plant to die and the other portions of the plant to remain healthy (Rangahau, 2003). The symptoms often occur on one side of the plant or on individual leaves (Powelson and Rowe, 1993). A light brown vascular discoloration in basal stem tissues is usually present in symptomatic plants. This symptom, however, is not diagnostic, as vascular discoloration can result from stress factors unrelated to the disease. As the disease develops within a field, widely scattered individual plants or groups of plants often show early symptoms, resulting in a nonuniform decline of the stand. In severe cases, plants across an entire field will die over a period of several weeks (Rowe and Powelson, 2002). Advanced symptoms usually do not occur until after flowering and may consist of decline of isolated plants or, in severe cases, early maturity of the entire crop (Krikun and Orion, 1979; Davis, 1981; Mace et al., 1981; Rowe, 1983).

ROLE OF BIOTIC FACTORS IN DISEASE DEVELOPMENT

Potato Early Dying is caused by soil-borne fungi and is a monocyclic disease, meaning that only one cycle of infection, pathogen growth, and reproduction occurs during each cropping season.

A major contributor to this syndrome is the soil-borne fungus *Verticillium*. Two species of the fungus, *Verticillium dahliae* and *Verticillium albo-atrum,* cause the disease. Both species naturally occur at low levels in soils and can attack susceptible crops when these are planted. *V. dahliae* grows better at slightly higher temperatures (25-28°C), whereas *V. albo-atrum* may be the dominant pathogen in cooler production areas of the most northern United States and southern Canada, where average soil temperature during the growing season rarely exceeds 25°C (Powelson and Rowe, 1993; Rowe et al., 1987). The two species of *Verticillium* differ from each other in the type of infective propagules on the dying potato plants. Microsclerotia are formed by *V. dahliae* (Figure 8.1), and melanized hyphae are formed by *V. albo-atrum* (Powelson, Johnson and Rowe, 1993). *V. dahliae* survives from season to season as microsclerotia in the soil, free or embedded in the plant debris, or as mycelium in the vascular ring of the tuber (Rowe and Powelson, 2002). These microsclerotia are stimulated to germinate in response to root exudates (Mace et al., 1981; Figure 8.2). The fungus infects newly planted potatoes and alternative hosts through young developing root tissues. As the disease progresses, *Verticillium* mycelium invades dying tissue, forming microsclerotia or melanized hyphae that are released back into the soil as the plant decays (Sturz and Clark, 2000). The species have similar vegetative hyphae and produce conidiophores that are septate with

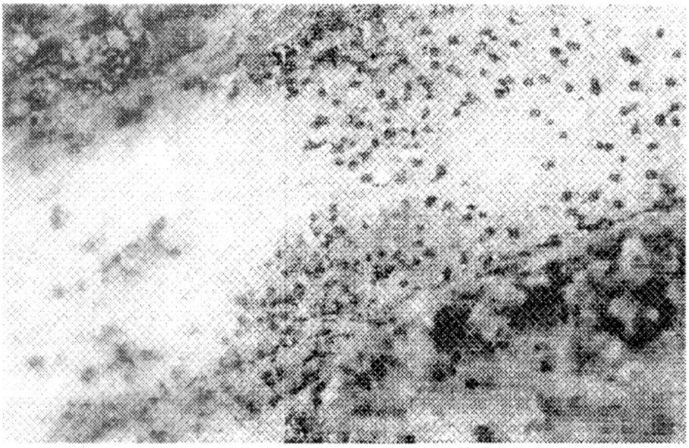

FIGURE 8.1. Microsclerotia of *Verticillium dahliae* formed on the surface of a dying potato stem (photograph courtesy of R.C. Rowe). Reprinted with permission of the American Phytopathological Society.

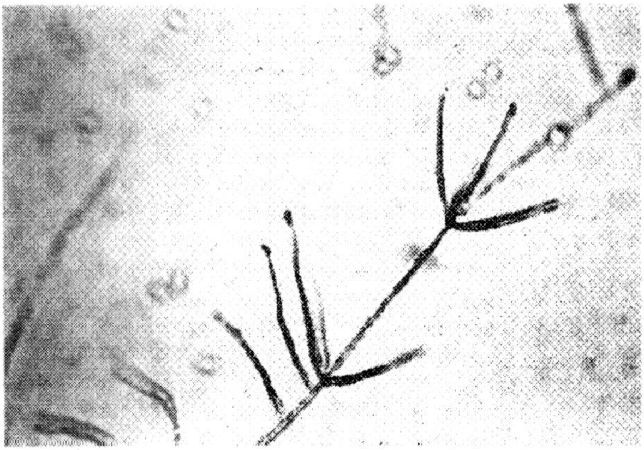

FIGURE 8.2. Conidiophores of *Verticillium dahliae* and *Verticillium albo-atrum* are septate with side branches, swollen at the base, and arranged in verticillate whorls (photograph courtesy of H. Torres). Reprinted with permission of The American Phytopathological Society.

side branches, swollen at the base, and arranged in verticilliate whorls (Figure 8.3). Although the disease is primarily caused by the vascular wilt pathogen *V. dahliae,* coinfection of potato by the root-lesion nematode *Pratylenchus* can increase both the severity of the disease and tuber loss (LaMondia, 1999; Back, Haydock and Jenkinson, 2002). Nematodes are microscopic wormlike animals ranging from 0.004 to 0.04 inches in length (Figure 8.4). Their life cycles, which include egg, juvenile, and adult stages, are completed in twenty to fifty days depending on the temperature (MacGuidwin, 1993). *Pratylenchus penetrans* is the most economically important species in eastern North America (Mai, Bloom and Chen, 1977).

Kimpinski (1998) reported that root-lesion nematodes (*Pratylenchus* spp.) are the most prevalent nematode parasites in roots and soils in potato fields in the Maritime region of Canada. These are migratory endoparasites that can enter and feed upon root tissues and also can move freely through soil from root to root (Rowe and Powelson, 2002; Figure 8.5).

Nematodes are thought to provide entry points for the fungus while feeding on potato roots, or they may reduce the natural host defense mechanism of the potato plant, allowing the fungus to attack roots and infect the host (Ranghau, 2003). However, the findings of

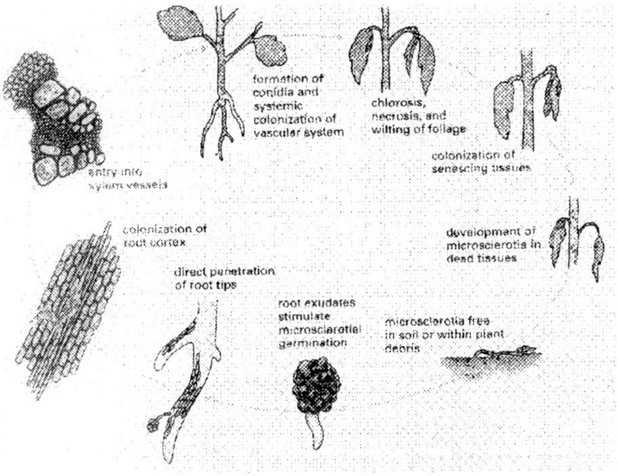

FIGURE 8.3. Potato Early Dying is a monocyclic disease with only one cycle of disease and inoculum production per season (courtesy of R. C. Rowe). Reprinted with permission of The American Phytopathological Society.

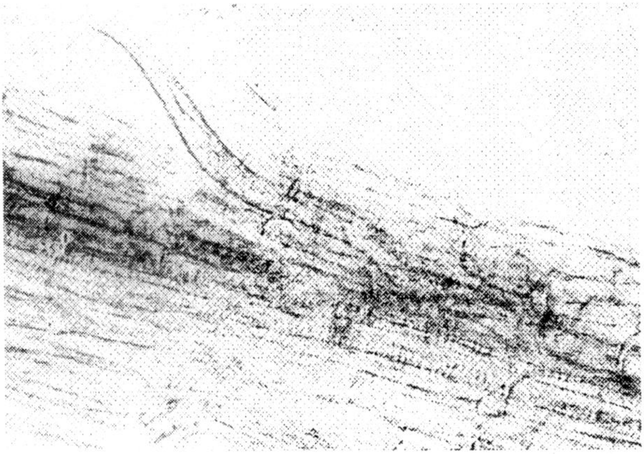

FIGURE 8.4. Necrosis caused by feeding by the root-lesion nematode *Pratylenchus penetrans* (photograph courtsey of J. Bowers). Reprinted with permission of The American Phytopathological Society.

Rotenberg et al. (2004) indicate that the role of the nematode in the fungus-host interaction is more than simply to facilitate extravascular and/or vascular entry of the fungus into potato roots. Other studies have shown that *V. dahliae* and root-lesion nematodes can interact

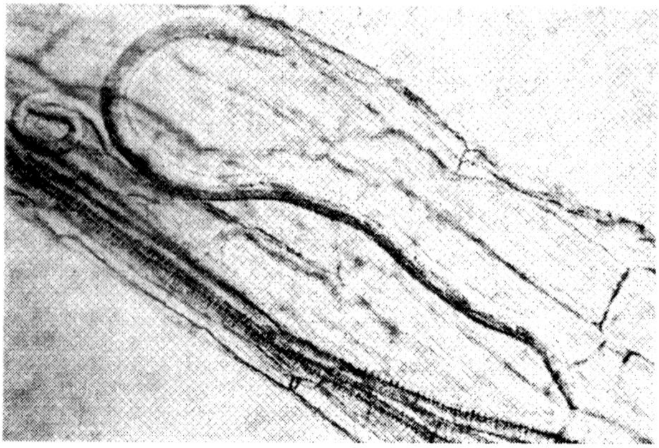

FIGURE 8.5. Adult and juvenile root-lesion nematode *(Pratylenchus penetrans)* inside root tissue (photograph courtesy of J. Bowers). Reprinted with permission of The American Phytopathological Society.

synergistically; together they cause severe symptom development and significant yield reduction at population levels that have little or no effect when each organism is present individually. The involvement of the root-lesion nematode *Pratylenchus* in Potato Early Dying is important because of its ability to activate relatively low populations of *V. dahliae* that would otherwise not be biologically significant (Martin, Riedel and Rowe, 1982; Bowers et al., 1996).

Root injury caused by nematode feeding has been considered a likely mechanism for the synergism of *V. dahliae* and *P. penetrans* in the disease. It is possible that wounded roots either stimulate the germination of dormant microsclerotia by increased root exudation or facilitate access for the fungus to the vascular cylinder; these actions may also occur together. Klinkenberg (1963) and Muller (1977) reported that the two species of *Pratylenchus (P. penetrans* and *Pratylenchus crenatus)* have similar feeding habits and reproduce well on potato roots (Riedel, Rowe and Martin, 1985), but only *P. penetrans* interacts with *V. dahliae* in this disease syndrome (Riedel, Rowe and Martin, 1985; Rowe, Riedel and Marin, 1985).

Similar results have been shown by Rowe et al. (1987), MacGuidwin and Rouse (1990), Botesas and Rowe (1994), Wheeler and Reidel (1994), and Bowers et al. (1996). Kotcon, Rouse and Mitchell (1985) found that the effects of the root-lesion nematode

P. penetrans in combination with *V. dahliae* on symptom expression were additive. They carried out an experiment to determine the effect of *V. dahliae, Colletotrichum coccodes, Rhizoctonia solani,* and *P. penetrans* singly and in all possible combinations with the 'Russet Burbank' variety of potato. *P. penetrans* caused stunting, chlorosis, and premature senescence but did not reduce yield. Plants in treatments containing *V. dahliae* had reduced root growth, foliage weight, and tuber yield. *C. coccodes* and *R. solani* had no effect on disease symptoms, plant growth, or yield. Interactions of *P. penetrans, C. coccodes,* or *R. solani* with *V. dahliae* did not result in significant yield reductions, whereas the effects of *P. penetrans* in combination with *V. dahliae* on symptom expression were additive. The effect of coinoculation by two fungi, *C. coccodes* and *V. dahliae,* has been studied by Tsror and Hazanovsky (2001), whose results showed that disease symptoms and occurrence of sclerotia were similar in plants inoculated with the combination and with a single pathogen. Compared with the effects of inoculation with either pathogen, simultaneous inoculation with both pathogens can, in some cultivars, increase the incidence of Potato Early Dying syndrome and thus severely decrease yields. As with nearly all investigations in science, many reports on disease complexes contradict one another.

Although some of these disparities might be explained by experimental procedure and accuracy, some findings highlight the specificity of certain disease complexes and the influence of biotic and abiotic factors on them. This is exemplified by studies on the *V. dahliae-Pratylenchus* complex of potato, where it has been found that the interaction between these organisms varies among different nematode species (Riedel, Rowe and Martin, 1985) and populations, as well as fungal genotypes (Botseas and Rowe, 1994). For example, Riedel, Rowe and Martin (1985) and Bowers et al. (1996) observed that Potato Early Dying disease was enhanced by populations of *P. penetrans* but not by *P. crenatus* or *Pratylenchus scribneri*.

Furthermore, greenhouse experiments undertaken by Hafez et al. (1999) demonstrated that populations of *Pratylenchus neglectus* collected from Ontario, Canada, interacted synergistically with *V. dahliae,* whereas populations of *P. neglectus* from Parma, Idaho, did not increase disease or yield loss any more than treatment with *V. dahliae* alone.

ENVIRONMENTAL FACTORS

It has long been understood that the development of disease symptoms is not determined solely by the pathogen(s) responsible but is dependent on the complex interrelationship between host, pathogen, and prevailing environmental conditions. Understanding the effects of environment on the severity of Potato Early Dying and associated yield reduction is hampered by the long time span of disease progression (Francl et al., 1990). The development of Potato Early Dying is affected by external abiotic factors, including temperature and moisture. The former has been reported to be the most important factor that influences the severity of Potato Early Dying (Rowe, Riedel and Marin, 1985; Johnson, 1988; Francl, 1990). Geographical distribution of the pathogens, development and severity of pathogens, development and severity of Potato Early Dying, and effects of the disease on yield are all related to temperature during the potato growing season. Disease severity in potatoes infected with *V. dahliae* tends to increase as the mean temperature rises from 20 to 28°C. Symptom development was arrested when the temperature was lowered from 20 to 13°C (Nnodu and Harrison, 1979). Similar results were reported by Sturz and Clark (2000) concerning the effect of dry summer conditions on disease severity. In a multilocational study in Ohio, yields were 24-37 percent lower for potatoes grown in fumigated microplots infested with *V. dahliae* when the July-August temperature was 20°C. However, *V. dahliae* in the same experimental design had little effect on tuber yield when the average July-August temperature was 20°C (Rowe, Riedel and Marin, 1985). In a detailed eight-year study in two locations, Francl et al. (1990) suggested that periods of high temperature have a negative effect on yield of infected plants.

Soil moisture is another factor that influences the development of Potato Early Dying. An increase in soil water was related to an increase in the disease severity (Cappaert et al., 1992). Similar results of positive correlation between soil moisture and disease severity were reported by Davis et al. (1990), McLean (1955), and Rowe et al. (1987).

As mentioned, the disease progression of Potato Early Dying is rather slow, and this hampers the study of the effect of environment on disease development and severity. Experimental research is required to arrive at definitive conclusions about the effects of environment on Potato Early Dying.

CONTROL/MANAGEMENT FACTORS

Selection of control strategies for Potato Early Dying must be based on an understanding of the biology of the pathogens and the effects of various agricultural practices (Rowe et al., 1987). Management practices are aimed at reducing the population of these pathogens in soil or changing the susceptibility of the host. Although there is interest in management of Potato Early Dying, simply confirming the presence of *V. dahliae* and *P. penetrans* in soil is not sufficient cause to implement expensive management strategies (Wheeler et al., 1994). No single practice will control Potato Early Dying. Therefore integrated management systems are emphasized in which several decisions are implemented throughout multiple-year rotational cropping programs. Cultural practices are an important component of integrated crop management strategies. Management practices that modify the rate of stem colonization by *V. dahliae* or increase the host's tolerance to colonization also may be useful control measures for this disease (Rowe et al., 1987).

Potential components of integrated systems are cultural methods such as crop rotation, green manures, fertilizers, irrigation, and vine removal, along with host resistance and biological controls (Rangahau, 2003). Management practices for the suppression of *Verticillium* wilt of 'Russet Burbank' potato include sanitation, optimized sprinkler irrigation practices, soil solarization, and an adequate soil fertility program (Davis et al., 1990).

Disease Resistance

Planting resistant cultivars is the most efficient economical and environmentally sound way of managing this disease. Studies from the United States have shown that planting resistant cultivars for five seasons can reduce pathogen levels by 60-70 percent compared with growing only susceptible cultivars (Selvanathan, 2003). Unfortunately, not many cultivars are resistant to this disease, and even the most resistant cultivars will be affected by the disease under favorable environmental conditions and when the soil inoculum levels are very high (Ranghau, 2003). Davis (1985) reported that market acceptance of these cultivars has not been widespread.

Sanitation and Soil Fumigation

Soil fumigation is the primary means of managing this disease, but cost may limit its use (Ohms, 1962). The burning of potato vines may be of value as a first line of defense by delaying the build up of soil-borne inoculum and improving control of the disease (Mpofu and Hall, 2002). Also, *Verticillium*-free seed may reduce the rate of inoculum increase. This is particularly true on new potato land and land that has been recently fumigated with an effective fungicide (e.g., chloropicrin, metam-sodium).

Irrigation

Maintaining plant vigor with careful irrigation is a good management strategy for controlling the disease. In 'Russet Burbank' potato, early dying can be suppressed if water is reduced before tuber initiation. If the potato cultivar has mechanisms at the root to tolerate mild moisture deficit stress, then some of these same mechanisms may be responsible for resisting infection to *V. dahliae*. Disease severity was greater throughout each season when plants grown in infested soil were irrigated excessively than when they received deficit irrigation. Increased disease severity under excessive irrigation was not related to nitrogen deficiency because all petiole nitrate-nitrogen concentrations exceeded critical levels (Arbogast et al., 1999).

In Idaho, the disease occurred earlier and symptoms were more severe during a growing season with wet soil conditions than in a dry season (Cappart et al., 1992). These studies showed that in irrigated regions, the amount of water applied can be manipulated by both frequency and duration of irrigation to reduce Potato Early Dying and associated yield loss. High soil-water content early in the season may be critical to the onset of infection of potato plants, whereas water stress in plants after infection may enhance symptom expression. Soil-water content must remain optimal after tuberization to prevent tuber malformation. Water management before tuberization, however, may be a viable strategy to manage this disease (Cappaert et al.,

1992). In Wisconsin, tuber yield was significantly lower in deficit than in moderate or excessive pre-tuber initiation irrigation regimes (Cappart et al., 1994). This microplot study showed that onset of senescence indicative of Potato Early Dying was earlier and severity was greater in plants irrigated in excess of estimated consumptive use before tuber initiation. Irrigation treatment after tuber initiation had no effect on rates of foliar senescence. In this study, plant senescence was suppressed when soil moisture levels were kept below 100 percent of estimated consumptive use before tuber initiation, whereas the amount of water applied after tuber initiation appeared to have no effect on the rate of plant senescence (Cappaert et al., 1994).

The volume and timing of irrigation have been used in Oregon as a management tool to suppress the disease. Where potatoes are grown under irrigation, the amount of water applied can be reduced early in the season, before tuber initiation, to minimize water use by the crop without sacrificing tuber yield or quality. Increased frequency of irrigation can improve potato root growth through the soil profile, increasing the probability of the roots coming into contact with *Verticillium* and hence of greater disease levels (Rangahau, 2003).

Crop Rotation

Considerable controversy can be found in the literature regarding the usefulness of rotation as a means of control (Busch, Smith and Njoh-Elango, 1978). Conflicting results have been reported and in many cases are difficult to explain (McKay, 1926; Nelson, 1950; Keyworth, 1952). After several crops, the level of soil-borne inoculum is generally sufficient to cause severe wilt (Davis et al., 1990). According to Davis (1981), in "old" potato ground the short rotation practices commonly used (two to three years) are ineffective in reducing inoculum levels. Davis and McDole (1979) showed that when land cropped for eight consecutive years with potatoes was compared with land cropped under rotations including either corn or barley for two consecutive years between potato crops during the same eight-year period, the soil-borne *V. dahliae* populations were not significantly different. Similarly, Huisman and Ashworth (1976) showed that rotations that included nonhost crops had little effect on *V. dahliae* populations. The longevity of *V. dahliae* microsclerotia and its wide host range, including many weed species, severely diminishes the effectiveness of crop rotation in controlling this disease. In contrast, crop rotation for suppression of *V. albo-atrum* has been effective. Talboys (1987) found that in England, where infested hop gardens were rotated to grass for a minimum of two years before being replanted with hops, *V. albo-atrum* was suppressed. The limited persistence of *V. albo-atrum* is a major factor in the reduction of inoculum for this method. In Ohio, potato-wheat rotation has been found to be useful as part of an overall control strategy (Joaquim, Smith and Rowe, 1988). Potato yield increases were observed with barley-corn rotation (Davis and McDole, 1979).

Fertility Program

According to Davis et al. (1990), the use of optimal soil fertility with sprinkler irrigation may currently provide one of the most viable solutions to the management of this disease. As nitrogen availability increases, colonization of plant tissue by *V. dahliae* decreases. Early Dying Complex in 'Russet Burbank' has been found to be most severe when nitrogen is deficient. However, this inverse relationship between available nitrogen and severity of the disease has not been found with 'Norgold Russet'. Phosphorus levels have significant effects on disease suppression. As phosphorus concentration in the soil increased to the optimum, the colonization of *V. dahliae* in potato stem-tissue decreased. The effects of potassium on *Verticillium* have not been documented for potato but have been reported for other crops (Davis et al., 1990). Addition of potassium to deficit soils not only reduced the incidence of wilt in cotton but also decreased symptom severity. Cultural factors involving nitrogen/phosphorus and soil electroconductivity measurements have accounted for as much as 71 percent of field variability in disease severity (Davis, 1986). Nitrogen concentration is commonly negatively related to the severity and disease incidence (Pennypacker, 1989), and the combined effects of nitrogen and phosphorus on wilt severity are even greater than their independent effects (Davis et al., 1994). Adequate nitrogen and phosphorus fertilization reduces disease incidence in 'Russet Burbank' and minimizes yield losses that occur with the intensive cropping of potato, and the effects of nitrogen and phosphorus on *Verticillium* wilt were often highly interdependent. Similar interactive effects have been observed by Pennypacker (1989).

Green Manure

Green manure has been shown to be effective in controlling this disease complex to a certain extent. Davis et al. (1996) reported that in Idaho, the disease was controlled after green manure treatments of either sudangrass or corn and an increase in yield was seen. Spent mushroom compost may be a means of reducing the effects of the disease on superior varieties of potato and increasing the yield when one or both pathogens are present (LaMondia, 1999). A study conducted in Washington showed that mustard green manure has the potential to replace the fumigant metam-sodium for the production of potatoes in some cropping systems. The practice can also improve water infiltration rates and provide substantial savings to farmers (McGuire, 2003a). Mustard green manuring is not an isolated practice. It must be integrated into a cropping system to produce the maximum benefits. Systems that reduce tillage, avoid compaction, rotate crops, and control erosion help maintain soil quality gains that come through green manure use. Good management of water and soil fertility ensure that gains in soil-borne pest control will not be lost to waterlogged soils or overfertilization (McGuire, 2003b).

Rangahau (2003) reported that incorporating organic material such as soymeal into soil a few months before planting potatoes to encourage naturally occurring antagonists in the individual paddock will reduce Potato Early Dying disease. Davis et al. (2001) reported that organic matter can be manipulated for suppression of *Verticillium* wilt without reducing the soil population of the pathogen.

FUTURE PROSPECTS

Studies on green manures should be performed for various rotations, longer- and short-season potatoes, and different soil types. The aim should be to select antagonists that inactivate the microsclerotia before these can affect the host plant. Disease-resistant varieties should be developed, and an integrated approach is required for the suitable management of this disease.

REFERENCES

Arbogast, M., M.L. Powelson, M.R. Cappaert and L.S. Watrud. (1999). Response of six potato cultivars to amount of applied water and *Verticillium dahliae. Phytopathology* 89: 782- 788.

Back, M.A., P.P.J. Haydock and P. Jenkinson. (2002). Disease complexes involving parasitic nematodes and soilborne pathogens. *Plant Pathology* 51: 683-697.

Botseas, D.D. and R.C. Rowe. (1994). Development of potato early dying response to infection by two pathotypes of *Verticillium dahliae* and co-infection by *Pratylenchus penetrans. Phytopathology* 84: 275-282.

Bowers, J.H., S.T. Nameth, R.M. Riedel and R.C. Rowe. (1996). Infection and colonization of potato roots by *Verticillium dahliae* as affected by *Pratylenchus penetrans* and *P. crenatus. Phytopathology* 86: 614-621.

Busch, L.V., A.E. Smith and F. Njoh-Elango. (1978). The effect of weeds on the value of rotation as a practical control for Verticillium wilt of potato. *Canadian Plant Disease Survey* 58: 61-64.

Cappaert, M.R., M.L. Powelson, N.W. Christensen and F.J. Crowe. (1992). Influence of irrigation on severity of potato early dying and tuber yield. *Phytopathology* 82: 1448-1453.

Cappaert, M.R., M.L. Powelson, N.W. Christensen, W.R. Stevenson and D.I. Rouse. (1994). Assessment of irrigation as a method of managing potato early dying. *Phytopathology* 84: 792-800.

Davis, J.R. (1981). Verticillium wilt of potato in southeastern Idaho. *University of Idaho Current Informatin Series. Number* 564.

Davis, J.R. (1985). Approaches to control of potato early dying caused by *Verticillium dahliae. American Potato Journal* 62: 177-185.

Davis, J.R. and D.O. Everson. (1986). Relation of *Verticillium dahliae* in soil and potato tissue, irrigation method and N-fertility to Verticillium wilt of potato. *Phytopathology* 76: 730-736.

Davis, J.R., O.C. Huisman, D.O. Everson and A.T. Schneider. (2001). Vertcillium wilt of potato: A model of key factors related to disease severity and tuber yield in Southeastern Idaho. *American Journal of Potato Research* 78: 291-300.

Davis, J.R., O.C. Huisman, D.T. Westerman, S.L. Hafez, D.O. Everson, L.H. Sorensen and A.T. Schneider. (1996). Effects of green manures on Verticillium wilt of potato. *Phytopathology* 86: 444-453.

Davis, J.R. and R.E. McDole. (1979). Influence of cropping sequences on soil-borne populations of *Verticillium dahliae* and *Rhizoctonia solani*. In *Soil-Borne Plant Pathogens,* B. Schipper and W. Gams, eds. Academic Press, London, pp. 399-405.

Davis, J.R., L.H. Sorensen, J.C. Stark and D.T. Westerman. (1990). Fertility and management practices to control Verticillium wilt of the Russet Burbank potato. *American Potato Journal* 67: 55-65.

Davis, J.R., J.C. Stark, L.H. Sorensen and A.T. Schneider. (1994). Interactive effects of nitrogen and phosphorus and Verticillium wilt of Russet Burbank of potato. *American Potato Journal* 71: 467-481.

Francl, L.J., L.V. Madden, R.C. Rowe and R.M. Riedel. (1990). Correlation of growing season environmental variables and the effect of early dying on potato yield. *Phytopathology* 80: 425-432.

Hafez, S.L., S. Al-Rehiyami, M. Thornton and P. Sundararaj. (1999). Differentiation of two geographically isolated populations of *Pratylenchus neglectus* based on their parasitism of potato interaction with *Verticillium dahliae. Nematropica* 29: 25-36.

Huisman,O.C. and L.C. Jr. Ashworth. (1976). Influence of crop rotation and survival of *Verticillium albo-atrum* in soils. *Phytopathology* 66: 978-981.

Joaquim, T.R. and R.C. Rowe. (1990). Reassessment of vegetative compatibility relationships among strains of *Verticillum dahliae* using nitrate nonutilizing mutants. *Phytopathology* 80: 1160-1166.

Joaquim, T.R., V.L. Smith and R.C. Rowe. (1988). Seasonal variation and effects of wheat rotation on populations of *Verticillium dahliae* Kleb in Ohio potato field soils. *American Potato Journal* 65: 439-447.

Johnson, K.B. (1988). Modelling the influences of plant infection rate and temperature on potato foliage and yield losses caused by *Verticillium dahliae. Phytopathology* 78: 1198-1205.

Keyworth, W.G. (1952). Verticillium wilt of potatoes in Connecticut in 1951. *Plant Disease Reporter* 36: 16-17.

Kimpinski, J., H.W. Platt, S. Perley and J.R. Walsh. (1998). *Pratylenchus* spp. and *Verticillium* spp. in New Brunswick potato field. *American Journal of Potato Research* 75: 87-91.

Klinkenberg, C.H. (1963). Observation on the feeding habits of *Rotylenchus uniformis, Pratylenchus crenatus, Pratylenchus penetrans, Tylenchorynchus dubius* and *Hemicycliophora similes. Nematologica* 9: 502-506.

Kotcon, J.B., D.I. Rouse and J.E. Mitchell. (1985). Interactions of *Verticillium dahliae, Colletotrichum coccodes, Rhizoctonia solani* and *Pratylenchus penetrans* in the early dying syndrome of Russet Burbank potatoes. *Phytopathology* 75: 68-74.

Krikun, J. and D. Orion. (1979). Verticillium wilt of potato: Importance and control. *Phytoparasitica* 7: 107-116.

LaMondia, J.A., M.P.N. Gent, F.J. Ferrandino, W.H. Elmer and K.A. Stoner. (1999). Effect of compost amendment or straw mulch on potato early dying disease. *Plant Disease* 83: 361-366.

Mace, M.E., A.A. Bell and C.H. Beckman, eds. (1981). *Fungal Wilt Diseases of Plants*. Academic Press, New York, 640p.

MacGuidwin, A.E. (1993). Management of nematodes. In *Potato Health Management,* R.C. Rowe, ed. American Phytopathological Society, St. Paul, MN, pp. 159-166.

MacGuidwin, A.E. and D.I. Rouse. (1990). Role of *Pratylenchus penetrans* in potato early dying disease of Russet Burbank potato. *Phytopathology* 80: 1077-1082.

Mai, W.F., J.R. Bloom and T.A. Chen. (1977). *Biology and Ecology of the Plant Parasitic Nematode Pratylenchus penetrans*. Bulletin 815 University Park, PA, USA: Pennsylvania State University, College of Agriculture.

Martin, A. (1965). *Introduction to Soil Microbiology.* John Wiley and Sons, New York.

Martin, M.J., R.M. Riedel and R.C. Rowe. (1982). *Pratylenchus penetrans* and *Verticillium dahliae*: Interactions in the early dying complex of potato in Ohio. *Phytopathology* 72: 640-644.

McGuire, A.M. (2003a). *Mustard Green Manures Replace Fumigant and Improve Infiltration in Potato Cropping System.* Online.Crop Management. doi:10.1094./CM-2003-0822-01-RS.

McGuire, A.M. (2003b). *Green Manuring with Mustard—Improving an Old Technology.* http://aenews.wsu.edu/June03AENews/June03AENews.htm#GreenManure.

McKay, M.B. (1926). Further studies of potato wilt caused by *Verticillium albo-atrum. Journal of Agricultural Research* 32: 437-470.

McLean, J.G. (1955). Selecting and breeding potatoes for field resistance to Verticillium wilt in Idaho. *Idaho Agricultural Research Station Bulletin* 30: 1-19.

Mpofu, S.I. and R. Hall. (2002). Effect of annual sequence of removing or flaming potato vines and fumigating soil on Verticillium wilt of potato. *American Journal of Potato Research* 79: 1-7.

Muller, J. (1977). Interactions between five *Pratylenchus* species and *Verticillium albo-atrum. Zeitschrift Pflanzenkrankenheit Pflanzenschutz* 84: 215-220.

Nelson, R. (1950). Verticillium wilt of peppermint. *Michigan State University, Agriculture Experiment Station Technical Bulletin 221.*

Nnodu, E.C. and M.D. Harrison. (1979). The relationship between *Verticillium albo-atrum* inoculum density and potato yield. *American Potato Journal* 56: 11-25.

Ohms, R.E. (1962). Producing the Idaho Potato. *University of Idaho College Agriculture Bulletin 367.*

Pennypacker, B.W. (1989). The role of mineral nutrition in the control of Verticillium wilt. In *Management of Diseases with Macro and Microelements,* A.W. Engelhard, ed. The American Phytopathological Society, St.Paul, MN, pp. 33-45.

Powelson, M.L., K.B. Johnson and R.C. Rowe. (1993). Management of diseases caused by soilborne pathogens. In *Potato Health Management,* R.C. Rowe, ed. American Phytopathological Society, St.Paul, MN, pp. 315-326.

Powelson, M.L. and R.C. Rowe. (1993). Biology and management of early dying of potatoes. *Annual Review of Phytopathology* 31: 111-126.

Rangahau, M.K. (2003). Potato early dying (Verticillium wilt)—Strategies for control crop and food research (Broad sheet) number 126. http://www.crop.cri.nz/psp/books/bradsheets.htm.

Riedel, R.M., R.C. Rowe and M.J. Martin. (1985). Differential interactions of *Pratylennchus crenatus,P.penetrans* and *P.scribneri* with *V.dahliae* in potato early dying disease. *Phytopathology* 75: 419-422.

Rotenberg, D., A.E. MacGuidwin, I.A.M. Saeed and D.I. Rouse. (2004). Interaction of spatially separated *Pratylenchus penetrans* and *Verticillium dahliae* on potato measured by impaired photosynthesis. *Plant Pathology* 53: 294-302.

Rowe, R.C. (1983). Early dying—East and west. *American Vegetable Grower* 31: 8-10.

Rowe, R.C., J.R. Davis, M.L. Powelson and D.J. Rouse. (1987). Potato early dying: Causal agents and management strategies. *Plant Disease* 71: 482-489.

Rowe, R.C. and M.R. Powelson. (2002). Potato early dying: Management challenges in a changing production environment. *Plant Disease Reporter* 60: 248-251.

Rowe, R.C., R.M. Riedel and M.J. Marin. (1985). Synergistic interactions between *Verticillium dahliae* and *Pratylenchus penetrans* in potato early dying disease. *Phytopathology* 75: 412-418.

Selvanathan, N. (2003). Potato early dying complex: Key factors related to disease development. In *Manitoba Agronomist Conference,* University of Manitoba, pp. 198-201.

Stevenson, W.R., R.M. Loria, G.D. Franc and D.P. Weingartner. (2001). *Compendium of Potato Diseases.* American Phytopathological Society, St. Paul, MN, 2nd ed., 106p.

Sturz, A.V. and M.M. Clark. (2000). *Vertcillium Wilt of Potatoes.* Agriculture and Forestry pp. 1-5. http://www.gov.pe.ca/af/agweb/index.php3?number=74050.

Talboys, P.W. (1987). Verticillium wilt in English hops: Retrospect and prospect. *Canadian Journal of Plant Pathology* 9: 68-77.

Talboys, P.W. and L.V. Busch. (1970). Pectic enzymes produced by *Verticillium* species. *Transactions of British Mycological Society* 55: 367-381.

Tsror, L. and M. Hazanovsky. (2001). Effect of coinoculation by *Verticillium dahliae* and *Colletotrichum coccodes* on disease symptoms and fungal colonization in four potato cultivars. *Plant Pathology* 50: 483-488.

Wheeler, T.A., L.V. Madden, R.M. Riedel and R.C. Rowe. (1994). Distribution and yield-loss relations of *Verticillium dahliae, Pratyelnchus penetrans, P. scribneri, P. crenatus* and *Meloidogyne hapla* in commercial potato fields. *Phytopathology* 84: 843-852.

Wheeler, T.A. and R.M. Riedel. (1994). Interactions among *Pratylenchus penetrans, P. scribneri* and *Verticillium dahliae* in the potato early dying disease complex. *Journal of Nematology* 26: 228-234.

Chapter 9

Nematode Resistance in Vegetable Crops

Philip A. Roberts

INTRODUCTION

In this chapter, the scope of resistance availability for controlling plant-parasitic nematodes in vegetable crops is reviewed, using profiles of several important resistance traits in vegetables of world significance. The opportunities for breeding resistant vegetable crops are discussed in terms of introgression of resistance from nondomesticated gene donor species, phenotype screens for nematode resistance, and development and use of molecular markers for marker-assisted selection (MAS). Approaches to implementing resistance for nematode management are given on the basis of how resistance affects vegetable crop yield and seasonal multiplication rates of nematode populations. Resistance specificity, selection for resistance-breaking or virulent nematode populations, and environmental factors such as resistance gene sensitivity to heat influence the effectiveness of resistant vegetable cultivars (Roberts, Matthews and Veremis, 1998). The integration of host plant resistance with other nematode management strategies is described, especially resistance use in crop rotations and also with nematicides and other cultural control tactics such as cover cropping and soil amendments.

Definitions of important terms used here are provided to clarify meaning, particularly because certain terms applied to nematology

Management of Nematode and Insect-Borne Plant Diseases

doi:10.1300/5754_09

do not have quite the same meaning as in classical plant pathology. These terms have been defined and described by several authorities in reviews and the reader is referred to those references for additional descriptions (Roberts, 1982; Cook and Evans, 1987; Cook, 1991; Trudgill, 1991; Shaner et al., 1992). *Resistance* is used to describe the ability of a plant to suppress the development or reproduction of the nematode. It can range from *low* or *moderate* (*partial* or *intermediate*) resistance to *high* resistance. The term "resistance" is also used to describe the capacity to suppress the disease, especially root-knot (Sasser, 1980). A highly resistant plant allows only trace amounts of nematode reproduction, whereas moderately resistant plants allow intermediate levels of reproduction. *Susceptibility* is used as the opposite of resistance; thus a *susceptible* plant allows normal nematode development to take place and the expression of any associated disease (Roberts, 2002).

Tolerance describes the ability of the plant to withstand the damage resulting from nematode infection. *Tolerant* plants grow and yield well despite nematode infection, whereas *intolerant* plants do not grow as well or die when infected. Most resistant plants are tolerant, and the majority of susceptible plants are injured to some extent by nematode infection. However, resistance and tolerance have been shown to be under separate genetic control in some plant-nematode interactions (Evans and Haydock, 1990; Trudgill, 1991). When resistant plants are injured by initial nematode infection, the lack of tolerance often results from tissue injury and reduced root function, owing to the hypersensitivity reaction in the resistance response, as found in the sweet potato's (*Ipomoea batatas* (L.) Lam) response to root-knot nematodes (Roberts and Scheuerman, 1984).

VEGETABLE CROPS

Resistance to plant-parasitic nematodes has been identified in numerous vegetable crops and their wild relative species. Some of these resistance traits have been bred into commercially acceptable cultivars and varieties of important vegetable crops, including common bean (*Phaseolus vulgaris* L.), lima bean (*Phaseolus lunatus* L.), carrot (*Daucus carota* L.), cowpea (*Vigna unguiculata* (L.) Walp.), peppers (*Capsicum* spp.), potato (*Solanum tuberosum* L.), sweet potato

(*I. batatas* Lam.), and tomato (*Lycopersicon esculentum* Mill.). In addition, nematode-resistant rootstocks have been developed and used for some vegetables, such as *Solanum torvum* Swartz rootstock for eggplant (*Solanum melongena* L.) production and pepper rootstocks for pepper production (Oka, Offenbach and Pivonia, 2004). Nematode resistance has also been identified for other crops, such as cucurbitaceous crops (melons, squash, cucumber), but the resistance resides in related germplasm that requires introgression into commercial cultivars. Still other important vegetables, such as lettuce (*Lactuca sativa* L.), do not have resistance available for introgression at this time. These might require transgenic resistance using effective genes cloned from other crops, such as the *Mi* gene from tomato. However, transgenic resistance using natural *R* genes from even related crops presents technological challenges because the genes do not function in many plant backgrounds, including different genetic backgrounds of the same crop, as found with the *Mi-1* gene in tomato (Goggin et al., 2004). The trait determinants may not be useful unless all the components in the signal transduction pathway or their functional equivalents are present.

MAJOR NEMATODE PESTS

Most of the resistance has been developed for managing the highly specialized nematode parasites, especially the root-knot nematodes (*Meloidogyne* spp.), cyst nematodes (*Heterodera* and *Globodera* spp.), the false root-knot nematode (*Naccobus aberrans* Thorne and Allen), and the reniform nematode (*Rotylenchulus reniformis* Linford and Oliveira). Some resistance has also been reported against the stem and bulb nematode (*Ditylenchus dipsaci* (Kühn) Filipjev) in beans and lesion nematode (*Pratylenchus* spp.) in various crops. These are the most important nematode pests of vegetables on a worldwide scale and, justifiably, most of the resistance development has been focused on them. However, many other nematodes are associated with vegetables in temperate, subtropical, and tropical regions (Luc, Sikora and Bridge, 1990; Evans, Trudgill and Webster, 1993). In almost all these other cases, resistance has not been identified or developed for affected crops, but it is likely that local crop selection and improvement efforts on infested fields over time may have selected for some tolerance to damaging nematodes in local production programs (Bridge, 1996). Most of the detailed accounts of nematode resistance in vegetable crops are based on *Meloidogyne* spp. Resistance to root-knot nematodes is the most diverse to any nematode group, occurring in a range of crops from different plant taxa. The genus *Meloidogyne* contains numerous species, of which several have wide host ranges and are major nematode pests in world agriculture (Sasser and Freckman, 1987).

STRATEGIES FOR BREEDING RESISTANT CULTIVARS

Nematode pests are often not targeted for resistance breeding in vegetable crops because their damaging effects are often not recognized. With the exception of root-knot nematodes and cyst nematodes, whose root galling and visible presence as females on roots, respectively, provide a clear diagnosis, most nematode infections do not generate diagnostic symptoms; typically infected plants do not thrive and may yellow and wilt and look stunted, but these are symptoms usually attributed to poor soil conditions, low fertility, or water stress. More effort is needed to engage trained nematologists to determine the presence of nematode parasitism as a yield-limiting factor and then to work with the plant breeder to develop resistant crops for managing the target nematode.

Resistance Phenotype Screening

Detailed descriptions of phenotype screening for resistance to the important groups of plant-parasitic nematodes are provided in Starr, Cook and Bridge (2002). Effective screening procedures are vital for selecting and advancing resistant genotypes in a breeding program, whether the approach is pedigree breeding or population selection. Biochemical and molecular markers, such as acid phosphatase isozymes and polymerase chain reaction (PCR)-based REX-1 (Williamson et al., 1994) for the gene *Mi-1* in tomato, are very useful for resistance selection using MAS. However, even when markers are available, phenotype screening for resistance needs to be made at certain stages of the breeding program. Initially, expression of the resistance in desired genetic backgrounds should be confirmed to ensure that the traits of interest are appropriate to advance within the breeding lines. Second, the phenotype of selected lines based on marker screening should be confirmed to ensure the linkage is maintained. The tighter the marker linkage to the resistance trait, the less likely that marker-positive lines without the resistance will be selected, but for most markers, some level of recombination is expected. The recombination rates can be predicted from the linkage data generated during the marker development (Young and Mudge, 2002). In some cases, the initial crosses may result in a complete series of breeding materials in which the marker linkage has been broken through selection of recombinants between marker and resistance locus. This occurred for the *Aps-1* marker for the gene *Mi-1* in tomato, where the Hawaiian tomato breeding program was founded on recombinant lines in which the *Aps-1* marker variant was uncoupled from the *Mi-1* gene (Medina-Filho and Tanksley, 1983).

In many breeding programs, markers for *R* genes are not available at present; therefore the entire screening and selection process is dependent upon an efficient and reliable phenotype screening protocol. Fortunately, relatively straightforward, scorable phenotypes for the most important nematode groups are available. The root-knot nematodes induce characteristic root-galling symptoms in susceptible plants, symptoms that are suppressed in resistant plants (Hussey and Jansen, 2002). The cyst nematodes produce visible adult females and cysts (the egg-filled dead females) on the root surface of susceptible plants that can be enumerated, whereas in resistant plants females and cysts are either absent or greatly reduced in number (highly resistant genotypes) or partially reduced in number (moderate resistance) (Cook and Noel, 2002). Nevertheless, these nematode screens may require a more detailed quantification of infection potential more accurately to measure resistance, based on enumeration of root-knot nematode juveniles, egg masses or eggs produced by females, or numbers of cyst nematode eggs in cysts. For most other nematodes for which resistance is available, infected plant tissues must be processed to extract the nematode life stages (eggs, juveniles, and/or adults), which must then be enumerated.

In all these cases, the measure of infection and nematode reproduction as an index of resistance must be related to responses of known susceptible and resistant genotype control plants tested in the same screening (Starr, Cook and Bridge, 2002). Often the parent genotypes in the breeding programs are used, including plants of the susceptible recipient genotype and the resistance gene donor line. Using the infection or reproduction scores of the control and the tester genotypes, a relative index of resistance/susceptibility can be made. This is often calculated as a ratio between the nematode infection on the tester and that on the susceptible control line, to give a reproduction factor (RF) or multiplication rate (MR). The RF or MR is calculated as the ratio (P_f/P_i) between the final population density of nematodes, P_f, assessed at test termination, to the initial population density, P_i, typically the inoculum level used in the test. Experience gained with expected responses for the resistance phenotype often lead to a simpler process, such as a root-galling index for *Meloidogyne* resistance (Bridge and Page, 1980) or a female index for cyst nematodes (Cook and Noel, 2002).

Even with these simplifications, continuing expense is associated with the labor, materials, and facilities required for repeated screening at different generations of resistance advancement during the breeding effort. Many of the early-generation phenotype screens are conducted in controlled greenhouse pot tests, growth pouch screens in greenhouses or controlled environment chambers, or small field microplots (Ehlers et al., 2000; Starr, Cook and Bridge, 2002). Nematode-plant interactions are environmentally sensitive and the test conditions can have large effects on the results of the phenotype screen. Temperature, day length, light intensity, watering regime, and plant nutrient status can each influence the nematode reproductive potential via the effect mediated through the host plant (Starr, Cook and Bridge, 2002). Verification of the practical utility of resistance in advanced breeding lines is important. This can be achieved only by screening in inoculated or naturally infested field plots. A uniform distribution of inoculum is essential to make accurate assessments of the resistance performance.

Screening for Tolerance

Measurements of growth and yield are important to assess tolerance to infection of resistant genotypes. The most informative assessment of tolerance is based on a four-way comparison of the resistant breeding line with a standard equivalent susceptible line (preferably a near-isogenic or sister line) tested in split-plots of infested (or inoculated) and noninfested (nematicide-treated or noninoculated) treatments. These comparisons have proven to be highly informative for gauging the value of resistance in improving crop tolerance to nematode attack under normal growing conditions (Roberts, 1982, 1992; Roberts and May, 1986). Where resources are limited for conducting such detailed comparisons, a minimum test protocol should include a comparison of resistant line growth or yield performance to that of a standard susceptible line or cultivar in infested field plots.

Marker-Assisted Selection

An intensive research effort is being directed toward development of molecular markers of resistance genes for indirect selection of resistance using MAS (Young and Mudge, 2002; Collard et al., 2005). MAS has important proven benefits for genetic improvement of crops based on selection of numerous agronomic and other traits, and major effort is being directed at resistance genes for pests and diseases (Collard et al., 2005). Resistance traits are especially good targets for application of MAS as they are often conferred by simply inherited major genes. Such genetic markers are based on defined deoxyribonucleic acid (DNA) sequences, from which specific DNA primer pairs can be designed. The primers amplify only the marker sequence when used in PCR reactions with DNA template prepared from whole-DNA samples extracted from tissues (typically leaves) of the resistant plant. Several common advantages in using markers for indirect selection of the resistance trait include (1) being able to test only one plant or line from advanced materials that are fixed (nonsegregating for the target gene); (2) nondestructive assays of the test plant; (3) testing at a young (seedling) stage, saving both time and resources (no need to grow out unwanted breeding lines); and (4) avoiding challenges associated with environmental sensitivity in phenotype screens.

In addition, more specific advantages exist for indirect selection of resistance: (1) scaling back the nematode phenotype screening effort, with savings on space (greenhouse, growth chamber), time, and resources (labor especially); (2) ability to screen in locations where the nematode pest culture is unavailable or excluded for quarantine reasons; and, (3) being able to screen for multiple genes where gene combinations are a goal of breeding. In the latter case, combining new genes with ones that have been defeated (by virulent nematode populations) may be important. In addition, MAS has potential for selecting polygenically inherited resistance based on two or more genes determined by QTLs (quantitative trait loci). Here the problems associated with advancing lines with the full complement of genes necessary for high levels of resistance can be tackled by indirect selection of the loci with markers. This type of quantitative nematode resistance is present in some crops, for example against potato cyst nematode *(Globodera pallida)* in potato and root-knot nematode *(Meloidogyne trifoliophila)* in clover (Jones, 1985; Mercer, Hussain and Moore, 2004). Breeding QTL-based nematode resistance into vegetable crops has potential, and multiple gene systems of resistance in crops such as peppers (the *Me* genes; Dijian-Caporalino et al., 2001) and tomatoes (the *Mi* genes; Williamson, 1998; Veremis and Roberts, 2000; Amaraju et al., 2003) with resistance to *Meloidogyne* spp., require this approach in the future. Different suites of genes will be required in a cultivar to provide the desired breadth and level of resistance to diverse *Meloidogyne* species such as *Meloidogyne incognita, Meloidogyne javanica, Meloidogyne hapla, Meloidogyne chitwoodi,* and *Meloidogyne fallax,* as well as their avirulent and virulent populations.

EXAMPLES OF RESISTANCE IN VEGETABLE CROPS

Meloidogyne Resistance in Tomato

The Mi-1 *Gene*

The best known and most widely utilized resistance to *Meloidogyne* spp. is the *Mi-1* gene in tomato (Gilbert and McGuire, 1956; Roberts, 1992; Milligan et al., 1998; Williamson, 1998). This gene has been used in tomato cultivars for approximately fifty years, since its introgression into *L. esculentum* from the wild species relative *Lycopersicon peruvianum* (L.) Mill. via embryo rescue (Smith, 1944). Today it is used in numerous fresh-market and processing tomato cultivars around the world, and it is effective in controlling three of the most important species of root-knot nematode: *Meloidogyne arenaria* (Neal) Chitwood, *M. incognita* (Kofoid and White) Chitwood, and *M. javanica* (Treub.) Chitwood (Roberts and Thomason, 1986). The *Mi-1* gene does not confer resistance to *M. hapla* Chitwood, a common species in cooler tomato-growing areas, or to *Meloidogyne mayaguensis* Rammah and Hirschmann, a species recently identified in Puerto Rico, Florida, West Africa, and other subtropical vegetable production areas (Brito et al., 2004). However, it does confer resistance to aphids and whiteflies (Rossi et al., 1998). *Mi-1*-based resistance has proven to confer a high level of tolerance to incompatible root-knot nematode populations. In field experiments in which resistant and susceptible cultivars of machine-harvested processing tomatoes were compared on *M. incognita*-infested and preplant nematicide-fumigated plots, the yield of the resistant cultivars was not reduced in the infetsed plots, whereas the susceptible cultivar yields were reduced in infested plots by approximately 50 percent compared with the yield in fumigated plots (Roberts and May, 1986). These results are consistent with numerous comparisons of resistant and susceptible varieties in field and greenhouse settings. The resistance reaction varies slightly among incompatible isolates of *M. incognita* and *M. javanica,* but the reaction is very effective in limiting the nematode reproduction to less than about 2 percent of that on equivalent susceptible tomatoes (Roberts and Thomason, 1986).

The *Mi-1* gene is easy to transfer between tomato genetic backgrounds in conventional breeding programs, a process greatly aided by the availability of genetic markers for indirect selection. The *Mi-1* gene has been cloned and sequenced and is a focus of research for its application as a transgenic source of resistance in tomato breeding. However, the apparent instability of *Mi-1* as a transgene in tomato after two or more selfing generations (Goggin et al., 2004), indicates that this transgenic approach to gene transfer even within the crop species may prove to be complex and to require much additional research.

The *Mi-1* gene has notable limitations despite its widespread documented utility. First, the gene is sensitive to high temperature, weakening in expression above 28-30°C. This creates problems of resistance breakdown in hot tomato-growing areas, such as Florida (Noling, 2000), in many subtropical and tropical regions, and also in glasshouse or plastic house production systems. Second, the resistance is not completely durable. Populations of *Meloidogyne* spp. with virulence to *Mi-1* have been found to occur both naturally, in the apparent absence of selection pressure from resistant tomato culture, and after selection for virulence by exposure to resistant tomato plantings (Roberts et al., 1990; Castagnone-Sereno et al., 1994; Roberts, 1995; Williamson, 1998). Virulent populations pose a threat in intensive production systems where rotation with other crops or use of nematicide treatments is not practiced or is not feasible.

Novel Resistance Genes

In light of the problems associated with heat sensitivity and virulent nematode populations, other genes for resistance have been found in the wild tomato relative *L. peruvianum*. These include several genes (*Mi-2* to *Mi-9*) that are typically dominant major genes and possess highly useful characteristics (Veremis and Roberts, 2000; Amiraju et al., 2003). Some of these genes are heat stable and remain effective at soil temperatures of 32-34°C, such as *Mi-2* and *Mi-9* (Cap, Roberts and Thomason, 1993; Veremis and Roberts, 1996a, b, c; Veremis, van Heusden and Roberts, 1999; Veremis and Roberts, 2000; Amiraju et al., 2003). These heat-stable genes have not been transferred to commercial tomato cultivars, although several breeding programs are working toward this goal. As with the original introgression into edible tomato of *Mi-1* from *L. peruvianum,* problems associated with both self- and cross-incompatibility and linkage of the resistance genes to any undesirable traits in the *L. peruvianum* donor genome must be overcome. Some of these novel genes are effective against both natural and selected *Mi-1*-virulent *Meloidogyne* populations (Roberts et al., 1990), such as *Mi-3,* which was mapped to chromosome 12 in tomato (Yaghoobi et al., 1995), and *Mi-7* and *Mi-8,* which reside in other regions of the tomato genome (Veremis and Roberts, 1996b). These novel genes are excellent candidates for genetic improvement of tomato for use in nematode management programs.

Meloidogyne Resistance in Peppers

Peppers (*Capsicum* spp.) are among the most widely grown vegetables worldwide, and they are highly susceptible to several common species of root-knot nematodes. Cultivars of both *Capsicum annuum* L. and *Capsicum frutescens* L., peppers suffer severe yield losses owing to infection by *M. incognita, M. javanica, M. arenaria, M. hapla,* and *M. mayaguensis* (Lindsey and Clayshutte, 1982, Thies and Fery, 2000). As a result, both preplant soil fumigation with nematicides and resistant pepper cultivars are used to manage the root-knot nematode problem (Thomas, Murray and Cardenas, 1995; Thies and Fery, 2002). The effectiveness of the two management strategies depends on the type of nematicide treatment and an incompatible match of the resistance gene(s) with the specific nematode species and population present in a given field or production area (Thies, Mueller and Fery, 1997; Thies and Fery, 2002). High yields of resistant peppers in heavily infested, nonfumigated soils have been documented, and even cultivars with only moderate resistance appear to be tolerant of root-knot infection (Thies and Fery, 2002). Resistant peppers have also been shown to be effective in suppressing nematode soil population densities when used as a rotational crop (Thies, Mueller and Fery, 1998), although cultivars carrying only moderate resistance, such as the bell pepper cv. Keystone Resistant Giant, allow considerable nematode reproduction and soil population increase.

Diverse sources of highly effective resistance to *Meloidogyne* have been identified and characterized in *Capsicum* spp. Several distinct genes for resistance have been determined to control the resistance in *C. annuum* and *C. frutescens*. The *N* gene was identified in *C. frutescens* approximately fifty years ago and was found to be resistant to some but not all of the major *Meloidogyne* spp. following transfer into a range of commercial pepper cultivars, including those of *C. annuum* (Hare, 1956; Thies and Fery, 2000). Hendy, Pochard and Dalmasso (1985) described a series of at least five major genes in the lines "PM687" and "PM217" derived from Indian and Central American germplasm sources, respectively, with the genes acting in a dominant fashion in gene-for-gene-type interactions with different nematode populations. Dijian-Caporalino et al. (1999) determined that the *Me3* gene in this material had broad-spectrum effectiveness against *M. incognita, M. javanica, M. arenaria,* and some but not all tested isolates of *M. hapla*. They also determined that the resistance was heat stable at temperatures as high as 42°C (Dijian-Caporalino et al., 1999). Sources of resistance such as this that have a broad spectrum of utility and can be used in hot production areas are especially valuable. The *Me3* gene and another linked heat stable gene, *Me4,* were further characterized in the *C. annuum* genome by high-resolution mapping (Dijian-Caporalino et al., 2001). These genes were determined to be approximately 10 cM apart and mapped to a linkage group that may have homology with chromosome 7 or 12 of tomato and chromosome XII of potato (Dijian-Caporalino et al., 2001). These putative genome relationships within related solanaceous crop species provide some indication of common resistance gene evolution, as these linkage groups each carry nematode resistance genes.

Meloidoygne Resistance in Carrot

Current Status

Carrots are extremely sensitive to root-knot infection because the invading second-stage juveniles induce galling and ramification or forking of the taproot. Although this does not suppress plant growth at low nematode population densities, the "cosmetic" injury to roots renders the carrots unmarketable, resulting in economic loss (Roberts, 1987). In most carrot production systems, the ground is preplant fumigated with a nematicide. For example, most of the carrot fields in California are treated with shank-injected 1,3-dichloropropene or metam-sodium (a methyl isothiocyanate liberator), applied through the irrigation system before planting (Roberts et al. 1988). The future availability of these and other nematicides is in doubt, and development of commercially suitable carrots with resistance to *M. incognita* and *M. javanica* are in the final stages of development in several breeding programs.

New Resistance

The primary resistance in carrot is conferred by the *Mj-1* locus, a major dominant gene that shows some allelic dosage effect (Boiteux et al., 2000; Simon, Matthews and Roberts, 2000). The resistance is highly effective in blocking both the root-galling of the taproot and fibrous roots and nematode reproduction on roots (Simon, Matthews and Roberts, 2000). The resistance has been advanced in breeding lines to develop fresh-market carrots suitable for California and elsewhere. The source of the resistance is the carrot variety 'Brasilia', initially developed in Brazil (Huang, 1986; Charchar and Viera, 1994). Recent breeding efforts in the United States have used Brasilia in crosses with elite USDA inbred breeding lines, followed by several rounds of advancement, sib-mating, and selection for resistance and agronomic traits. Advanced resistant breeder lines have been released to commercial seed companies for use in crossing with their elite carrot types to produce a range of commercially acceptable resistant carrot cultivars.

Mj-1 resistance is not as effective in the heterozygous condition as in the homozygous condition, with heterozygous plants supporting slightly more root-galling and nematode reproduction than the homozygous resistant plants (Simon, Matthews and Roberts, 2000; Boiteux et al., 2004). This effect is exacerbated by high-temperature growing conditions. The *Mj-1* gene is heat sensitive, with resistance weakened at temperatures above 30°C and heterozygous plants showing a more pronounced decrease in resistance at high temperatures (Matthews, Simon and Roberts, 1999). Carrot breeders typically develop hybrid carrots as commercial cultivars. However, to optimize the *Mj-1* resistance, they are being advised to develop homozygous resistant cultivars by ensuring that both parents used to generate the hybrid planting seed are homozygous for the *Mj-1* gene.

Mj-1 resistance is more variable to *M. incognita* than to *M. javanica,* depending on the nematode population or isolate tested. The reaction appears to differ more on heterozygous plants, and the indications are that factors other than *Mj-1* may be involved in the resistance reaction to *M. incognita* (Boiteux et al., 2000). However, in the California carrot breeding program, we have found that most *M. incognita* populations, including those from *M. incognita* race 1 and race 3, are effectively controlled by *Mj-1* resistance. In addition, the *Mj-1* gene does not resist *M. hapla,* a root-knot problem on carrots in cooler production areas. Sources of *M. hapla* resistance in carrot have been reported (Wang and Goldman, 1996), and the opportunity to combine the recessive *M. hapla* resistance genes with *Mj-1*-conferred *M. incognita* and *M. javanica* resistance could lead to a broad-based root-knot resistance in commercial carrots. The relationship between *M. hapla* resistance and the *Mj-1* locus has not been studied.

Markers for Resistance in Carrot

Because carrots are difficult to phenotype for resistance in a breeding program and to do so accurately requires considerable investment, molecular markers for the *Mj-1* locus were developed for use in MAS for resistance (Boiteux et al., 2000). Using bulked segregant analysis and an F_2 population of more than 400 plants segregating for *Mj*-1, PCR-based random amplified polymorphic DNA (RAPD) markers were developed that were closely linked to thc locus. Two of these markers were converted to STS (sequence tag site) markers for ease of use in MAS, providing a pair of codominant flanking markers, one completely cosegregating in the F_2 and the second within 1 cM of the *Mj-1* gene (Boiteux et al., 2004). The markers were highly correlated with several resistance phenotype parameters, including fibrous root-galling indices, total nematode eggs per root system, and numbers of eggs per gram of root. These markers are being applied to our breeding populations to select for *Mj-1* resistance. The markers are effective in distinguishing homozygous from heterozygous resistant plants and for developing nonsegregating resistant carrot populations. As with most genetic markers, they require assessment for their identifiable polymorphism in each set of parents used in a given breeding program. The linkage to the resistance is present, but the susceptible parent genotype must differ from that of the resistant parent for the marker. In our studies, the coupling-phase-linked RAPD markers from which the PCR-based STS markers were developed were polymorphic in a broad range of carrot germplasm lines (Boiteux et al., 2000). Only 5 (all resistant and related to 'Brasilia') out of 121 accessions of carrot germplasm entries had the full complement of closely linked markers, indicating that the markers should be highly valuable for incorporating the *Mj-1*-based resistance into different elite carrot germplasm (Boiteux et al., 2000, 2004). Nevertheless, the markers need to be tested initially for their utility in each specific cross and may not be useful in some breeding programs, depending on the desired carrot background into which the resistance is being incorporated.

Meloidogyne Resistance in Beans

Several classes of beans are used as vegetables, and most are highly susceptible to a range of nematodes, including the common *Meloidogyne* spp. In particular, common beans *(P. vulgaris)* and cowpeas *(V. unguiculata)* used as green pod snap beans, lima bean *(P. lunatus)* as a fresh or frozen vegetable, and faba or broadbeans (*Vicia faba* L.) are all root-knot nematode susceptible (Luc, Sikora and Bridge, 1990; Evans, Trudgill and Webster, 1993). In the first three cases, breeding for resistance to root-knot nematodes has proved successful, with many examples of effective commercial cultivars of these crops having been developed and used for nematode management around the world (Omwega et al., 1989; Roberts, 1992; Hall et al., 2003; Helms et al., 2004).

Common Bean

For common beans, several sources of root-knot nematode resistance have been identified and in some cases bred into commercial cultivars, dating back to a first report by Isbell (1931). These include recessive resistance from *P. vulgaris* cv. Alabama No. 1, which was used to breed *M. incognita* resistance into pole beans such as 'Manoa Wonder' (Blazey et al., 1964). This resistance is effective against some but not all populations of *M. incognita* and *M. hapla,* but not against *M. arenaria* or *M. javanica* (Omwega et al., 1989; Omwega and Roberts, 1992). Two other important sources of resistance in *P. vulgaris* provide a complementary set of resistance genes that together are effective against the four common *Meloidogyne* spp. (Omwega and Roberts, 1992). One set of materials (landraces G2618 and G1805 and their derived breeding lines such as A315 and A445) contains a single dominant gene, *Me1,* with resistance to *M. javanica* and *M. arenaria* and possibly some but not most *M. incognita* isolates (Omwega et al., 1989; Omwega, Thomason and Roberts, 1990a). In a second set of resistant genotypes, including PI165426, PI 165435, and in part 'Alabama No. 1', resistance genes *Me2* (dominant), *me3* (recessive), and possibly other genes control *M. incognita,* but not *M. javanica* and *M. arenaria.* To optimize the level and spectrum of resistance in bean cultivars, developing genotypes with the combination of the *Me1, Me2,* and *me3* genes would be a desirable and achievable goal. These complementary resistance gene systems in common beans are excellent candidates for gene pyramiding using indirect selection with molecular markers. However, to date, useful linked markers have not been developed, and the two types of resistance have not been combined into a genotype in bean breeding programs. In addition, a good source of resistance to all common *Meloidogyne* species was found in the tepary bean *Phaseolus acutifolius* A. Gray accession PI310606 (Omwega et al., 1989); however, transfer to common bean would be difficult owing to genetic incompatibility between *P. acutifolius* and *P. vulgaris*.

An important feature of the common bean resistance genes is their expression response to various temperature regimes. Omwega, Thomason and Roberts (1990b) determined that the two gene systems had different responses to temperature when challenged with incompatible *Meloidogyne* species isolates. Resistance conferred by recessive genes (*me3* in PI165426 and other recessive genes in 'Alabama No. 1' and PI165435) was effective at or below 26°C but not at 28°C and higher temperatures. In contrast, the dominant genes *Me1* (in system 2) and *Me2* (in system 1) were weakened but not broken at or above 28°C, being semidominant or neutral on the basis of an allelic dosage effect at the higher temperatures (Omwega and Roberts, 1992). An analysis of nematode development and reproduction in these resistant genotypes at different temperatures confirmed these findings and indicated that rates of development rather than root penetration rates were important in determining the reproduction potential of root-knot nematodes on these resistant bean genotypes (Sydenham, McSorley and Dunn, 1996).

A common pattern of resistance gene specificity is apparent from analysis of resistance systems with multiple genes in different crops. As found for gene *Mi-1* in tomato, when specific resistance genotypes are challenged with multiple isolates or populations of a root-knot nematode species, differential nematode–plant interactions can be identified. In common bean, not only do the two gene systems differentiate root-knot nematodes at the species level, but isolates within a species vary in their response to the resistance genes. For example, screens with several *M. hapla* isolates on PI 165426 revealed clear-cut differences in the (a)virulence profiles of the nematodes (Chen and Roberts, 2003a). Some isolates were avirulent to this resistance, being unable to reproduce significantly on resistant plants, whereas other isolates were highly virulent and reproduced abundantly (Figure 9.1). Such differential reactions are typical of gene-for-gene systems in which nematode avirulence genes are matched or not with specific host resistance genes.

Genetic analysis of these avirulent and virulent *M. hapla* isolates by controlled crossing revealed a pattern of simple inheritance of avirulence in the nematode being conferred by a single gene, with avirulence dominant to virulence (Chen and Roberts, 2003b). The results with *M. hapla* were similar to the genetic control of avirulence in the potato cyst nematode *G. rostochiensis* matching the resistance gene *H1* in potato (Janssen, Bakker and Gommers, 1991). The practical implications of these specific nematode-plant interactions support the need to broaden the genetic base for resistance by identifying and introgressing different resistance genes. Genetic markers for novel resistance genes should expedite their application in crop improvement and nematode management.

Lima Beans

For lima beans, several sources of root-knot nematode resistance have been identified, and in some cases they have been bred into

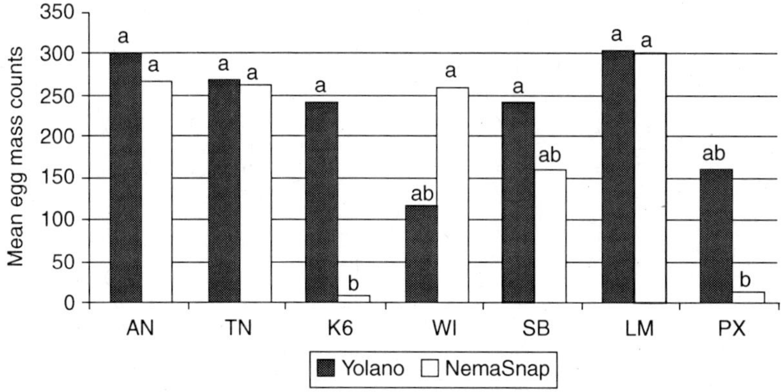

FIGURE 9.1. Reproduction of *Meloidogyne hapla* on resistant (Nemasnap) and susceptible (Yolano) common beans, differentiating virulent (K6, PX) from avirulent (AN, TN, WI, SB, LM) nematode populations (Chen and Roberts, 2003a).

commercial cultivars. Resistance to root-knot was identified in a range of *P. lunatus* germplasm representing the Central and South American gene pools of the species (Allard, 1954; McGuire, Allard and Harding, 1961). The resistance is conferred by multiple genes that differ in several respects. Some genes are effective against *M. incognita* only, whereas others also resist *M. javanica*. The genes identified so far are a combination of both recessive and dominant genes. The gene differences extend to their distinct effects in suppressing nematode reproduction in roots and nematode-induced root galling. For example, the resistance in large-seeded lima beans such as 'White Ventura N' and 'Maria' is effective only against *M. incognita,* but it suppresses both root galling and reproduction of this species (Helms et al., 2004). A more promising source of resistance from Puerto Rico, accession L-136, was found to contain three nonallelic genes that confer resistance to *M. incognita* reproduction, *M. incognita* root galling, and *M. javanica* root galling, respectively (Helms et al., 2004; Roberts and Matthews, unpublished data). This suite of genes was transferred by conventional breeding into determinate (bush-type) and indeterminate (vine-type) lima bean backgrounds, culminating in the release of the commercial small-seeded bush lima cv. Cariblanco N (Helms et al., 2004).

In a different resistance source, PI256874, others genes were identified that differentially control root galling and reproduction of both *M. incognita* and *M. javanica* (Matthews, Helms and Roberts, 2000). Thus lima bean has a rich resource of root-knot nematode resistance traits that are amenable to breeding into elite cultivars. Efforts are underway to determine the relative tolerance to nematode infection that these independent genes impart, using near-isogenic lima bean lines that differ for the specific resistance genes. This system provides an excellent model for examining the relative contribution to plant injury and yield loss by root-knot nematode-induced root galling and by nematode feeding and reproduction on infected roots. The development of molecular markers, including orthologs from *P. vulgaris* with homology to *P. lunatus* DNA sequences, will expedite the transfer of these important resistance genes into preferred lima bean cultivars and possibly into common bean as transgenes if they can be cloned.

Cowpea

Cowpea, also known as southern pea or blackeyed pea, is mostly used as a dry grain legume but is increasing in popularity as a fresh green pod snap bean vegetable (Timko, Ehlers and Roberts, 2006). Interest has also been growing in the use of root-knot nematode-resistant cowpea cultivars that have been bred for high biomass production as cover crops in vegetable production systems (Roberts, Matthews and Ehlers, 2005). Cowpea is highly susceptible to root-knot nematode, and growth and yield can be reduced by more than 50 percent on heavily infested fields (Roberts, Matthews and Ehlers, 2005). Several sources of root-knot nematode resistance have been identified in cowpea, and some have been bred into commercial cultivars. Most resistant cowpea cultivars have the single dominant gene *Rk* that confers resistance to *M. incognita, M. javanica, M. arenaria,* and *M. hapla* (Fery and Dukes, 1980). This resistance is highly effective against many root-knot populations, but some *M. javanica* populations are able to reproduce on plants with *Rk,* and recently *Rk*-virulent populations of *M. incognita* were identified from fields with a history of growing resistant cowpea (Roberts et al., 1995; Ehlers et al., 2003). The resistance is valuable because cowpeas are a relatively low cash value crop and do not justify the cost of nematicide treatments. Their increased value as a green pod vegetable may support the cost of nematicide treatment. Additional sources were identified that have a broader and higher level of resistance expressed to populations of *M. javanica* and *M. incognita,* which are not controlled by *Rk* (Roberts, Matthews and Ehlers, 1996; Ehlers et al., 2000). The Rk^2 gene was found in a breeding line from Nigeria and is either an allele at the *Rk* locus or tightly linked to *Rk* within a genetic distance of 0.17 cM (Roberts, Matthews and Ehlers, 1996). A third gene, *rk3,* is recessive and independent of the *Rk* locus; it confers only partial or moderate resistance alone, but acts additively in combination with the gene *Rk* to provide broad-based resistance (Ehlers et al., 2000). The *Rk* locus has been mapped on linkage group 1 in the cowpea genome (Ouédraogo et al., 2002), and efforts are underway to develop molecular markers for use in indirect selection for resistance in various cowpea breeding programs.

An important feature of nematode resistance in cowpea is that it helps to limit the severe plant damage and yield loss caused by the root-knot nematode Fusarium wilt *(Fusarium oxysporum* Schlechtend.: Fr.f sp. *tracheiphilum* (E. F. Sm.) W. C. Snyder and H. N. Hanson) disease complex, in which nematode and fungus joint infection can be devastating (Roberts et al., 1995). Although the nematode is not required for disease associated with Fusarium wilt infection to occur in cowpea, the wilt is exacerbated in the presence of the nematode. In field tests, the combined independent resistances to both the nematode and Fusarium wilt were found to be effective in managing the disease complex (Roberts et al., 1995).

Nematode Resistance in Potato

Potato *(S. tuberosum)* is highly susceptible to root-knot nematodes (Luc, Sikora and Bridge, 1990; Evans, Trudgill and Webster, 1993). Problems from infection are especially acute because nematodes penetrate into the edible tuber, resulting in blemishes and rot that render the crop unmarketable or of lower grade to the processor (Evans, Trudgill and Webster, 1993; Ferris, Carlson and Westerdahl, 1994). Potato has the added problem of susceptibility to the potato cyst nematodes, *G. pallida* and *G. rostochiensis,* that infest many of the world's potato production areas (Turner and Evans, 1998). Resistance to *Globodera* spp. in potato is highly effective in some cases, but resistance implementation and efficacy are confounded in part owing to the occurrence of numerous pathotypes within each species that have specific (a)virulence profiles on the identified resistance genes (Marks and Brodie, 1998). In addition, resistance to *G. pallida* populations is typically multigenic and quantitative, based on several major and minor genes, and as a result has proved especially challenging in potato breeding. In fact, after many years of effort, high levels of resistance to *G. pallida* have not been incorporated into commercial cultivars, although partial resistance and tolerance have been bred into some cultivars (Trudgill, 1991). Among the undesirable outcomes of implementing the highly effective *G. rostochiensis* resistance gene *H1* in British potato cultivars has been the inadvertent selection of *G. pallida* in many *G. rostochiensis*-infested fields in which *G. pallida* was present at initially undetectably low frequencies. This *H1*-resistance-induced shift in the cyst nematode species composition has left many fields with *G. pallida* infestations that are more difficult to control

(Minnis et al., 2002). Further challenges to using resistant cultivars in potato cyst nematode management result from the genetic potential in potato cyst nematodes for virulence selection on resistance genes in *Solanum* spp. (Turner and Fleming, 2002).

Resistance to *Meloidogyne* spp. in potato and its wild relatives has been identified in a diverse range of *Solanum* spp. that is effective against *M. chitwoodi, M. fallax, M. hapla,* and *M. incognita* (Brown, Mojtahedi and Santo, 1995, 1999; Janssen et al., 1996). Several breeding programs are attempting to transfer these resistance traits into commercial cultivars. For example, resistance to *M. chitwoodi* conferred by major dominant genes from *Solanum bulbocastanum* and *Solanum hougasii* has been transferred into *Solanum tubersosum* elite breeding lines (Brown, Mojtahedi and Santo, 1995, 1999). Resistance to *M. fallax* and *M. chitwoodi* is being pursued by European breeding programs, such as in the Netherlands, where the complex of these nematode species presents a serious pest management problem (Janssen et al., 1996).

Meloidogyne Resistance in Sweet Potato

In warm subtropical and tropical areas, sweet potato *(I. batatas)* is highly susceptible to root-knot nematodes (Luc, Sikora and Bridge, 1990). Root-knot infection causes cracking of the edible storage root and reduces yield (Roberts and Scheuerman, 1984; Luc, Sikora and Bridge, 1990). Sweet potato clones with resistance to *M. incognita, M. javanica,* and *M. arenaria* have been identified (Jatala and Russell, 1972; Dukes et al., 1978; Roberts and Scheuerman, 1984; Bridge, 1996) and incorporated into commercial cultivars in some cases. In subsistence agriculture in New Guinea, sweet potatoes with *M. incognita* resistance or tolerance have been selected over many years by local farmers and are used in rotation schemes that maintain a sustainable level of root-knot nematode management (Bridge, 1996). Field evaluations of resistant sweet potato clones in *M. incognita*-infested fields in plots with and without preplant fumigant nematicide treatment revealed that resistant sweet potatoes are quite intolerant of infection and sustain significant yield loss in infested soils (Roberts and Scheuerman, 1984). The resistance was effective in protecting the harvested storage roots from cracking, but the plant growth and yield weights were reduced owing to plant response to nematode invasion and feeding. Therefore, the resistance requires integration with other management tactics, including nematicide treatment and crop rotation to lower nematode population densities in soil.

RESISTANCE IN NEMATODE MANAGEMENT PROGRAMS

Two primary characteristics of nematode-resistant crops underpin their utility in nematode management programs: (1) most resistant cultivars are more tolerant of nematode infection than comparable susceptible cultivars of the host crop and as such are protected against yield reductions caused by nematode infection. Sometimes this is called "self-protection" (Roberts, 2002). (2) Resistant plants suppress nematode reproduction compared with susceptible plants, resulting in reduced seasonal multiplication rates of nematode populations in soil. These two attributes are discussed further in relation to nematode management programs.

Tolerance and Yield of Resistant Plants

The impact of resistance and tolerance traits on crop yield can be considered in terms of critical point models or damage functions that relate relative plant growth or crop yield to the initial nematode population density in soil at planting time (Roberts, 1982; Ferris, 1985a,b; Trudgill, 1991; McSorley, 1998). These damage functions include simple linear regression models and also more biologically descriptive sigmoidal functions such as that developed by Seinhorst (1965). These have been described for a range of vegetable crops (Ferris et al., 1986; Duncan and Noling, 1998). For susceptible, intolerant crops, this relationship is linear except at very low or very high population densities. Both the position and the slope of the curve are governed by the relative tolerance of the particular cultivar (Figure 9.2). The incorporation of resistance and its associated tolerance shifts the function curve to the right, indicating a higher initial population density required to cause detectable yield reduction and reduce the function slope because less damage per nematode results in increased relative yields at high population densities. Knowledge of these damage

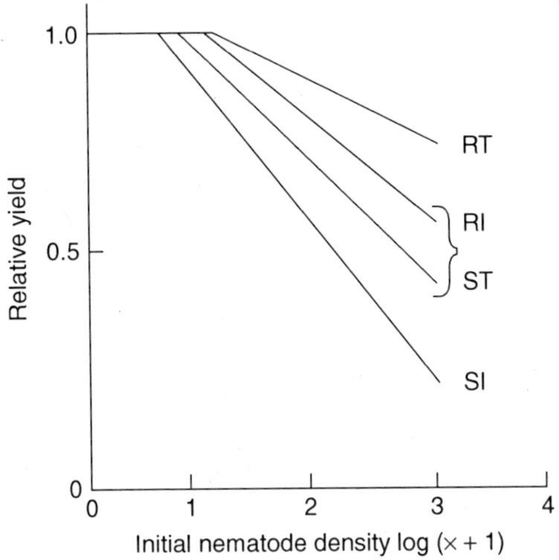

FIGURE 9.2. Hypothetical damage functions (relationship between yield and initial nematode density) for crop cultivars with different nematode resistance and tolerance traits. RT = resistant, tolerant; RI = resistant, intolerant; ST = susceptible, tolerant; SI = susceptible, intolerant (Roberts, 1982).

function curves is desirable for successful nematode management planning on the basis of nematode sampling and assay procedures (Duncan and Noling, 1998). Even simple estimates of nematode damage potential on common cultivars in a given region can provide useful guidance for nematode management decisions. In subsistence farming in developing countries, experienced field observation by local growers can help provide a ranking of crop cultivar performance for local variety selection and planting decisions (Bridge, 1996).

Yield comparisons of near-isogenic or closely related resistant and susceptible cultivars on infested and treated field plots have revealed a range of yield responses to nematode infection, indicating that resistant crops vary in their relative tolerance. Often these tests have been made at extremely high infestation levels, whereas in commercial farming the nematode population levels encountered are often less damaging and more variable, with patchy distributions. Tolerance levels in different cyst and root-knot nematode-resistant crops were found to range from 15 to 48 percent reduction in normal yield, whereas comparable susceptible cultivars had 49-90 percent reductions (Roberts,

1982). Comparisons of tolerance to heavy infestations of *M. incognita* in California fields ranged from no measurable yield reduction in highly tolerant resistant processing tomatoes (approximately 50 percent reduction in susceptible cultivars) (Roberts and May, 1986) to approximately 30 percent yield reduction in intolerant resistant sweet potatoes (Roberts and Scheuerman, 1984). Similar ranges of tolerance in potatoes resistant to potato cyst nematodes have been reported (Evans and Haydock, 1990; Trudgill, 1991). Even partial tolerance is useful because it will increase relative yield and may be all that is necessary for managing low to moderate nematode population levels (Alphey, Phillips and Trudgill, 1988). Tolerance combined with resistance is desirable because the large, healthier root systems of tolerant susceptible plants allow nematode populations to increase. However, in well-managed cropping systems where strategies to control nematodes are integrated or combined, tolerant susceptible crops can be very important (Roberts, 1993, 2002).

Effects of Resistance on Nematode Population Dynamics

Suppression of nematode multiplication during the growing season using nematode-resistant cultivars results in lower residual or final population densities in soil that could impact the next crop to be planted. In vegetable cropping systems, where one to several crops per year may be grown on the same ground, nematode problems can be managed by including crops with different levels of resistance, either singly or in combination (Roberts, 1993). Susceptible crops allow large increases in nematode populations, even from low initial densities, although the rate of population increase declines at higher initial densities owing to density-dependent effects on infected root systems (Ferris, 1985b). At high initial densities competition for feeding sites and food reserves increases and root systems of intolerant plants are smaller as a result of nematode injury (Ferris, 1985b; McSorley, 1998).

Nematode multiplication rates are lower on resistant plants than on susceptible plants but vary depending on the level of resistance. The relationship between P_f/P_i over a range of P_i values can be described by a negative exponential function (Ferris, 1985b). Highly resistant crops have a $P_f/P_i < 1$, and although moderately resistant plantings may generate a $P_f/P_i > 1$, the P_f is typically much lower than that resulting from planting a susceptible crop (Roberts, 2002). Relationships between initial and final *M. incognita* population densities on resistant and susceptible cotton from field experiments in California demonstrate these important resistance effects (Figure 9.3). Similar low P_f values and $P_f/P_i < 1$ were found for *M. incognita* populations in fields with resistant tomatoes carrying the *Mi-1* gene (Roberts and May, 1986). Suppression of nematode multiplication varies with the nematode virulence status and by any weakening of the resistance at high soil temperatures, as found for the *Mi-1* gene in tomato (Tzortzakakis, Trudgill and Phillips, 1998; Noling, 2000).

Integration of Nematode Resistance with Other Control Tactics

Protection of subsequent crops in a rotation by growing resistant cash or cover crops that suppress nematode multiplication is a powerful management strategy that can improve crop yield and economic returns (Rhoades, 1984; Roberts, 1993). Integration of resistance and crop rotation is commonly used in California, where highly resistant processing tomato cultivars with the *Mi-1* gene are used to manage *M. incognita* in rotation with susceptible cotton, bean, and other field

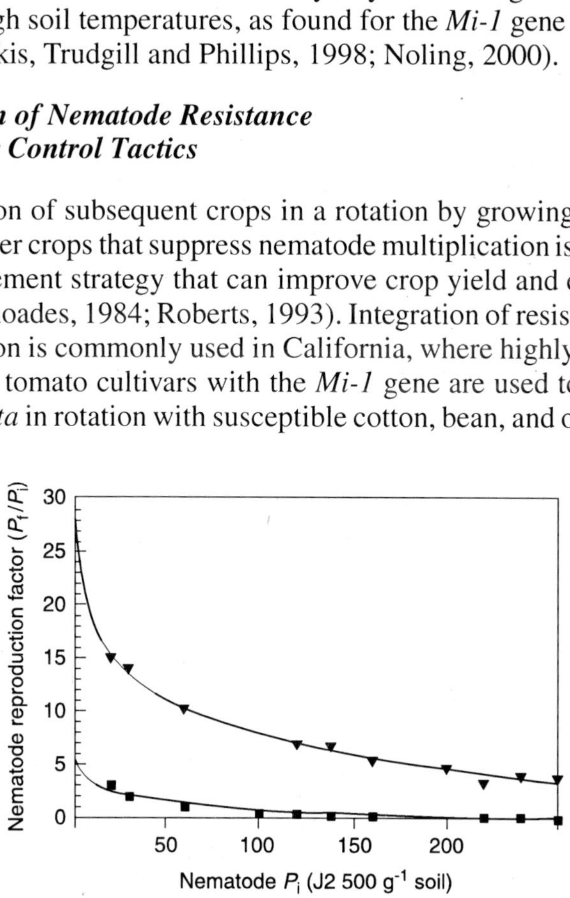

FIGURE 9.3. Varying initial densities (P_i) *of M. incognita and nematode reproduction factors* (P_f/P_i) on plots planted to resistant NemX ( ) and/or susceptible Maxxa ( ) cotton in rotations. The curves represent a logarithmic function. J2 = second-stage juveniles (from Ogallo et al., 1999).

and vegetable crops (Roberts and May, 1986). When susceptible crops are planted after resistant tomato, they require no nematicide treatment and yield normally because of low population densities after resistant tomato. In some cases, a highly resistant crop can provide at least two years of nematode control benefit, but soil sampling to estimate the nematode P_i before planting the next crop is advisable. In California, other resistant crops have been included in the annual rotations to strengthen the nematode management program. New cotton cultivars are available with resistance to *M. incognita* that suppress nematode multiplication, as shown in Figure 9.4A (Ogallo et

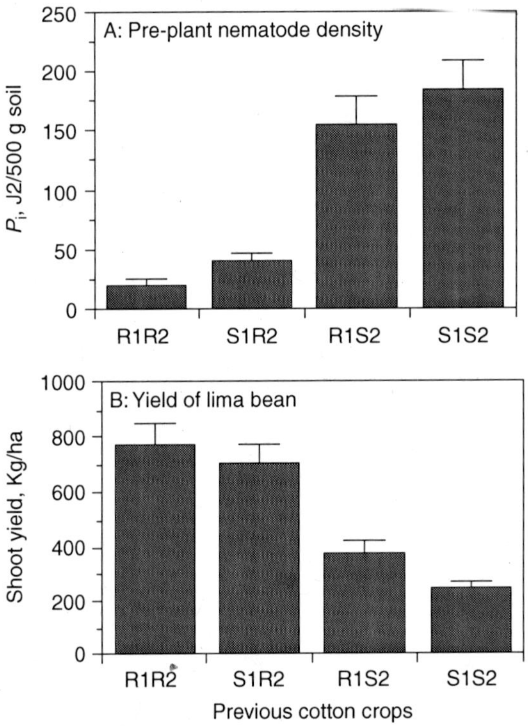

FIGURE 9.4. (A) Initial (preplant) population densities (P_i) of *Meloidogyne incognita* and (B) total shoot yields of susceptible lima bean planted in year 3 after resistant NemX (R) and (or) susceptible Maxxa (S) cotton were planted in year 1 and year 2 of a 3-year rotation. J2 = second-stage juveniles (from Ogallo et al., 1999).

al., 1999). The population suppression results in higher yields of susceptible lima bean and other crops grown after resistant cotton. The effect on lima bean shoot growth after one or two years of resistant or susceptible cotton is shown in Figure 9.4B, in which lima bean yield more than doubled following one or two years of resistant cotton (Ogallo et al., 1999).

Resistant Cover Crops

A variation on the strategy of combining resistance and crop rotation involves growing a resistant cover crop before planting the primary susceptible cash crop. The principle is the same, but the resistance is in a noncrop planting used to reduce nematode populations and improve soil health and structure, often by incorporating the cover crop top growth as a soil amendment (Bridge, 1996; Wang, McSorley and Gallaher, 2003; Roberts, Matthews and Ehlers, 2005). In tomato and other vegetable production systems in California and Florida, root-knot nematode-resistant cowpea carrying the *Rk* and Rk^2 genes has been shown to be highly beneficial in reducing nematode populations and protecting the following crop, especially when the cover crop is soil-incorporated (Wang, McSorley and Gallaher, 2003; Roberts, Matthews and Ehlers, 2005). Trap crops can also be used for reducing nematode population densities in soil before planting the primary cash crop, but the feasibility of these strategies must be determined for each farming operation (Noe, 1998).

Integration of resistance with other nematode control tactics is also used to some extent. Alphey, Phillips and Trudgill (1988) demonstrated that the application of small amounts of nematicide combined with planting a partially resistant potato cultivar was an effective approach to managing potato cyst nematode. Combining partial or high levels of resistance with other tactics, including biological control agents, manipulation of planting and harvest dates, and other cultural management approaches, all hold potential for integration (Roberts, 1993). Examples of integrated approaches are few, however, and require complex multifactorial experimentation to determine their true benefits. An integrated approach involving resistant crops can have the added benefit of extending the durability of the resistance, assuming that it reduces the virulence selection pressure of growing resistant plants by having to grow them less frequently or on reduced population levels. As new resistant vegetable crops become available, opportunities to develop a more integrated nematode management approach will increase.

REFERENCES

Allard, R.W. (1954). Sources of root-knot nematode resistance in lima beans. *Phytopathology* 44: 1-4.

Alphey, T.J.W., M.S. Phillips and D.L. Trudgill. (1988). Integrated control of potato cyst nematodes using small amounts of nematicide and potatoes with partial resistance. *Annals of Applied Biology* 113: 545-552.

Ammiraju, J.S.S., J.C. Veremis, X. Huang, P.A. Roberts and I. Kaloshian. (2003). The heat-stable root-knot nematode resistance gene *Mi-9* from *Lycopersicon peruvianum* is localized on the short arm of chromosome 6. *Theoretical and Applied Genetics* 106: 478-484.

Blazey, D.A., P.G. Smith, A.G. Gentile and S.T. Muyagawa. (1964). Nematode resistance in the common bean. *Journal of Heredity* 55: 20-22.

Boiteux, L.S., J.G. Belter, P.A. Roberts and P.W. Simon. (2000). RAPD linkage map of the genomic region encompassing the root-knot nematode (Meloidogyne javanica) resistance locus in carrot. *Theoretical and Applied Genetics* 100: 439-446.

Boiteux, L.S., J.R. Hayman, I.C. Bach, M.E.N. Fonseca, W.C. Matthews, P.A. Roberts and P.W. Simon. (2004). Employment of flanking codominant STS markers to estimate allelic substitution effects of a nematode resistance locus in carrot. *Euphytica* 136: 37-44.

Bridge, J. (1996). Nematode management in sustainable and subsistence agriculture. *Annual Review of Phytopathology* 34: 201-225.

Bridge, J. and S.L.J. Page. (1980). Estimation of root-knot nematode infestation levels on roots using a rating chart. *Tropical Pest Management* 2: 296-298.

Brito, J., T.O. Powers, P.G. Mullin, R.N. Inserra and D.W. Dickson. (2004). Morphological and molecular characterization of *Meloidogyne mayaguensis* isolates from Florida. *Journal of Nematology* 36: 232-240.

Brown, C.R., H. Mojtahedi and G.S. Santo. (1995). Introgression of resistance to Columbia and Northern root-knot nematodes from *Solanum bulbocastanum* into cultivated potato. *Euphytica* 83: 71-78.

Brown, C.R., H. Mojtahedi and G.S. Santo. (1999). Genetic analysis of resistance to *Meloidogyne chitwoodi* introgressed from *Solanum hougasii* in cultivated potato. *Journal of Nematology* 31: 264-271.

Cap, G.B., P.A. Roberts and I.J. Thomason. (1993). Inheritance of heat-stable resistance to *Meloidogyne incognita* in *Lycopersicon peruvianum* and its relationship to the *Mi* gene. *Theoretical and Applied Genetics* 85: 777-783.

Castagnone-Sereno, P., E. Wajnberg, M. Bongiovanni, F. Leroy and A. Dalmasso. (1994). Genetic variation in *Meloidogyne incognita* virulence against the tomato

Mi gene: Evidence from isofemale line selection studies. *Theoretical and Applied Genetics* 88: 749-753.

Charchar, J.M. and J.V. Viera. (1994). Selecao de cenoura com resistancia a nematoides de galhas (*Meloidogyne* spp.). *Horticalis Brasiliensis* 12: 144-148.

Chen, P. and P.A. Roberts. (2003a). Virulence in *Meloidogyne hapla* differentiated by resistance in common bean (*Phaseolus vulgaris*). *Nematology* 5: 39-47.

Chen, P. and P.A. Roberts. (2003b). Genetic analysis of (a)virulence in *Meloidogyne hapla* to resistance in bean (*Phaseolus vulgaris*). *Nematology* 5: 687-697.

Collard B.C.Y., M.Z.Z. Jahufer, J.B. Brouwer and E.C.K. Pang. (2005). An introduction to markers, quantitative trait loci (QTL) mapping and marker-assisted selection for crop improvement: The basic concepts. *Euphytica* 142: 169-196.

Cook, R. (1991). Resistance in plants to cyst and root-knot nematodes. *Agricultural Zoology Reviews* 4: 213-240.

Cook, R. and K. Evans. (1987). Resistance and tolerance. In *Principles and Practice of Nematode Control in Crops,* R.H. Brown and B.R. Kerry, eds. Academic Press, Sydney, Australia, pp. 179-231.

Cook, R. and G.R. Noel. (2002). Cyst nematodes: *Globodera and Heterodera* species. In *Plant Resistance to Parasitic Nematodes,* J.L. Starr, R. Cook and J. Bridge, eds. CABI Publishing, Wallingford, UK, pp. 71-105.

Djian-Caporalino, C., L. Pijarowski, A. Fanuel, V. Lefebre, A. Daubeze, A. Palloix, A. Dalmasso and P. Abad. (1999). Spectrum of resistance to root-knot nematodes and inheritance of heat-stable resistance in pepper (*Capsicum annuum* L.). *Theoretical and Applied Genetics* 99: 496-502.

Djian-Caporalino, C., L. Pijarowski, A. Fazari, M. Samson, L. Gaveau, C. O'Byrne, V. Lefebre, C. Caranta, A. Palloix and P. Abad. (2001). High resolution mapping of the pepper (*Capsicum annuum* L.) resistance loci Me_3 and Me_4 conferring heat-stable resistance to root-knot nematodes (*Meloidogyne* spp.). *Theoretical and Applied Genetics* 103: 592-600.

Dukes, P.D., A. Jones, F.P. Cuthbert and M.G. Hamilton. (1978). W-51 root-knot resistant sweet potato germplasm. *HortScience* 13: 201-202.

Duncan, L.W. and J.W. Noling. (1998). Agricultural sustainability and nematode integrated pest management. In *Plant Nematode Interactions,* K.R. Barker, G.A. Pederson and G.L. Windham, eds. American Society of Agronomy, Madison, Wisconsin, USA, pp. 251-287.

Ehlers, J.D., W.C. Matthews, A.E. Hall and P.A. Roberts. (2000). Inheritance of a broad-based form of root-knot nematode resistance in cowpea. *Crop Science* 40: 611-618.

Ehlers, J.D., W.C. Matthews, A.E. Hall and P.A. Roberts. (2003). Breeding and evaluation of cowpeas with high levels of broad-based resistance to root-knot nematodes. In *Challenges and Opportunities for Enhancing Sustainable Cowpea Production,* C.A. Fatokun, S.A. Tarawali, B.B. Singh, P.M. Kormawa and M. Tamo, eds. International Institute Tropical Agriculture, Ibadan, Nigeria, pp. 41-51.

Evans, K. and P.P.J. Haydock. (1990). A review of tolerance by potato plants of cyst nematode attack, with consideration of what factors may confer tolerance and methods of assaying and improving it in crops. *Annals of Applied Biology* 117: 703-740.

Evans, K., D.L. Trudgill and J.M. Webster, eds. (1993). *Plant Parasitic Nematodes in Temperate Agriculture.* CAB International, Wallingford, UK, 648p.

Ferris, H. (1985a). Population assessment and management strategies for plant-parasitic nematodes. *Agricultural, Ecosystems and Environment* 12: 285-299.

Ferris, H. (1985b). Density-dependent nematode seasonal multiplication rates and overwinter survivorship: A critical point model. *Journal of Nematology* 17: 3-100.

Ferris, H., D.A. Ball, L.W. Beem and L.A. Gudmundson. (1986). Using nematode count data in crop management decisions. *California Agriculture* 40: 12-14.

Ferris, H., H.D. Carlson and B.B. Westerdahl. (1994). Nematode population changes under crop rotation sequences: Consequences for potato production. *Agronomy Journal* 86: 340-348.

Fery, R.L. and P.D. Dukes. (1980). Inheritance of root-knot resistance in the cowpea (*Vigna unguiculata* (L.) Walp.). *Journal of the American Society for Horticultural Science* 105: 671-674.

Gilbert, J. C. and D.C. McGuire. (1956). Inheritance of resistance to severe root-knot from *Meloidogyne incognita* in commercial type tomatoes. *Proceedings of the American Society for Horticultural Science* 68: 437-442.

Goggin, F.L., S. Hebert, G. Shah, V.M. Williamson and D.E. Ullman. (2004). Instability of Mi-mediated resistance in transgenic tomato plants. In *Proceedings of Plant and Animal Genome Conference XII,* San Diego, California, USA, 98p.

Hare, W.W. (1956). Resistance in pepper to *Meloidogyne incognita acrita. Phytopathology* 46: 98-104.

Hall, A.E., N. Cisse, S. Thiaw, H.O.A. Elawad, J.D. Ehlers, A.M. Ismail, R.L. Fery, et al. (2003). Development of cowpea cultivars and germplasm. *Field Crops Research* 82: 103-134.

Helms, D.M., W.C. Matthews, S.R. Temple and P.A. Roberts. (2004). Registration of 'Cariblanco N' Lima bean. *Crop Science* 44: 352-353.

Hendy, H., E. Pochard and A. Dalmasso. (1985). Transmission héréditaire de la résistance aux *Meloidogyne* portée par deux lignées de *Capsicum annuum*: études de descendance d'homozygotes issues d'androgénèse. *Agronomie* 5: 93-100.

Huang, S.P. (1986). Penetration, development, reproduction and sex ratio of *Meloidogyne javanica* in three carrot cultivars. *Journal of Nematology* 18: 408-412.

Hussey, R.S. and G.J.W. Janssen. (2002). Root-knot nematodes: *Meloidogyne* species. In *Plant Resistance to Parasitic Nematodes,* J.L. Starr, R. Cook and J. Bridge, eds. CABI Publishing, Wallingford, UK, pp. 43-70.

Isbell, C.L. (1931). Nematode resistance studies with pole snap-beans. *Journal of Heredity* 22: 191-198.

Janssen, G.J.W., A. van Norel, B. Verkerk-Bakker and R. Janssen. (1996). Resistance to *Meloidogyne chitwoodi, M. fallax* and *M. hapla* in wild tuber-bearing *Solanum* spp. *Euphytica* 92: 287-294.

Janssen, R., J. Bakker and F.J. Gommers. (1991). Mendelian proof for a gene-forgene relationship between virulence of *Globodera rostochiensis* and the *H1* resistance gene in *Solanum tuberosum* ssp. *andigena* CPC 1673. *Revue de Nématologie* 14: 213-219.

Jatala, P. and C.C. Russell. (1972). Nature of sweet potato resistance to *Meloidogyne incognita* and the effects of temperature on parasitism. *Journal of Nematology* 4: 1-7.

Jones, F.G.W. (1985). Modeling multigenic resistance to potato cyst nematodes. OEPP/EPPO *Bulletin* 15: 155-166.

Lindsey, D.L. and M.S. Clayshulte. (1982). Influence of initial population densities of *Meloidogyne incognita* on three chili cultivars. *Journal of Nematology* 14: 353-358.

Luc, M., R.A. Sikora and J. Bridge, eds. (1990). *Plant Parasitic Nematodes in Tropical and Subtropical Agriculture.* CAB International, Wallingford, UK, 629p.

Marks, R.J. and B.B. Brodie, eds. (1998). *Potato Cyst Nematodes: Biology, Distribution and Control.* CABI Publishing, Wallingford, UK, 432p.

Matthews, W.C., D.M. Helms and P.A. Roberts. (2000). Evidence for independent genetic control of reproduction and galling to *Meloidogyne javanica* in lima bean. *Journal of Nematology* 32: 444 (Abstr.).

Matthews, W.C., P.W. Simon and P.A. Roberts. (1999). Influence of temperature on expression of resistance in carrot to *Meloidogyne javanica. Journal of Nematology* 31: 553 (Abstr.).

McGuire, D.C., R.W. Allard and J.A. Harding. (1961). Inheritance of root-knot nematode resistance in lima beans. *Journal of American Society of Horticultural Science* 78: 302-307.

McSorley, R. (1998). Population dynamics. In *Plant Nematode Interactions,* K.R. Barker, G.A. Pederson and G.L. Windham, eds. American Society of Agronomy, Madison, Wisconsin, USA, pp. 109-133.

Medina-Filho, H.P. and S.D. Tanksley. (1983). Breeding for nematode resistance. In *Handbook of Plant Cell Culture, Vol. 1,* D.A. Evans, W.R. Sharp, P.V. Ammirato and Y. Yamada, eds. Macmillan Publishing Company, London, UK, pp. 904-923.

Mercer, C.F., S.W. Hussain and K.K. Moore. (2004). Resistance reactions to *Meloidogyne trifoliophila* in *Trifolium repens* and *T. semipilosum. Journal of Nematology* 36: 499-504.

Milligan S.B., J. Bodeau, J. Yaghoobi, I. Kaloshian, P. Zabel and V.M. Williamson. (1998). The root knot nematode resistance gene *Mi* from tomato is a member of the leucine zipper, nucleotide binding, leucine-rich repeat family of plant genes. *Plant Cell* 10: 1307-1320.

Minnis, S.T., P.P.J. Haydock, S.K. Ibrahim, I.G. Grove, K. Evans and M.D. Russell. (2002). Potato cyst nematodes in England and Wales—Occurrence and distribution. *Annals of Applied Biology* 140: 187-195.

Noe, J.P. (1998). Crop- and nematode-management systems. In *Plant Nematode Interactions,* K.R. Barker, G.A. Pederson and G.L. Windham, eds. American Society of Agronomy, Madison, Wisconsin, USA, pp.159-171.

Noling, J.W. (2000). Effects of continuous culture of a resistant tomato cultivar on *Meloidogyne incognita* soil population densities and pathogenicity. *Journal of Nematology* 32: 452 (Abstr.).

Ogallo, J.L., P.B. Goodell, J. Eckert and P.A. Roberts. (1999). Management of root-knot nematodes with resistant cotton cv. NemX. *Crop Science* 39: 418-421.

Oka, Y., R. Offenbach and S. Pivonia. (2004). Pepper rootstock graft compatibility and response to *Meloidogyne javanica* and *M. incognita*. *Journal of Nematology* 36: 137-141.

Omwega, C.O. and P.A. Roberts. (1992). Inheritance of resistance to *Meloidogyne* spp. in common bean and the genetic basis of its sensitivity to temperature. *Theoretical and Applied Genetics* 83: 720-726.

Omwega, C.O., I.J. Thomason and P.A. Roberts. (1990a). A single dominant gene in common bean conferring resistance to three root-knot nematode species. *Phytopathology* 80: 745-748.

Omwega, C.O., I.J. Thomason and P.A. Roberts. (1990b). Effect of temperature on expression of resistance to *Meloidogyne* spp. in common bean (*Phaseolus vulgaris*). *Journal of Nematology* 22: 446-451.

Omwega, C.O., I.J. Thomason, P.A. Roberts and J.G. Waines. (1989). Identification of new sources of resistance to root-knot nematodes in *Phaseolus*. *Crop Science* 29: 1463-1468.

Ouédraogo, J.T., B.S. Gowda, M. Jean, T.J. Close, J.D. Ehlers, A.E. Hall, A.G. Gillaspie, P.A. Roberts, A.M. Ismail, G. Bruening et al. (2002). An improved genetic linkage map for cowpea (*Vigna unguiculata* L.) combining AFLP, RFLP, RAPD, biochemical markers and biological resistance traits. *Genome* 45: 175-188.

Rhoades, H.L. (1984). Effects of fallowing, summer cover crops and fenamiphos on nematode populations and yields in a cabbage-field corn rotation in Florida. *Nematropica* 14: 131-138.

Roberts, P.A. (1982). Plant resistance in nematode pest management. *Journal of Nematology* 14: 24-33.

Roberts, P.A. (1987). The influence of date of planting of carrot on *Meloidogyne incognita* reproduction and injury to roots. *Nematologica* 33: 335-342.

Roberts, P.A. (1992). Current status of the availability, development and use of host plant resistance to nematodes. *Journal of Nematology* 24: 213-227.

Roberts, P.A. (1993). The future of nematology: Integration of new and improved management strategies. *Journal of Nematology* 25: 383-394.

Roberts, P.A. (1995). Conceptual and practical aspects of variability in root-knot nematodes related to host plant resistance. *Annual Review of Phytopathology* 33: 199-221.

Roberts, P.A. (2002). Concepts and consequences of resistance. In *Plant Resistance to Parasitic Nematodes,* J.L. Starr, R. Cook and J. Bridge, eds. CABI Publishing, Wallingford, UK, pp. 23-41.

Roberts, P.A., A. Dalmasso, G.B. Cap and P. Castagnone-Sereno. (1990). Resistance in *Lycopersicon peruvianum* to isolates of *Mi* gene-compatible *Meloidogyne* populations. *Journal of Nematology* 22: 585-589.

Roberts, P.A., C.A. Frate, W.C. Matthews and P.P. Osterli. (1995). Interactions of virulent *Meloidogyne incognita* and Fusarium Wilt on resistant cowpea genotypes. *Phytopathology* 85: 1288-1295.

Roberts, P.A., A.C. Magyarosy, W.C. Matthews and D.M. May. (1988). Effects of metam-sodium applied by drip irrigation on root-knot nematodes, *Pythium ultimum* and *Fusarium* sp. in soil and on carrot and tomato roots. *Plant Disease* 72: 213-217.

Roberts, P.A., W.C. Matthews and J.D. Ehlers. (1996). New resistance to virulent root-knot nematodes linked to the *Rk* locus of cowpea. *Crop Science* 36: 889-894.

Roberts, P. A., W.C. Matthews and J.D. Ehlers. (2005). Root-knot nematode resistant cowpea cover crops in tomato production systems. *Agronomy Journal* 97: 1626-1635.

Roberts, P.A., W.C. Matthews and J.C. Veremis. (1998). Genetic mechanisms of host plant resistance to nematodes. In *Plant Nematode Interactions,* K.R. Barker, G.A. Pederson and G.L. Windham, eds. American Society of Agronomy, Madison, WI, pp. 209-238.

Roberts, P.A. and D.M. May. (1986). *Meloidogyne incognita* resistance characteristics in tomato genotypes developed for processing. *Journal of Nematology* 18: 353-359.

Roberts, P.A. and R.W. Scheuerman. (1984). Field evaluation of sweet potato clones and cultivars for reaction to root-knot and stubby root nematodes in California. *Hort Science* 19: 270-273.

Roberts, P.A. and I.J. Thomason. (1986). Variability in reproduction of isolates of *Meloidogyne incognita* and *M. javanica* on resistant tomato genotypes. *Plant Disease* 70: 547-551.

Rossi, M., F.L. Goggin, S.B. Milligan, I. Kaloshian, D.E. Ullman and V.M. Williamson. (1998). The nematode resistance gene *Mi* of tomato confers resistance against the potato aphid. *Proceedings of the National Academy of Sciences USA* 95: 9750-9754.

Sasser, J.N. (1980). Root-knot nematodes: A global menace to crop production. *Plant Disease* 64: 36-41.

Sasser, J.N. and D.W. Freckman. (1987). A world perspective on nematology: the role of the society. In *Vistas on Nematology,* J.A. Veech and D.W. Dickson, eds. Society of Nematologists Inc., Hyattsville, MD, pp. 7-14.

Seinhorst, J.W. (1965). The relationship between nematode density and damage to plants. *Nematologica* 11: 137-154.

Shaner G, G.H. Lacy, E.L. Stromberg, K.R. Barker and T.P. Pirone. (1992). Nomenclature and concepts of pathogenicity and virulence. *Annual Review of Phytopathology* 30: 47-66.

Simon, P.W., W.C. Matthews and P.A. Roberts. (2000). Evidence for a simply inherited dominant resistance to *Meloidogyne javanica* in carrot. *Theoretical and Applied Genetics* 100: 735-742.

Smith, P.G. (1944). Embryo culture of a tomato species hybrid. *Proceedings of the American Society for Horticultural Science* 44: 413-416.

Starr, J.L., R. Cook and J. Bridge, eds. (2002). *Plant Resistance to Parasitic Nematodes.* CABI Publishing, Wallingford, UK, 258p.

Sydenham, G.A., R. McSorley and R.A. Dunn. (1996). Effects of resistance in *Phaseolus vulgaris* on development of *Meloidogyne* species. *Journal of Nematology* 28: 485-491.

Thies, J.A. and R.L. Fery. (2000). Characterization of resistance conferred by the *N* gene to *Meloidogyne arenaria* races 1 and 2, *M. hapla* and *M. javanica* in two sets of isogenic lines of *Capsicum annuum. Journal of American Society of Horticultural Science* 125: 71-75.

Thies, J.A. and R.L. Fery. (2002). Host plant resistance as an alternative to methyl bromide for managing *Meloidogyne incognita* in pepper. *Journal of Nematology* 34: 374-377.

Thies, J.A., J.D. Mueller and R.L. Fery. (1997). Effectiveness of resistance to southern root-knot nematode in 'Carolina Cayenne' pepper in greenhouse, microplot and field tests. *Journal of American Society of Horticultural Science* 122: 200-204.

Thies, J.A., J.D. Mueller and R.L. Fery. (1998). Use of a resistant pepper as a rotational crop to manage southern root-knot nematode. *HortScience* 18: 353-359.

Thomas, S.H., L.W. Murray and M. Cardenas. (1995). Relationship of preplant population densities of *Meloidogyne incognita* to damage in three chile pepper cultivars. *Plant Disease* 79: 557-559.

Timko, M.P., J.D. Ehlers and P.A. Roberts. (2006). Cowpea. In *Genome Mapping and Molecular Breeding in Plants,* C. Kole, ed. Springer, Heidelberg, Germany, p. 3.

Trudgill, D.L. (1991). Resistance to and tolerance of plant parasitic nematodes in plants. *Annual Review of Phytopathology* 29: 167-92.

Turner, S.J. and K. Evans. (1998). The origins, global distribution and biology of potato cyst nematodes (*Globodera rostochiensis* (Woll.) and *G. pallida* Stone). In *Potato Cyst Nematodes: Biology, Distribution and Control,* R.J. Marks and B.B. Brodie, eds. CABI Publishing, Wallingford, UK, pp. 7-26.

Turner, S.J. and C.C. Fleming. (2002). Selection of potato cyst nematode *Globodera pallida* virulence on a range of potato species I. Serial selection on *Solanum*-hybrids. *European Journal of Plant Pathology* 108: 461-467.

Tzortzakakis, E.A., D.L. Trudgill and M. Phillips. (1998). Evidence for a dosage effect of *Mi* gene on partially virulent isolates of *Meloidogyne javanica. Journal of Nematology* 30: 76-80.

Veremis, J.C. and P.A. Roberts. (1996a). Differentiation of *Meloidogyne incognita* and M. *arenaria* novel resistance phenotypes in *Lycopersicon peruvianum* and derived bridge-lines. *Theoretical and Applied Genetics* 93: 960-967.

Veremis, J.C. and P.A. Roberts. (1996b). Relationships between *Meloidogyne incognita* resistance genes in *Lycopersicon peruvianum* differentiated by heat sensitivity and nematode virulence. *Theoretical and Applied Genetics* 93: 950-959.

Veremis, J.C. and P.A. Roberts. (1996c). Identification of resistance to *Meloidogyne javanica* in the *Lycopersicon peruvianum* complex. *Theoretical and Applied Genetics* 93: 894-901.

Veremis, J.C. and P.A. Roberts. (2000). Diversity of heat-stable resistance to *Meloidogyne* in Maranon races of *Lycopersicon peruvianum* complex. *Euphytica* 111: 9-16.

Veremis, J.C., A.W. van Heusden and P.A. Roberts. (1999). Mapping a novel heat-stable resistance to *Meloidogyne in Lycopersicon peruvianum. Theoretical and Applied Genetics* 98: 274-280.

Wang, M. and I.L. Goldman. (1996). Resistance to root-knot nematode (*Meloidogyne hapla* Chitwood) in carrot is controlled by two recessive genes. *Journal of Heredity* 87: 119-123.

Wang, K.H., R.M. McSorley and R.N. Gallaher. (2003). Host status and amendment effects of cowpea on *Meloidogyne incognita* in vegetable production systems. *Nematropica* 33: 215-224.

Williamson, V.M. (1998). Root-knot nematode resistance genes in tomato and their potential for future use. *Annual Review of Phytopathology* 36: 277-293.

Williamson, V.M., J.Y. Ho, F.F. Wu, N. Miller and I. Kaloshian. (1994). A PCR-based marker tightly linked to the nematode resistance gene, *Mi,* in tomato. *Theoretical and Applied Genetics* 87: 757-763.

Yaghoobi, J., I. Kaloshian, Y. Wen and V.M. Williamson. (1995). Mapping a new nematode resistance locus in *Lycopersicon peruvianum. Theoretical and Applied Genetics* 91: 757-763.

Young, N. and J. Mudge. (2002). Marker-assisted selection for soybean cyst nematode resistance. In *Plant Resistance to Parasitic Nematodes,* J.L. Starr, R. Cook and J. Bridge, eds. CABI Publishing, Wallingford, UK, pp. 241-252.

Chapter 10

RNAi: A Novel Approach to the Management of Insect- and Nematode-Borne Diseases

Tony Grace
Ginny Antony

INTRODUCTION

According to the central dogma, proteins, which are responsible for the bulk of cellular functions, are made in two steps. The first step, transcription, copies genes from double-stranded deoxyribonucleic acid (DNA) molecules to single-stranded ribonucleic acid (RNA) molecules called messenger RNA (mRNA) in the nucleus. The mRNA then exits the nucleus, and in the second step, called translation, the mRNA is converted into protein in the cytoplasm. Although this seems to be simple and straightforward, the real story is far more complex, with many different proteins and RNAs involved in the transcription and translational machinery. In the past, significant discoveries from around the world soon established that not all RNAs that are transcribed end up as protein. These novel classes of RNAs, known as noncoding RNAs, direct an astonishing diversity of regulatory pathways and have established a new paradigm for understanding eukaryotic gene regulation. One such class of RNAs directs sequence-specific degradation of mRNA by the process called RNA interference (RNAi). Since RNAi was first discovered in the late 1990s, laboratories around

Management of Nematode and Insect-Borne Plant Diseases

doi:10.1300/5754_10

the world—both academic labs and biotech firms—have rapidly adopted this technology and are competing to harness its potential. Intensive studies have focused on developing RNAi technologies for treating human diseases. Yet the application of RNAi in the management of insect- and nematode-vectored diseases of plants is still in its infancy. This chapter will provide some insights into the historical background of RNAi, its biology, and its functions in the eukaryotic system. Current concepts and strategies for the management of insect- and nematode-borne diseases of crop plants using RNAi will also be discussed.

HISTORICAL BACKGROUND

Jorgenson et al. (1996) stumbled upon the first evidence for an alternative role for RNA in plants. In an attempt to deepen the purple color of petunia plants, they introduced extra copies of the pigment-producing gene chalcone synthase (*chs*). In addition to producing plants with purple flowers that were no different from nontransgenic plants, a small proportion of the transgenic plants produced either white flowers or variegated flowers. Although *chs* mRNA levels in white and variegated tissues were dramatically reduced, the transcription rates of the introduced as well as endogenous *chs* were no different from those in nontransgenic purple plants (Napoli, Lemieux and Jorgensen, 1990; Jorgensen et al., 1996). This phenomenon was termed cosuppression. Later experiments demonstrated that homologous transcripts are produced but are rapidly degraded in these transgenic plants (Metzlaff et al., 1997).

The real breakthrough in understanding the mechanism of silencing by RNA came from the work done on the free-living nematode *Caenorhabditis elegans*. Guo and Kemphures (1995) attempted to use RNA complementary to *C. elegans par1* mRNA to block *par1* gene expression. To their surprise, they found that both antisense and control sense RNA preparations induced silencing. Subsequently, Fire et al. (1998) discovered that double-stranded RNA (dsRNA) was substantially more effective at producing interference than either strand individually. Only a few molecules of injected dsRNA were required per cell to bring about specific interference, suggesting the possibility of an amplification component in this process. Another important result from their experiment was that this knockout phenotype was "transmissible" and was observed in F_2 and F_3 progeny.

Consequently, it was found that it was not necessary to inject dsRNA into the nematode *C. elegans* because similar phenocopies can also be produced by feeding nematodes on bacteria expressing dsRNA or even by soaking nematodes in dsRNA (Timmons and Fire, 1998; Sijen et al., 2001). However, this silencing effect was achieved only with a transcribed gene segment, and nontranscribed regions such as introns or promoters were less effective. This potent and highly specific silencing induced by dsRNA was RNAi. Further experiments suggested that RNAi targets a posttranscriptional event in gene expression and leads to sequence-specific degradation of mRNA. This results in phenotypes that are identical to genetic null mutations.

Scientists working on developing virus-resistant plants using viral genes (pathogen-derived resistance) observed a similar silencing mechanism that operates by targeted RNA degradation (Dougherty and Parks, 1995). This phenomenon was termed post-transcriptional gene silencing (PTGS). Important insights into the PTGS mechanism came from the discovery of short RNA species of approximately 22-25 nucleotides, corresponding to both the sense and antisense sequence of the targeted mRNA in silenced plants (Hamilton and Baulcombe, 1999). These RNA species were designated siRNAs (small interfering dsRNA) and were found consistently in a broad spectrum of plants undergoing PTGS.

Analogous to PTGS in plants, RNAi in *C. elegans* and *Drosophila* was also found to be associated with the formation of small RNAs (Elbashir, Lendeckel and Tuschl, 2001). The presence of siRNAs in association with the phenomenon of RNAi in *C. elegans* was detected by injecting radiolabeled dsRNA into the gonad of the nematode. Thus, siRNAs became recognized as the signature component in silencing triggered by transgenes, microinjected RNAs, viruses, and transposons. This enabled scientists to conclude that PTGS (in plants) and RNAi (in animals) involve siRNAs and have the same underlying mechanism. To date, researchers from all over the world have shown that RNAi is evolutionarily conserved and occurs in a wide range of eukaryotic organisms, including nematodes, flies, fungi, plants, and animals.

MECHANISM OF RNAi

The realization that RNAi is a conserved biological phenomenon led to much excitement in the biological world. Over the past few years much research has been conducted to unravel the RNAi phenomenon. The emerging view from these studies is that the initiator in this silencing pathway is dsRNA, which acts as a trigger and/or intermediate in the process. Various sources of dsRNA in the cell are summarized in Figure 10.1.

Several lines of evidence also suggest that transgenes can trigger silencing through mechanisms in which dsRNA is not the initial stimulus. Plants, fungi, and *C. elegans* have genes that encode putative cellular RNA-dependent RNA polymerases (RdRPs), which have been found to be involved in the RNA silencing machinery (Cogoni and Macino, 1999, 2000). These cellular RdRPs are required for transgene-mediated cosuppression in plants but are not essential for virus-induced silencing, presumably because the virus provides its own viral RNA polymerase (Dalmay et al., 2000). This means that the transgene or its single-stranded mRNA can also activate the RNAi machinery. It is hypothesized that RdRPs somehow recognize the transgene products as abnormal or aberrant and subsequently convert them to dsRNA. In this case the dsRNA is an intermediate in the pathway rather than a trigger. But how the cell recognizes the mRNA as aberrant is still not known. The sequence of dsRNA need not be identical to the target mRNA to effect the degradation: dsRNAs that were 88 percent identical to target mRNA triggered RNAi, but these triggers still contained significant stretches that perfectly complemented mRNA. dsRNA is then cleaved into siRNA of 21-25 nucleotides in length in the cytoplasm, by a protein complex containing the dsRNA processing enzyme Dicer (Bernstein et al., 2001). Dicer is an RNase III-like enzyme that shows high specificity for dsRNAs. This enzyme has four distinct domains: a dsRNA-binding domain, a helicase domain, a PAZ domain, and dual RNAse III motifs (Tabara et al., 1999). Dicer is evolutionarily well conserved, with homologs present in fungi, plants, and animals including *C. elegans, Drosophila,* and mammals. These siRNAs are then incorporated into an RNA-induced silencing complex (RISC). Within this complex, the strands of short dsRNA are separated, and one of the strands is used as a guide to rec-

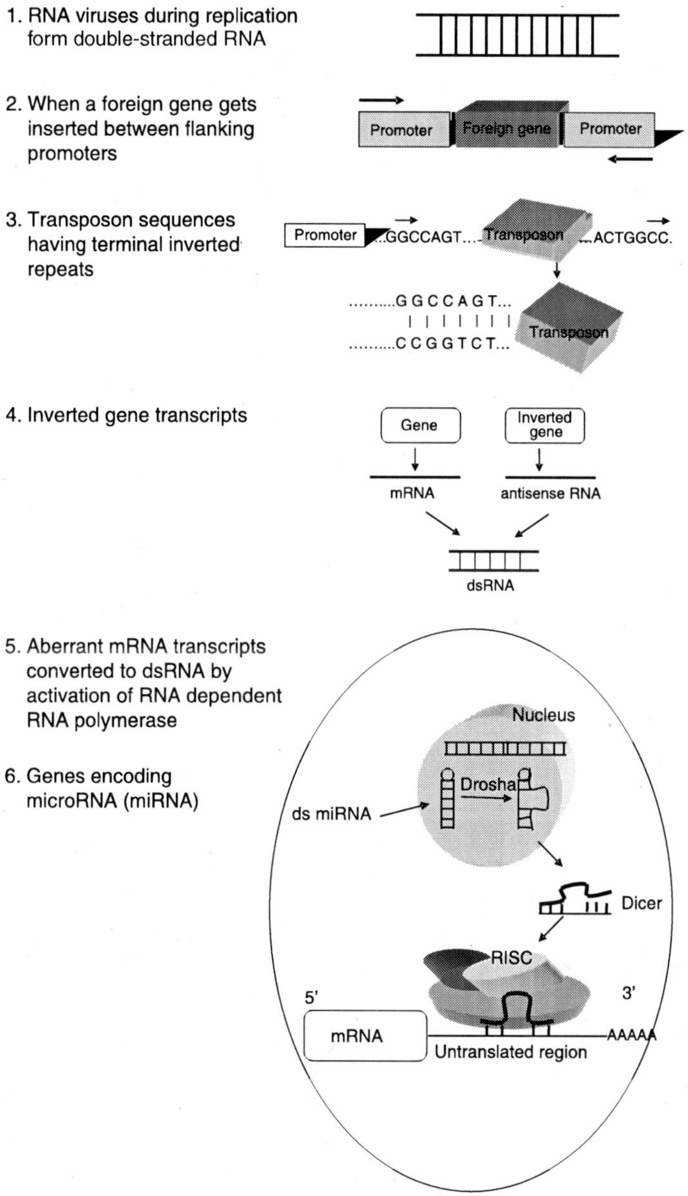

FIGURE 10.1. Sources of double-stranded RNA (dsRNA). mRNA, messenger RNA; RISC; RNA-induced silencing complex.

ognize single-stranded RNA (target mRNA) with complementary sequences (Figure 10.2). During the purification of RISC, discrete 25-nucleotide-long RNA species were found to cofractionate with RISC (Hammond et al., 2000).

The complex then cleaves the target mRNA approximately in the middle of the guide sequence. In plants and *C. elegans,* the effects of

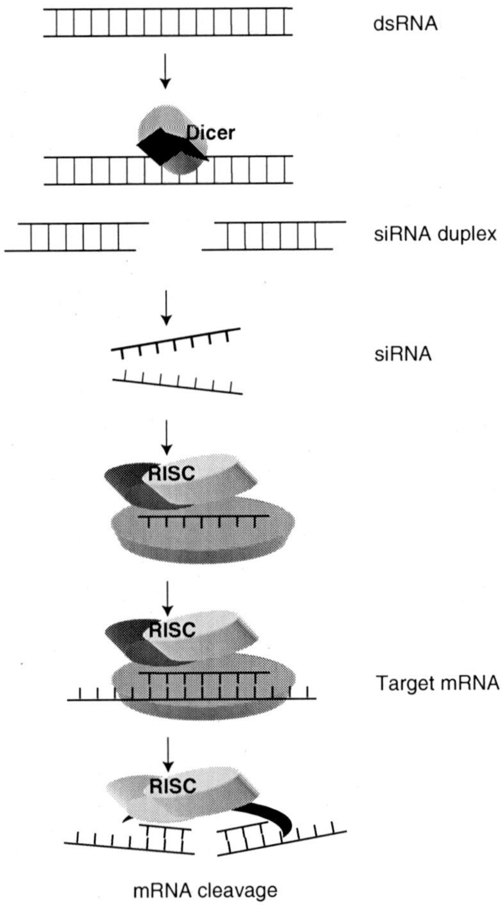

FIGURE 10.2. Mechanism of gene silencing induced by double-stranded RNA (dsRNA). mRNA, messenger RNA; RISC, RNA-induced silencing complex; siRNA, small interfering dsRNA.

silencing extend beyond those cells in which the silencing is initiated and can spread systemically (Fire et al., 1998). In plants, it has been shown that the silencing can move across a graft union. The transgene in a GUS (reporter gene)-expressing scion grafted onto a GUS-silenced rootstock became silenced, and short RNAs were detected in both scion and rootstock (Palauqui et al., 1997). This has led to speculation concerning a mobile silencing signal that can move from cell to cell and even across graft unions. It has been proposed that long-distance movement of the signal occurs through the phloem and cell-to-cell movement occurs through the plasmodesmata (Fagard et al., 2000). However, the nature of this mobile signal is still a mystery.

The RNA silencing machinery is also able to amplify the degradation signal. This is evident in invertebrates where siRNAs function as primers that are extended on the targeted mRNA by an RdRP, resulting in production of dsRNA and thus more siRNAs (Fire et al., 1998). This amplification component has not yet been observed in mammalian systems.

In addition to siRNAs, another class of small RNAs that regulate gene expression have been identified. These are termed micro RNAs (miRNA) and originate from indigenous transcripts that contain complementary or near-complementary inverted repeats, which fold back to form dsRNA hairpins. These long primary transcripts are processed by the enzyme Drosha in the nucleus. The miRNA precursor is then exported to cytoplasm, where Dicer further processes it. miRNA then forms a complex with ribonucleoproteins to form miRNPs (Lee et al., 2002). This miRNP complex targets mRNA and prevents translation in animals and mRNA cleavage in plants. Initially, miRNAs were discovered in *C. elegans* to be regulating developmental genes (Lee, Feinbaum and Ambros, 1993). However, the range of targets for miRNA in plants is not restricted to developmental genes (Baehrecke, 2003). miRNAs with regulatory roles in developmental timing, cell death, and cell proliferation have been identified in animals.

Hundreds of miRNA genes have been found in diverse organisms, and many of these are phylogenetically conserved (Lagos-Quintana et al., 2001). Plant and metazoan miRNA pathways are more or less the same. Plant miRNAs are more perfectly paired to their target RNA and result in RNA cleavage rather than translational repression. Animal miRNAs normally target the 3′ UTR (untranslated region) of mRNA, whereas plant miRNAs have targets in the coding sequence or even in the 5′-UTR of mRNA. The silencing pathways are further complicated by the fact that some miRNAs can act as siRNAs and some siRNAs can act as miRNAs (Doench, Petersen and Sharp, 2003; Saxena, Jonsson and Dutta, 2003). A complete understanding of the numerous branching and converging silencing pathways that appear to exist in diverse organisms will no doubt take years of research. Not all transgenes/transgenic events induce RNA silencing in plants and animals, because novel traits can be introduced using transgenesis.

Several models have been proposed to explain the cause of RNA silencing in plants, and in most cases these models can be extended to animal systems (Waterhouse, Wang and Finnegan, 1999).

Threshold Induction of Gene Silencing

According to the model of threshold induction, a surveillance system operates within the cell that detects mRNAs expressed above an acceptable level. This model was proposed to explain the varied responses of transgenic plants to virus inoculation: susceptible, resistant, and showing a recovery phenotype. Examination of these plants revealed that the resistant plants contained high numbers of integrated transgene copies compared with the susceptible and recovery phenotypes. The resistant plants, owing to the presence of several transgenes in their genome, produce large amounts of transgene mRNA. The high level of transcripts triggers the RNAi machinery. In recovery phenotype plants, the transgene mRNA does not accumulate above the threshold amount required to trigger the silencing pathway but will do so when combined with the RNA from the virus during infection.

Induction by Aberrant RNA

It has been observed that gene silencing is not always associated with actively transcribed genes. It can be induced even with promoterless constructs. In this case it is suggested that a particular feature of the transgene mRNA rather than its abundance induces silencing. The aberrant features that have been suggested include untranslatability, lack of introns, transcriptional truncation, and self-complementarity. The aberrant RNA with self-complementarity may be produced from repeat integrations of the transgene that might produce cruciform structures or by transcriptional readthrough of inverted repeats. Truncated RNAs might be produced by premature transcriptional and translational termination of mRNA. The aberrant RNAs form the template for RdRPs, and subsequently dsRNAs are produced.

BIOLOGICAL FUNCTIONS

Protection Against Invading Nucleic Acids

The RNA silencing mechanism was first recognized as an antiviral mechanism that protects organisms from RNA viruses or that prevents the random integration of transposable elements. Plants have been engineered to achieve virus resistance by transforming a susceptible plant with genes derived from the pathogen itself (Dasgupta, Malathi and Mukherjee, 2003). This form of resistance was termed "pathogen-derived resistance" and was known long before the discovery of RNAi. Most plant viruses have genomes composed of RNA. dsRNA is produced as an intermediate during virus replication in the cell. This acts as an effective trigger for RNAi. Even DNA viruses may produce dsRNA as a by-product of bidirectional transcription from their genome. Strong evidence supports the view that the RNA silencing mechanism evolved as an intrinsic defense mechanism owing to the ongoing battle between host and virus. Plants are naturally resistant to most viruses. This may be because plants are able to initiate RNA silencing against most viruses.

Viral Suppressors of RNAi

Certain plant viruses produce proteins that counteract RNA-mediated virus resistance. Helper component proteinase (HCPro) from potyvirus and 2b proteins of the cucumber mosaic virus (CMV) function as effective suppressors of PTGS (Brigneti et al., 1998; Kassachau et al., 2003). No siRNAs were observed in plants where silencing was suppressed by HCPro. Potex viruses express a protein, p25, that prevents the spread of the mobile silencing signal but does not inhibit the existing PTGS. Grafting experiments showed that systemic silencing was absent in the presence of functional p25. The use of the RNA silencing mechanism as a defense against viral infection is not limited to plants. Flock house virus (FHV), which can infect both vertebrate and invertebrate hosts, is both an inducer and target of RNA silencing. It also encodes a protein, B2, that acts as an RNAi suppressor (Li, Li and Ding, 2002). The silencing effect has also been shown against dengue virus in mosquitoes (Adelman et al., 2001). These results suggest a role of RNA silencing in antiviral defense in invertebrates. Recently it has been found that siRNAs can block replication of several different types of mammalian viruses. In short, viruses act as inducers, suppressors, and targets of the RNA silencing pathway.

Maintenance of Genome Integrity

It has been observed in *C. elegans* and *Chlamydomonas reinhardtii* that mutations in genes required for RNA silencing resulted in increased rates of transposition, implicating RNAi in the control of transposons (Dernburg et al., 2000). Transposons are mobile sequences of DNA that can move around in the genome of a single cell, and this process is termed transposition. Their movement can cause mutations. As excessive transposition can result in serious consequences to the cell machinery, the cell has developed its own countermeasures. Recently this defense mechanism was identified as RNAi. Retrotransposon-derived siRNAs have been detected in *Trypanosoma brucei* (Djikeng et al., 2001). RNAi provides a mechanism for programmed sequence-specific silencing by means of heterochromatic modifications that result in silencing of transposons and other repeat regions. siRNA-mediated DNA methylation and transcriptional gene silencing (TGS) are proposed to be responsible for minimizing the transposition frequency.

Regulation of Gene Expression

The control and maintenance of gene expression is critical for cell development and differentiation. Independent studies from around the world indicate that RNAi regulates endogenous gene expression through chromosomal modification, PTGS, and translational blocking. The role of RNA silencing in the regulation of endogenous gene expression became apparent only when it was realized that plants and animals have genes that encode foldback dsRNA (precursor of miRNA). The *Drosophila, C. elegans,* and human genomes contain more than 600 different miRNA genes. As discussed, yet another function of RNA silencing is to regulate the genes, which produces mRNAs above a certain threshold level. Any mutations within the promoter or the transcriptional factors can lead to excessive expression. Another associated function might be to detect and remove aberrant nonfunctional RNAs.

The RNAi Process in the Heterochromatization of Nuclear DNA

Heterochromatin is composed of DNA sequences with little or no coding potential, repeated thousands of times and silenced by covalent modifications such as methylation, acetylation, and phosphorylation. Heterochromatic regions of the chromosomes are mostly found near the centromeres and telomeres. The past few years of research have revealed that siRNAs not only work at the posttranscriptional stage but also affect transcription activity through chromatin modifications. RNAi has been found to play a central role in DNA methylation, heterochromatin formation, and programmed elimination of DNA. Plants showing PTGS often have their homologous genomic DNA regions methylated at almost all cytosine residues and the promoter regions transcriptionally silent (Wang et al., 2001). This process is termed RNA-dependent DNA methylation (Aufsatz et al., 2002). DNA methylation at the promoter region induces TGS, which, unlike PTGS, is stable and heritable.

Evidence of a direct correlation between PTGS and TGS has been obtained from *Arabidopsis thaliana* mutants. Two types of mutants, ddm1 (deficient in DNA methylation) and met1 (methyl transferase), showed a reduction in genome methylation. Both these mutants also exhibited a reduction in PTGS activity (Aufsatz et al., 2002). In *C. elegans,* the polycomb group of proteins is responsible for controlling gene expression by altering the chromatin conformation of target genes. Recently it has been found that some of the polycomb proteins are required for RNAi (Dykxhoorn, Novina and Sharp, 2003). Also in *Arabidopsis* and *Drosophila,* certain components of RNAi machinery are found to be directly involved in heterochromatin formation (Volpe et al., 2003; Zilberman, Cao and Jacobsen, 2003). The connection between RNAi and chromatin modifications has been strengthened by the fact that small RNAs complementary to centromeric repeats and long terminal repeats of retrotransposons have been identified in yeast. This suggests that RNAi-induced heterochromatin formation is a general method of regulating gene expression in yeast (Reinhart and Barlet, 2002). Small RNAs are also involved in guiding DNA sequence elimination of *Tetrahymena pyriformis*. *T. pyriformis* has two nuclei: the micronucleus and macronucleus. During conjugation approximately 15 percent of genomic micro DNA is eliminated. How this happens was a mystery to the scientific world until recently. In wild-type cells of this organism, siRNAs of approximately 26-31 nucleotides are produced, which then hybridize to the micronuclear genomic DNA. These siRNAs are named scan RNAs and are believed to guide the nuclear equivalents of RISC to cognate DNA for elimination (Mochizuki et al., 2002). Many aspects of the RNAi-mediated heterochromatic silencing mechanism are still unclear, but this role of RNAi widens the scope of RNA-mediated genome alterations.

RNAi FOR INSECT-VECTORED VIRAL DISEASES OF PLANTS

Plant viral diseases are the major limiting factors in crop production in many areas around the world. The most common way that the virus moves from plant to plant is via another organism known as a vector. These vectors include insects, nematodes, and certain fungi. By far, both in the number of viruses transmitted and in the damage caused, insects are the most important vector. At least 70 percent of all viruses are transmitted by an insect vector.

Transmission of viruses from plant to plant by a vector can occur in two different ways. The first is the noncirculative manner, in which the virus does not enter the circulatory system of the vector but remains attached to the outer tissues of the vector. In most cases, plant viruses are transmitted in the noncirculative fashion. In the circulative manner, a virus enters the circulatory system of the vector and eventually ends up in the salivary glands. The vector transmits the virus into plants while feeding. Occasionally the virus may replicate inside the vector.

In addition to being a tool for intensive basic research, RNAi holds the key to future pest management strategies. The past few years of research have provided compelling evidence supporting RNA silencing as a natural process behind antiviral immunity in insects.

A double-genomic sindbis virus that was engineered to transcribe an antisense RNA complementary to the luciferase gene (reporter gene) efficiently suppressed luciferase expression in the salivary glands of mosquitoes (Gaines et al., 1996). In another experiment, a dsSindbis virus system was used to transduce female mosquitoes *(Aedes aegypti)* with the antisense premembrane (prM) RNA of dengue type 2 (DEN-2) viruses (Blair, Adelman and Olson, 2000). The transduced mosquitoes were unable to support replication of the DEN-2 virus and were not competent to transmit the DEN-2 virus. Sense RNAs derived from DEN-2 viruses were just as effective as antisense RNA at blocking the homologous virus replication in *A. aegypti* (Adelman et al., 2001). Furthermore, mosquito cells transfected with a plasmid designed to express a hairpin RNA derived from the dengue virus genome were resistant to virus infection and also accumulated signature molecules of RNAi: si RNAs (Adelman et al., 2002).

Important evidence supporting the antiviral role of RNAi in insects was also obtained from studies on FHV. Fruit fly cells transfected with dsRNA corresponding to part of the RNA genome of FHV became resistant to this virus (Li, Li and Ding, 2002). FHV infection in *Drosophila* cells triggers strong virus RNA silencing and results in the accumulation of siRNAs specific for the viral genome. This indicates that the invertebrate Dicer ribonuclease is able to detect the infecting viruses. Increased accumulation of FHV RNAs was observed in *Drosophila* cells defective for the RNA silencing pathway.

Interestingly, the FHV virus encodes a protein, B2, that acts as an efficient suppressor of the RNAi machinery. This protein shares key features with the 2b protein of CMV, which is a suppressor of RNA silencing (Li, Li and Ding, 2002). The FHV B2 protein can functionally replace the 2b protein of CMV. These findings provide direct evidence for a role of RNA silencing as an adaptive antiviral defense in invertebrates. The existence of a functional RNA silencing pathway in insects opens up the possibility of using this pathway to reduce vector competence to transmit pathogens. However, the use of RNAi-based immunity to viruses for virus control is feasible only in the case of viruses that are transmitted in a circulative manner.

The release of sterile males is an important pest control strategy that has been successfully used in many areas around the world for the eradication of different pests. But this method is costly and, to ensure success, vast numbers of sterile insects must be released. This has led researchers to look at alternative strategies to make insects sterile. Hairpin constructs producing siRNAs corresponding to *boule* gene essential for fertility were introduced into *Drosophila*. The construct was linked to a genetic switch that is turned off by an antibiotic. To create normal males, the flies are fed with food laced with the antibiotic. When the flies are taken off the antibiotic and fed normal food, they produce sterile males. This genetic trick could be extended to other farm pests.

GENETIC ENGINEERING OF PLANTS FOR VIRUS RESISTANCE

Another approach to controlling diseases transmitted by vectors could be to make the plant resistant to the transmitted diseases using RNAi technology. Virus-resistant transgenes have been developed in many crops by introducing either the viral coat protein gene or the replicase gene of the virus. Among these experiments, the most successful was transgenic papaya resistant to papaya ring spot virus (PRSV). Papaya transformed with the coat protein gene of PRSV remained virus free for several months (Ferreira et al., 2002). The biochemical mechanism behind this was poorly known and was long thought to be protein mediated. Dougherty and Lindbo were the first to recognize that it was not the protein but the RNA from the transgene that was conferring resistance. They were the first to propose that the transgene mRNA in virus resistant plants induces degradation of RNAs with homologous sequences (Lindbo et al., 1993).

The degradation takes place exclusively in the cytoplasm: evidence shows that PTGS does not affect the full-length transcript levels in the nucleus. Gene-silencing-defective mutants of plants show increased sensitivity to viral infections (Mourrain et al., 2000). dsRNAs derived from viral sequences were shown to interfere with virus infection in a sequence-specific manner by delivering dsRNAs to leaf cells by mechanical coinoculation with the virus or by an *Agrobacterium*-mediated transient expression approach. The salient feature of RNA-mediated virus resistance (PTGS) is that even small gene segments (~230 base pairs) provide complete immunity against viruses (Pang, Jan and Gonsalves, 1997).

Studies on virus-resistant transgenic plants showed that the transgene-derived transcript and the viral genome are specifically eliminated from the cell when they show more than 90 percent similarity. A virus showing less than 80 percent similarity to the transgene replicates normally (Smith et al., 1994).

As crops are frequently infected with more than one virus under field conditions, the benefits from the transgenic approach to virus resistance can be fully realized if multiple virus resistance can be developed. Transforming plants with full-length viral genes for multiple resistances appears to be difficult; moreover, it provides narrow-spectrum protection. PTGS is a powerful, specific, and intracellular RNA degradation mechanism that can be used to prevent multiple viral infections if conserved small viral segments from different viruses are transcriptionally fused (Jan et al., 2000; Antony, Mishra and Praveen, 2005).

Viral genes show high levels of variability. Lack of proofreading function in the viral replicases and high recombination rates of viral genomes during the process of infection are the probable reasons behind this variability. To develop stable, broader, and multiple resistances, it is appropriate to use viral sequences that are conserved among the isolates/viruses. The use of small viral gene segments instead of the complete gene will reduce the possible risk of complementation, transencapsidation, and recombination with other viral genomes. This strategy will also reduce the translational load on plant expression systems. PTGS for virus-resistant endogenous gene silencing does not always require a high level of transcription of the inducing transgenes. It has been proposed that an aberrant quality of transgene mRNA and not always its abundance triggers silencing (Waterhouse, Wang and Finnegan, 1999). The aberrant features can be introduced through transcriptional fusion of different viral segments, which lacks translatability (Antony, Mishra and Praveen, 2005).

RNAi FOR FUNCTIONAL GENOMICS

Insecticides provide a convenient and economically feasible method of controlling pests. However, there are adverse biological and economic consequences related to their use, such as development of resistance, resurgence of pests, elimination of natural enemies, impact on nontarget species, and biomagnification. Most often, these consequences are due to insufficient knowledge about the mode of action of insecticides. Development of an ideal pesticide often requires detailed knowledge of the complex interactions between the insect and the host plant as well as the insect's physiology. Gene silencing through RNAi is a powerful tool for the elucidation of gene functions. dsRNA technology is now being used for the systematic exploration of the functions of genes in a variety of organisms. The most interesting aspect of this technology is the ability to knock out gene expression in a manner that is highly specific as well as technologically easy and cheap.

Cultured *Drosophila* cells transfected with specific dsRNAs can cause loss-of-function phenotypes (Hammond et al., 2000). Injection of lacZ dsRNA as well as hairpin RNA corresponding to lacZ into *Drosophila* embryos expressing lacZ resulted in interference with lacZ expression. The majority of the treated insects exhibited complete inhibition of lacZ expression. A transgene under the transcriptional control of the yeast UAS element of lacZ that can synthesize hairpin dsRNA was introduced into the *Drosophila* germline by transformation, and the transformed flies were crossed with flies carrying the yeast Gal 4 gene. The Gal 4 gene product activates the transcription of genes with the UAS regulatory element. The F_1 hybrid flies exhibited a disrupted pattern of lacZ expression (Kennerdell and Carthew, 1998). In *Plodia interpunctella* (Indian meal moth), a lepidopteran insect that infests stored products, injection of tryptophan oxygenase dsRNA into the embryos resulted in larvae with complete knockout of eye color pigmentation (Fabrick, Kanost and Baker, 2004).

Haemolin is a bacteria-inducible immune gene in *Hyalophora cecropia.* To investigate its role in embryonic development, dsRNA corresponding to this gene was injected into the hemocoel of diapausing pupae. The double-stranded hemolin pupae developed into normal moths, and the males and females mated normally, but none of their eggs hatched. This demonstrates the ability of RNAi to act across the ecdysis. The results also confirmed the dual roles of hemolin in immunity and development (Bettencourt, Terenius and Faye, 2002).

Arakane et al. (2005) performed a series of RNAi experiments in the red flour beetle, *Tribolium castaneum,* to test the involvement of phenoloxidases in cuticle tanning. dsRNAs corresponding to different phenoloxidases were injected individually into penultimate-instar larvae, last-instar larvae, and prepupal stages. Phenotypes obtained were similar to those for buffer-injected control for all phenoloxidases tested with the exception of laccase 2. Two isoforms of the enzyme laccase 2 were found to be indispensable for cuticle tanning because dsRNA-mediated knockdown of their transcripts at the late larval stage resulted in incomplete tanning of the larval, pupal, and adult cuticles and, subsequently, death. Table 10.1 summarizes RNAi-related events so far described in insects. These studies indicate the emergence of RNAi as a powerful tool for screening the insect system for candidate genes that can be used as targets for the development of future pesticides as well as biorational products.

RNAi FOR NEMATODE MANAGEMENT

The annual loss to world agriculture caused by plant-parasitic nematodes is estimated at US$100 billion (Sasser and Freckman, 1987). Cyst and root-knot nematodes are the most damaging plant-pathogenic nematode species. Juvenile nematodes invade the roots of the plant and persist in the root system for many weeks. Once established, they can cause deleterious effects on plant health and productivity. Moreover, a large number of nematodes act as vectors of crop diseases. Currently the control of plant-parasitic nematodes is inadequate and is heavily dependent on a few chemicals available in the market.

The demonstration by Fire et al. (1998) that RNAi in *C. elegans* can be initiated by soaking the worms in a solution containing dsRNA or by feeding the worms with *Escherichia coli* that express dsRNAs triggered a new line of thinking. What would happen if the plant cells in the potential region of infestation (roots in this case) were engineered to produce dsRNAs of a nematode gene? Would it be as successful as a virus-resistant transgenic plant? Root-specific promoters can be used for the production of dsRNAs in the root. These dsRNAs may then be processed into siRNAs by root cells. On infestation of the plant roots, nematodes may take up either dsRNAs or siRNAs through the skin while feeding on root cells. At Kansas State University transgenic soybeans resistant to the soybean cyst nematode have been developed. These transgenic soybeans, which express a hairpin construct corresponding to a major sperm protein of

TABLE 10.1. RNAi in insects.

Species	Common name	Initiator	Delivery	Reference
Anophelus gambie	Mosquito	dsRNA Transgene	Adult injection Transformation	Billecocq et al., 2000; Gaines et al., 1996
Apis mellifera	Honey bee	dsRNA	Embryo and pupal injection	Amdam et al., 2003; Beye et al., 2002
Bombyx mori	Silkworm moth	Viral RNA	Larval injection	Uhlirova et al., 2003
Ceratitis capitata	Mediterranean fruit fly	dsRNA	Embryo injection	Pane et al., 2002
Drosophila melanogaster	Fruit fly	dsRNA Transgene	Embryo injection Transformation	Kennerdell and Carthew, 1998, 2000
Hyalophora cecropia	Giant silkmoth	dsRNA	Embryo and pupal injection	Bettencourt, Terenius and Faye, 2002
Manduca sexta	Tobacco horn worm	dsRNA	Injection	Vermehren, Qazi and Trimmer, 2001
Periplaneta americana	Cockroach	dsRNA	Larval injection	Marie, Bacon and Blagburn, 2000
Plodia interpunctella	Indian mealmoth	dsRNA	Embryo injection	Fabrick, Kanost and Baker, 2004
Spodoptera litura	Army worm	dsRNA	Larval injection	Rajagopal et al., 2002
Tribolium castaneum	Red flour beetle	dsRNA	Larval and pupal injection	Bucher, Scholten and Klingler, 2002; Arakane et al., 2005
Trichoplusia ni	Cabbage looper	dsRNA	Transfection	Beck and Strand 2003

the soybean cyst nematode, are currently under evaluation. This major sperm protein is crucial for soybean cyst nematode male fertility. Preliminary results show a 50.6 percent reduction in cysts in transgenic plants inoculated with soybean cyst nematode compared with nontransgenic control.

DESIGNING SILENCING TRIGGERS FOR RNAi

Designing constructs that can act as point triggers is the key to the successful application of RNAi. For any RNAi-based stratergy for controlling insect- and nematode-borne diseases in plants, selection of the ideal target gene is most important. Gene silencing can be achieved by transformation of plants with sequences homologous to the target gene in the pathogen. siRNAs synthesized against different regions of the same target mRNA have shown different silencing efficiencies (Holen et al., 2002). Several groups have analyzed a number of parameters for optimizing siRNA-based gene silencing; however, no consensus has evolved. Usually 3′ regions of the gene segments are chosen. The length of the gene segments chosen is generally in the range 150-300 base pairs. Several software programs are available on the Web that offer helpful guidelines to select potential siRNA sequences and to determine whether these selected sequences match off-target mRNA sequences (e.g., Ambion: http://www.ambion.com/techlib/misc/siRNA_finder.html; Dharmacon: http://www.dharmacon.com/Default.aspx; Oligoengine: http://www.oligoengine.com/Home/mid_prodSirna.html#sirna_tool; SFold: http://sfold.wadsworth.org/sirna.pl; SiRNA target finder: http://www.genscript.com/ssl-bin/app/rnai). As dsRNAs act as the central player in initiating RNAi, two approaches have been developed for generating siRNAs in vivo. In the first approach, sequences homologous to the target mRNA are cloned in sense and antisense, each transcribed by individual promoters. In the second method, sense and antisense sequences are interrupted by a small sequence, usually 30-100 base pairs long, called a loop and are transcribed by a single promoter, giving rise to hairpin or foldback stem-loop structures. Hairpin constructs have been found to be highly efficient and can be used while targeting specific genes. For example, while selecting nematode genes for RNAi-mediated suppression in crop plants, species specific genes that have functions in fertility and reproduction should be targeted. Gene sequences with high homology (>80 percent) with nontarget organisms and humans should be avoided. In the case of viruses, simple viral gene segments in either sense or antisense are as good as hairpin structures in initiating RNAi in plants. Fusing different gene segments derived from different viruses will be a better stratergy than using hairpin constructs, as this will give multiple-virus resistance. Cloning sequences in hairpin form has been made easy by the generation of specialized plasmid vectors such as the pHANNIBAL, pHELLSGATE, and pANDA series (Helliwell and Waterhouse, 2003). In these vectors, the sequences derived from a target gene can be inserted in sense and antisense separated by a linker sequence in a single shot using a gateway-directed recombination system. pANDA vectors are available in two different versions, with either the CAMV35S promoter or the maize ubiquitin promoter. Suppression of mRNA expression was observed in more than 90 percent of transgenic plants transformed with a pANDA vector expressing hairpin RNA derived from different rice genes (Miki and Shimamoto, 2004). siRNA, indicative of RNA silencing, was detected in each of the silenced plants. Plants can be transformed either by the use of Agrobacterium or by Biolistics. Once transformed plants are obtained, expression of siRNAs can be detected by northern blotting. Resistant phenotypes can be evaluated using appropriate inoculation methods.

CONCLUSION

In the past few years, we have come a long way toward understanding the mechanism of RNAi through a combination of biochemical and genetic studies. Nevertheless, RNAi is still a maturing technology, with its true potential yet to be known. As this technology offers great potential in understanding the gene functions of hosts and pests, it will enable us to utilize the knowledge generated to improve crop production as well as the quality of the produce. If judiciously used, this technology will go a long way in the generation of disease-, insect-, nematode-, and virus-resistant crops. RNAi has great advantages over conventional methods of pest control owing to its specificity, potency, and lack of side effects. Tissue- or organ-specific RNAi can be achieved with minimal interference to the normal plant life cycle. Future dissection of this pathway will facilitate greater efficiency in fine-tuning RNAi machinery in plant cells for greater benefits to humans. It is only a matter of time before we see the fruits of RNAi research in the farmer's field.

REFERENCES

Adelman, Z.N., C.D. Blair, J.O. Carlson, B.J. Beaty and K.E. Olson. (2001). Sindbis virus-induced silencing of dengue viruses in mosquitoes. *Insect Molecular Biology* 10: 265-273.

Adelman, Z.N., I. Sanchez-Vargas, E.A. Travanty, J.O. Carlson, B.J. Beaty, C.D. Blair and K.E. Olson. (2002). RNA silencing of dengue virus type 2 replication in transformed C6/36 mosquito cells transcribing an inverted-repeat RNA derived from the virus genome. *Journal of Virology* 76: 12925-12933.

Amdam, G.V., Z.L. Simoes, K.R. Guidugli, K. Norberg and S.W. Omholt. (2003). Disruption of vitellogenin gene function in adult honeybees by intra-abdominal injection of double-stranded RNA. *BioMed Central Biotechnology* 3: 1.

Antony, G., A.K. Mishra and S. Praveen. (2005). A single chimeric transgene derived from two distinct viruses for multiple resistance. *Journal of Plant Biochemistry and Biotechnology* 14: 101-105.

Arakane,Y., S. Muthukrishnan, R.W. Beeman, M.R. Kanost and K.J. Kramer. (2005). *Laccase 2* is the phenoloxidase gene required for beetle cuticle tanning. *PNAS* 102: 11337-11342.

Aufsatz, W., M.F. Mette, J. Van der Winden, A.J. Matzke and M. Matzke. (2002). RNA-directed DNA methylation in *Arabidopsis. Proceedings of the National Academy of Sciences USA* 99: 16499-16506.

Baehrecke, E.H. (2003). miRNAs: Micro managers of programmed cell death. *Current Biology* 13: R473-R475.

Beck, M. and M.R. Strand. (2003). RNA interference silences *Microplitis demolitor* bracovirus genes and implicates glc1.8 in disruption of adhesion in infected host cells. *Virology* 314: 521-535.

Bernstein, E., A.A. Caudy, S.M. Hammond and G.J. Hannon. (2001). Role for a bidentate ribonuclease in the initiation step of RNA interference. *Nature* 409: 363-366.

Bettencourt, R., O. Terenius and I. Faye. (2002). Hemolin gene silencing by ds-RNA injected into *Cecropia* pupae is lethal to next generation embryos. *Insect Molecular Biology* 11: 267-271.

Beye, M., S. Hartel, A. Hagen, M. Hasselmann and S.W. Omholt. (2002). Specific developmental gene silencing in the honey bee using a homeobox motif. *Insect Molecular Biology* 11: 527-532.

Billecocq, A., M. Vazeille-Falcoz, F. Rodhain and M. Bouloy. (2000). Pathogen-specific resistance to Rift Valley fever virus infection is induced in mosquito cells by expression of the recombinant nucleoprotein but not NSs non-structural protein sequences. *Journal of General Virology* 81: 2161-2166.

Blair, C.D., Z.N. Adelman and K.E. Olson. (2000). Molecular strategies for interrupting arthropod-borne virus transmission by mosquitoes. *Clinical Microbiology Reviews* 13: 651-661.

Brigneti, G., O. Voinnet, W.X. Li, L.H. Ji, S.W. Ding and D.C. Baulcombe. (1998). Viral pathogenicity determinants are suppressors of transgene silencing in

Nicotiana benthamiana. *European Molecular Biology Organisation Journal* 17: 6739-6746.

Bucher, G., J. Scholten and M. Klingler. (2002). Parental RNAi in *Tribolium* (Coleoptera). *Current Biology* 12: R85-R86.

Cogoni, C. and G. Macino. (1999). Gene silencing in *Neurospora crassa* requires a protein homologous to RNA-dependent RNA polymerase. *Nature* 399: 166-169.

Cogoni, C. and G. Macino. (2000). Post-transcriptional gene silencing across kingdoms. *Current Opinions on Genetics and Development* 10: 638-643.

Dalmay, T., A. Hamilton, S. Rudd, S. Angell and D.C. Baulcombe. (2000). An RNA-dependent RNA polymerase gene in *Arabidopsis* is required for post transcriptional gene silencing mediated by a transgene but not by a virus. *Cell* 101: 543-553.

Dasgupta, I., V.G. Malathi and S.K. Mukherjee. (2003). Genetic engineering for virus resistance. *Current Science* 84: 341-354.

Dernburg, A.F., J. Zalevsky, M.P. Calaiacovo and A.M. Villeneuve. (2000). Transgene-mediated co-suppression in the *C. elegans* germ line. *Genes and Development* 14: 1578-1583.

Djikeng, A., H. Shi, C. Tschudi and E. Ullu. (2001). RNA interference in *Trypanosoma bruci*: Cloning of small interfering RNA provides evidence for retroposon-derived 24-26 nucleotide RNA. *RNA* 7: 1522-1530.

Doench, J.G., C.P. Petersen and P.A. Sharp. (2003). siRNAs can function as miRNAs. *Genes and Development* 17: 438-442.

Dougherty, W.G. and T.D. Parks. (1995). Transgenes and gene suppression: Telling us something new? *Current Opinions in Cell Biology* 7: 399-405.

Dykxhoorn, D.M., C.D. Novina and P.A. Sharp. (2003). Killing the messenger: short RNAs that silence gene expression. *Nature Review Molecular Cell Biology* 4: 457-467.

Elbashir, S.M., W. Lendeckel and T. Tuschl. (2001). RNA interference is mediated by 21- and 22-nucleotide RNAs. *Genes and Development* 15: 188-200.

Fabrick, J.A., M.R. Kanost and J.E. Baker. (2004). RNAi induced silencing of embryonic tryptophan oxygenase in the pyralid moth, *Plodia interpunctella*. *Journal of Insect Science* 4: 1-9.

Fagard, M., S. Boutet, J.B. Morel, C. Bellini and H. Vaucheret. (2000). AGO1, QDE-2 and RDE-1 are related proteins required for post-transcriptional gene silencing in plants, quelling in fungi and RNA interference in animals. *Proceedings of the National Academy of Sciences USA* 97: 11650-11654.

Ferreira, S.A., K.Y. Pitz, R. Manshardt, F. Zee, M. Fitch and D. Gonsalves. (2002). Virus coat protein transgenic papaya provides practical control of Papaya ring spot virus in Hawaii. *Plant Disease* 86: 101-105.

Fire, A., S. Xu, M.K. Montgomery, S.A. Kostas, S.E. Driver and C.C. Mello. (1998). Potent and specific genetic interference by double-stranded RNA in *C. elegans*. *Nature* 391: 806-811.

Gaines, P.J., K.E. Olson, S. Higgs, A.M. Powers, B.J. Beaty and C.D. Blair. (1996). Pathogen-derived resistance to dengue type 2 virus in mosquito cells by expression of the premembrane coding region of the viral genome. *Journal of Virology* 70: 2132-2137.

Guo, S. and K. Kemphures. (1995). *Par-1,* a gene required for establishing polarity in *C. elegans* embryos, encodes a putative Ser/Thr kinase that is asymmetrically distributed. *Cell* 81: 611-620.

Hamilton, A.J. and D.C. Baulcombe. (1999). A species of small antisense RNA in posttranscriptional gene silencing in plants. *Science* 286: 950-952.

Hammond, S., M.E. Berstein, D. Beach and G.J. Hannon. (2000). An RNA-directed nuclease mediates post-transcriptional gene silencing in *Drosophila* cells. *Nature* 404: 293-296.

Helliwell, C. and P. Waterhouse. (2003). Constructs and methods for high-throughput gene silencing in plants. *Methods* 30: 289-295.

Holen, T., M. Amarzguioui, M.T. Wiiger, E. Babaie and H. Prydz. (2002). Positional effects of short interfering RNAs targetting the human coagulation trigger tissue factor. *Nucleic Acid Research* 30: 1757-1766.

Jan F.J., C. Fagoago, S.Z. Pang and D. Gonsalves. (2000). A single chimeric transgene derived from two distinct viruses confers multivirus resistance in transgeneic plants through homology-dependent gene silencing. *Journal of General Virology* 81: 2103-2109.

Jorgensen, R., P.D. Cluster, J. English, Q. Que and C.A. Napoli. (1996). Chalcone synthase co-suppression phenotypes in petunia flowers: Comparison of sense Vs antisense constructs and single copy vs complex T-DNA sequences. *Plant Molecular Biology* 31: 957-973.

Kasschau, K.D., Z. Xie, E. Allen, C. Llave, E.J. Chapman, K.A. Krizan and J.C. Carrington. (2003). P1/HC-Pro, a viral suppressor of RNA silencing, interferes with *Arabidopsis* development and miRNA function. *Developmental Cell* 4: 205-217.

Kennerdell, J.R. and R.W. Carthew. (1998). Use of dsRNA-mediated genetic interference to demonstrate that frizzled and frizzled 2 act in the wingless pathway. *Cell* 95: 1017-1026.

Kennerdell, J.R. and R.W. Carthew. (2000). Heritable gene silencing in *Drosophila* using double-stranded RNA. *Nature Biotechnology* 18: 896-898.

Lagos-Quintana, M., R. Rauhut, W. Lendeckel and T. Tuschl. (2001). Identification of novel genes coding for small expressed RNAs. *Science* 294: 853-858.

Lee, R.C., L. Feinbaum and V. Ambros. (1993). The *C. elegans* heterochronic Gene lin-4 encodes small RNAs with antisense complementarity to lin-14. *Cell* 75: 843-854.

Lee, Y., K. Jeon, J.T. Lee, S. Kim and V.N. Kim. (2002). MicroRNA maturation: Stepwise processing and subcellular localization. *European Molecular Biology Organisation Journal* 21: 4663-4670.

Li, H.W., W.X. Li and S.W. Ding. (2002). Induction and suppression of RNA silencing by an animal virus. *Science* 296: 1319-1321.

Lindbo, J.A., L. Silva-Rosales, W.M. Proebsting and W.G. Dougherty. (1993). Induction of a highly specific antiviral state in transgenic plants: Implications for regulation of gene expression and virus resistance. *Plant Cell* 5: 1749-1759.

Marie, B., J.P. Bacon and J.M. Blagburn. (2000). Double-stranded RNA interference shows that Engrailed controls the synaptic specificity of identified sensory neurons. *Current Biology* 10: 289-292.

Metzlaff, M., M. ODell, P.D. Cluster and R.B. Flavell. (1997). RNA-mediated RNA degradation and chalcone synthase A silencing in *Petunia*. *Cell* 88: 845-854.

Miki, D and K. Shiamamoto. (2004). Simple RNAi vectors for stable and transient suppression of gene function in rice. *Plant and Cell Physiology* 45: 490-495.

Mochizuki, K., N.A. Fine, T. Fujisawa and M.A. Gorovsky. (2002). Analysis of a piwi-related gene implicates small RNAs in genome rearrangement in *Tetrahymena*. *Cell* 110: 689-699.

Mourrain, P., C. Beclin, T. Elmayan, F. Feuerbach, C. Godon, J.B. Morel, D. Jouette et al. (2000). *Arabidopsis* SGS2 and SGS3 genes are required for posttranscriptional gene silencing and natural virus resistance. *Cell* 101: 533-542.

Napoli, C., C. Lemieux and R. Jorgensen. (1990). Introduction of chimeric chalcone synthase gene into *Petunia* results in reversible co-suppression of homologous genes in trans. *Plant Cell* 2: 279-289.

Palauqui, J.C., T. Elmayan, J.M. Pollien and H. Vaucheret. (1997). Systemic acquired silencing: Transgene specific posttranscriptional gene silencing is transmitted by grafting from silenced stocks to non-silenced scions. *European Molecular Biology Organisation Journal* 16: 4738-4745.

Pane, A., M. Salvemini, P. Delli Bovi, C. Polito and G. Saccone. (2002). The transformer gene in *Ceratitis capitata* provides a genetic basis for selecting and remembering the sexual fate. *Development* 129: 3715-3725.

Pang, S., Z. Jan and D. Gonsalves. (1997). Non target DNA sequence reduce the transgene length necessary for RNA mediated tospovirus resistance in transgenic plants. *Proceedings of the National Academy of Sciences USA* 94: 8261-8266.

Rajagopal, R., S. Sivakumar, N. Agrawal, P. Malhotra and R.K. Bhatnagar. (2002). Silencing of midgut aminopeptidase N of *Spodoptera litura* by double-stranded RNA establishes its role as *Bacillus thuringiensis* toxin receptor. *Journal of Biological Chemistry* 277: 46849-46851.

Reinhart, B.J. and D.P. Bartel. (2002). Small RNAs correspond to centromere heterochromatic repeats. *Science* 297: 1831.

Sasser, J.N. and D.W. Freckman. (1987). A world perspective on nematology: The role of society. In *Vistas on Nematology,* eds. J.A. Veech and D.W. Dickerson, Society of Nematologists, U.S.A. Marceline, MO, pp. 7-14.

Saxena, S., Z.O. Jonsson and A. Dutta. (2003). Small RNAs with imperfect match to endogenous mRNA repress translation: Implications for off-target activity of siRNA in mammalian cells. *Journal of Biological Chemistry* 278: 44312-44319.

Sijen, T., J. Fleenor, F. Simmer, K.L. Thijssen, S. Parrish, L. Timmons, R.H. Plasterk and A. Fire. (2001). On the role of RNA amplification in dsRNA-triggered gene silencing. *Cell* 107: 465-476.

Smith, H.A., S.L. Swaney, T.D. Parks, E.A. Wernsman and W.G. Dougherty. (1994). Transgenic plant virus resistance mediated by untranslatable sense RNAs: Expression, regulation and fate of nonessential RNAs. *Plant Cell* 6: 1441-1453.

Tabara, H., M. Sarkissian, W.G. Kelly, J. Fleenor, A. Grishok, L. Timmons, A. Fire and C.C. Mello. (1999). The *rde-1* gene, RNA interference and transposon silencing in *C. elegans*. *Cell* 99: 123-132.

Timmons, L. and A. Fire. (1998). Specific interference by ingested dsRNA. *Nature* 395: 854.

Uhlirova, M., B.D. Foy, B.J. Beaty, K.E. Olson, L.M. Riddiford and M. Jindra. (2003). Use of Sindbis virus-mediated RNA interference to demonstrate a conserved role of Broad-Complex in insect metamorphosis. *Proceedings of the National Academy of Sciences USA* 100: 15607-15612.

Vermehren, A., S. Qazi and B.A. Trimmer. (2001). The nicotinic subunit MARA1 is necessary for cholinergic evoked calcium transients in *Manduca* neurons. *Neuroscience Letters* 313: 113-116.

Volpe, T.A., C. Kidner, I.M. Hall, G. Teng, S.L.S. Grewal, and R.A. Martienssen. (2003). Regulation of heterochromatic silencing and histone H3 Lysine-9 methylation by RNAi. *Science* 297: 1833-1837.

Wang, M.B., S.V. Wesley, E.J. Finnegan, N.A. Smith and P.M. Waterhouse. (2001). Replicating satellite RNA induces sequence specific DNA methylation and truncated transcripts in plants. *RNA* 7: 16-28.

Waterhouse, P.M., M.B. Wang and E.J. Finnegan. (1999). Virus resistance and gene silencing: Killing the messenger. *Trends in Plant Sciences* 6: 297-301.

Zilberman, D., X. Cao and S.E. Jacobsen. (2003). ARGONAUTE4 control of locus-specific siRNA accumulation and DNA and histone methylation. *Science* 299: 716-719. http://www.newscientist.com/article.ns?id=dn3465.

Index

Management of Nematode and Insect-Borne Plant Diseases

doi:10.1300/5754_11